Multivariate Analysis

STATISTICS

Textbooks and Monographs

EDITED BY

D. B. OWEN

Department of Statistics
Southern Methodist University
Dallas, Texas

VOLUME 1: The Generalized Jackknife Statistic, *H. L. Gray and W. R. Schucany*

VOLUME 2: Multivariate Analysis, *Anant M. Kshirsagar*

OTHER VOLUMES IN PREPARATION

Multivariate Analysis

ANANT M. KSHIRSAGAR
Institute of Statistics
Texas A & M University
College Station, Texas

1972

MARCEL DEKKER, INC. New York

MARCEL DEKKER, INC.

95 Madison Avenue, New York, New York 10016

LIBRARY OF CONGRESS CATALOG CARD NUMBER 78-182214

ISBN 0-8247-1386-9

PRINTED IN THE UNITED STATES OF AMERICA

PREFACE

For students and research workers interested in the theory of multivariate analysis, there appeared, to me at least, to be a need for a book which takes account of the developments in the subject during the last fifteen years or so. Multivariate analysis has made tremendous progress during the past few years. Distribution theory associated with multivariate analysis can be simplified considerably by elegant matrix transformations and "random orthogonal transformations," first introduced by Bowker and Wijsman. In the whole theory of multivariate analysis, regression analysis plays a key role; it is a unifying central theme underlying all the techniques, and it is essential that this aspect be brought forth explicitly while teaching multivariate analysis. This book is written with this idea in mind.

This book is not intended as an elementary textbook. If some real depth and understanding of multivariate methods is to be achieved, it is essential that the reader be not only familiar but rather well trained in the theory of matrices, vector spaces, linear models, quadratic forms, and univariate analysis of variance and covariance. This book is written assuming these prerequisites and also familiarity with most of the univariate statistical tests and techniques. Otherwise, it would have been necessary to spend a substantial portion of the book on these, and then perhaps it would not be possible to develop the subject proper of multivariate analysis as much as I had in mind. Fortunately excellent books on matrix algebra and statistical theory are available.

Even assuming all these prerequisites, some topics like factor analysis could not be included in the book, in order to limit the book to a reasonable size.

The aim was to restrict the book to the derivation and presentation of the theoretical results only. The readers will have to supplement this book, therefore, by other books given in the references at the end of each chapter, for illustrative examples. Already it is complained by many that the same examples recur again and again in multivariate analysis. This is thus avoided; however reference to classical examples in the literature is made in general terms in the context of particular techniques.

It is hoped that this book will be useful as a textbook for a one or two semester course in multivariate analysis. It is also likely to be helpful to students interested in doing theoretical research in multivariate analysis.

If the readers find this book useful, the credit is entirely due to my able teachers like Professor M. S. Bartlett and Professor M. C. Chakrabarti, who created in me a liking for this subject by their masterly presentation and deep insight. I am grateful to Professor Donald B. Owen for his valuable advice and constant encouragement to undertake this task.

Professor John T. Webster, Professor W. B. Smith, and Dr. John C. Young read the manuscript of the book and suggested many improvements. I am indebted to them. I am also thankful to Professor H. O. Hartley, Professor Paul D. Minton, and Dr. W. R. Schucany and to a number of my past and present students for their encouragement. I also wish to express my gratitude to Ms. Barbara Bemben, who copy-edited the book and improved it a lot and to Mrs. Karen Holder who typed the manuscript with great care and diligence.

A. M. Kshirsagar

College Station, Texas

CONTENTS

NOTATION AND ABBREVIATIONS

Theorems, lemmas and tables in each chapter are numbered serially. Equations and expressions are numbered serially in each section of a chapter, the first number indicating the section number and the second number indicating the equation number; the two numbers are separated by a dot. If a reference is made to, say equation number (3.2) in a chapter, it means equation 2 of section 3 of that chapter, but when a reference is made to (5.3.2), it means equation 2 of section 3 of chapter 5. Vectors are in general column vectors and are underscored; e.g. $\underline{x}$. For a partitioned matrix, the dimensions of the different parts are indicated outside the matrix. Thus for example,

$$\begin{array}{cc} \left[\begin{array}{c|c} A & B \\ \hline C & D \end{array}\right] & \begin{array}{l} p \\ q \end{array} . \\ \begin{array}{cc} p & \;\; q \end{array} & \end{array}$$

By a Jacobian of transformation, it is understood that the absolute value is considered. The following abbreviations are frequently used:

[1] Probability density function: p.d.f.

[2] Cumulative distribution function: c.d.f.

[3] Degrees of freedom: d.f.

[4] Trace: tr

[5] Variance-covariance matrix: V

[6] Expectation: E

[7] Transpose of A: A'

[8] Determinant of A: $|A|$

[9] A is positive definite: $A > 0$

[10] Kronecker product of A and B: $A \otimes B$

[11] Jacobian of transformation from V to W: $J(V \to W)$

[12] A diagonal matrix, with elements $a_1, \ldots, a_p$: $\mathrm{diag}(a_1, a_2, \ldots, a_p)$

[13] A p x q matrix of all unit elements: E_{pq}

[14] Lower triangular matrix: l.t.

[15] Exponential function: exp.

[16] Maximum likelihood estimate: m.l.e.

[17] Normal independent variables with zero mean and unit variance: NI(0, 1)

[18] Sum of squares and sum of products: s.s. and s.p.

Multivariate Analysis

CHAPTER 1

REGRESSION AND CORRELATION AMONG SEVERAL VARIABLES

1. MULTIVARIATE ANALYSIS

Multivariate analysis is the analysis of observations on several correlated random variables, for a number of individuals. Such analysis becomes necessary in Anthropology, Biology, Psychology, Education, Medicine, Agriculture, and Economics, when one deals with several variables simultaneously. A series of univariate statistical analyses carried out separately for each of the variables is, in general, not adequate as it ignores the correlations among the variables. It may even be misleading sometimes. On the contrary, multivariate analysis can throw light on the relationships, interdependence, and relative importance of the characteristics involved and yield more meaningful information, in general.

The theory of multivariate analysis developed so far almost invariably assumes that the joint distribution of the random variables is a multivariate normal distribution. There are other multivariate distributions too, such as the multinomial, multivariate gamma, multivariate beta, etc., but the mathematics based on the assumption of a multinormal distribution is tractable and further the analysis runs almost parallel to the corresponding univariate analysis based on a normal distribution. The similarity becomes more marked if matrix notation is used. The distribution of linear functions of normal variables is normal, and so most of the theory of multivariate analysis is based on linear transformations of original variables. The aim of the statistician undertaking multivariate analysis is to reduce the number of variables by employing suitable linear transformations and to choose a very limited number of the resulting linear combinations in some optimum manner, disregarding the remaining linear combinations in the hope that they do not contain much significant information. The statistician thus reduces the dimensionality of the problem.

The variables are assumed to be correlated, but the observations on different individuals are assumed to be independent. This is what is meant by a random sample of individuals. Analysis of data, where this assumption is not satisfied, is also possible and occurs in multivariate time series and stochastic processes, but this type of analysis will not be considered in this book. Similarly, nonparametric multivariate methods, i.e., methods which do not make use of the assumption of any particular distribution of the variables under consideration, will not be considered in this book. Occassionally, analysis of qualitative multivariate data is considered in the book, but for the most part, we shall be dealing with the analysis of independent observations on several variables having a joint normal distribution only. This may sound very restrictive but the methods and techniques developed under these assumptions are often found to be satisfactory for all practical purposes, and many times these techniques are quite robust and thus useful even when the underlying assumptions are not strictly satisfied.

Just to give an idea of the types of multivariate problems discussed in the statistical literature, we shall briefly mention a few of them. Determining the sex, from various measurements on a skull, is a very commonly occurring problem in anthropology. Whether time alone is useful in accounting for the differences among four series of Egyptian skulls was a problem considered by Barnard [1]. Diagnosis of a disease based on body temperature, blood pressure, and other medical factors involves multivariate analysis. The decision whether or not to admit a candidate to a professional school, such as medical, engineering, or architectural, on the basis of marks obtained in different subjects, can be made with the help of multivariate analysis. Assignment of optimum scores and representation on a linear scale of blood serological reactions was achieved by Fisher [5] and Bartlett [3] by using multivariate methods. A study of the relationship among various economic factors such as consumption, price, cost of production, and income of consumers was carried out by Bartlett [2] to derive a "demand" relation, with the help of canonical variables in multivariate analysis. The problem of identifying the neurotic state of an individual subjected to psychiatric examination is described by Rao [6]. Discrimination between two species of flowers, growth curves, determination of common and specific factors, determining the authorship of a literary work are some more problems discussed in the statistical literature with the help of multivariate methods.

Discriminant analysis, regression, multivariate analysis of variance and covariance, canonical variables, and principal components analysis are some of the important multivariate statistical techniques, and these will be considered in this book. However, regression is no doubt the most important technique in statistics, and all other techniques, in one way or another, are derivable from regression analysis. Regression analysis thus plays an important central role throughout. We shall, therefore, introduce this technique in this chapter.

2. MULTIVARIATE DISTRIBUTIONS

As in multivariate analysis one is interested in the relationship among several variates simultaneously, one has to consider their joint distribution and the various marginal and conditional distributions that can be obtained from it. Let $\underline{x}$ denote the column vector of p variates $x_1, x_2, \ldots, x_p$ having a continuous distribution, and let $f(x_1, x_2, \ldots, x_p)$ be the probability density function (p.d.f.) of this joint distribution. Then the p.d.f. of the marginal distribution of one or more of the x_i's, say of $x_1, x_2, \ldots, x_k$ $(k < p)$, can be obtained by integrating out the remaining variables. Thus it is

$$f_1(x_1, \ldots, x_k) = \int f(x_1, x_2, \ldots, x_p)\, dx_{k+1}\, dx_{k+2} \ldots dx_p, \tag{2.1}$$

where a single integral sign is used to denote the multiple integral and the integration is over the appropriate range of $x_{k+1}, \ldots, x_p$. The p.d.f. of the conditional distribution of $x_{k+1}, \ldots, x_p$, when $x_1, \ldots, x_k$ are fixed, will be denoted by $f_2(x_{k+1}, \ldots, x_p | x_1, \ldots, x_k)$ and is $f(x_1, \ldots, x_p)/f_1(x_1, \ldots, x_k)$. If the joint distribution is a discrete one, we shall be dealing with the joint probability function, and integration will be replaced by summation over the relevant variables, while finding the marginal distribution. We shall use E to denote the expectation operator, so that $E(\underline{x})$ will be a column vector with elements $E(x_i)$, $(i = 1, 2, \ldots, p)$. The symbols V and Cov will be used to denote variance and covariance, respectively. The symmetric matrix

$$\begin{bmatrix} V(x_1) & Cov(x_1, x_2) & \cdots & Cov(x_1, x_p) \\ Cov(x_2, x_1) & V(x_2) & \cdots & Cov(x_2, x_p) \\ \vdots & & & \\ Cov(x_p, x_1) & Cov(x_p, x_2) & \cdots & V(x_p) \end{bmatrix} \tag{2.2}$$

is known as the variance-covariance matrix of the vector $\underline{x}$ and shall be denoted by $V(\underline{x})$. From the definition of variances and covariances, it is obvious that

$$V(\underline{x}) = E\left\{\underline{x} - E(\underline{x})\right\}\left\{\underline{x} - E(\underline{x})\right\}'$$

$$= E(\underline{x}\,\underline{x}') - E(\underline{x})\,E(\underline{x}') \,. \tag{2.3}$$

If $\underline{x}$ and $\underline{y}$ are two vector variates of p and q components, respectively, the matrix

$$\begin{bmatrix} \mathrm{Cov}(x_1, y_1) & \mathrm{Cov}(x_1, y_2) & \cdots & \mathrm{Cov}(x_1, y_q) \\ \mathrm{Cov}(x_2, y_1) & \mathrm{Cov}(x_2, y_2) & \cdots & \mathrm{Cov}(x_2, y_q) \\ \vdots & & & \\ \mathrm{Cov}(x_p, y_1) & \mathrm{Cov}(x_p, y_2) & \cdots & \mathrm{Cov}(x_p, y_q) \end{bmatrix}, \tag{2.4}$$

where x_i is the ith component of $\underline{x}$ and y_j is the jth component of $\underline{y}$, is known as the covariance matrix of $\underline{x}$ and $\underline{y}$, and will be denoted by $\mathrm{Cov}(\underline{x}, \underline{y})$. It is not symmetric and $\mathrm{Cov}(\underline{y}, \underline{x})$ will be the transpose of $\mathrm{Cov}(\underline{x}, \underline{y})$. A transpose will be denoted by a prime. If A is some m x p matrix of constants, the variance-covariance matrix of the transformed variates $A\underline{x}$ is, by definition,

$$E\left\{A\underline{x} - E(A\underline{x})\right\}\left\{A\underline{x} - E(A\underline{x})\right\}'$$

and reduces to $AV(\underline{x})A'$. Similarly, the covariance matrix of $A\underline{x}$ and $B\underline{y}$, where B is some n x q matrix, is $A\,\mathrm{Cov}(\underline{x}, \underline{y})B'$.

$E(\underline{x}|\underline{z})$, $V(\underline{x}|\underline{z})$, $Cov(\underline{x}, \underline{y}|\underline{z})$ will denote, respectively, the conditional expectation of $\underline{x}$, conditional variance-covariance matrix of $\underline{x}$, and the conditional covariance matrix of $\underline{x}$, $\underline{y}$ when the variates $\underline{z}$ are fixed.

If the correlation coefficient between x_i and x_j is denoted by $\rho_{ij}(i, j = 1, 2, \ldots, p)$ and if ρ_{ii} is defined to be unity, the matrix $P = [\rho_{ij}]$ $(i, j = 1, 2, \ldots, p)$ is known as the correlation matrix of $\underline{x}$, and from (2.2) it follows that

$$V(\underline{x}) = DPD \tag{2.5}$$

where

$$D = \text{diag}\left\{[V(x_1)]^{1/2}, [V(x_2)]^{1/2}, \ldots, [V(x_p)]^{1/2}\right\}$$

and diag stands for a diagonal matrix, the elements of which are written in the adjoining bracket.

If X is a p x n matrix of variates x_{ir} $(i = 1, 2, \ldots, p;$ $r = 1, 2, \ldots, n)$, we shall use the symbol V(X) to denote the pn x pn variance-covariance matrix of the column vector, whose elements are $x_{11}, x_{12}, x_{13}, \ldots, x_{1n}, x_{21}, x_{22}, \ldots, x_{pn}$, i.e., they are the elements of the matrix X, taken row by row and written as a column vector. In many cases, this matrix V(X) is of the form

$$\begin{bmatrix} a_{11}B & a_{12}B & & a_{1p}B \\ a_{21}B & a_{22}B & & a_{2p}B \\ & & & \\ a_{p1}B & a_{p2}B & & a_{pp}B \end{bmatrix} \tag{2.6}$$

where $A = [a_{ij}]$ and $B = [b_{rs}]$ are, respectively, p x p and n x n matrices. The matrix (2.6) is known as the Kronecker product of A and B and will be denoted by $A \otimes B$. It is then easy to demonstrate that

$$V(PXQ') = (PAP') \otimes (QBQ') , \tag{2.7}$$

where P and Q are any two matrices of constants of order $\alpha \times p$ and $\beta \times n$ respectively.

The characteristic function $\phi(t)$ of the distribution of $\underline{x}$ is defined by

$$\phi(\underline{t}) = E(e^{\sqrt{-1}\ \underline{t}'\underline{x}}) ,$$

where $\underline{t}$ is the column vector of the real elements $t_1, t_2, \ldots, t_p$. The cumulant generating function of $\underline{x}$ is $\log_e \phi(\underline{t})$, and either this or the characteristic function $\phi(\underline{t})$ can be used for obtaining moments and product moments of the x_i's, by differentiating partially with respect to the appropriate t_i's an appropriate number of times and then setting $\underline{t} = \underline{0}$. Thus, for example, for $i \neq j$,

$$E(x_i x_j) = \left[\frac{-\partial^2 \log_e \phi(t)}{\partial t_i\ \partial t_j} \right]_{\underline{t} = \underline{0}} .$$

When $\underline{x}$ has a p.d.f., it is given by

$$\frac{1}{(2\pi)^p} \int_{-\infty}^{\infty} e^{-\sqrt{-1}\ \underline{t}'\underline{x}}\ \phi(\underline{t})\ dt_1\ dt_2\ \ldots\ dt_p . \tag{2.8}$$

This is the multivariate version of the inversion theorem for characteristic functions [4].

3. REGRESSION, PARTIAL AND MULTIPLE CORRELATION

Let us assume, for the sake of convenience, that $E(\underline{x}) = 0$, and also set $V(\underline{x}) = [\sigma_{ij}] = \Sigma$. The mean of the conditional distribution of x_1, when $x_2, x_3, \ldots, x_p$ are fixed, is known as the regression of x_1 on $x_2, x_3, \ldots, x_p$. When we consider this regression, x_1 is known as the dependent variable and $x_2, x_3, \ldots, x_p$ are called the independent variables. This regression $E(x_1 | x_2, x_3, \ldots, x_p)$ is used to predict x_1 from $x_2, x_3, \ldots, x_p$, because, among all functions of $x_2, \ldots, x_p$, it gives the closest fit to x_1 in the least squares sense. In other words, the minimum value of

$$u = E\left\{x_1 - g(x_2, x_3, \ldots, x_p)\right\}^2$$

over all possible functions g occurs when

$$g(x_2, \ldots, x_p) = E(x_1 | x_2, \ldots, x_p) . \tag{3.1}$$

This can be seen by writing u as

$$E\left\{E\left[\left(x_1 - g(x_2, \ldots, x_p)\right)^2 | x_2, x_3, \ldots, x_p\right]\right\}. \tag{3.2}$$

From a well-known result for univariate distributions, it then follows that the quantity inside the braces in (3.2) is minimum when (3.1) holds, and thus the result follows. However, in many practical situations, we do not try to seek the minimum of u from among all functions of $x_2, x_3, \ldots, x_p$ but from only linear functions of $x_2, x_3, \ldots, x_p$. If so, we minimize

$$v = E\left\{x_1 - \beta_2 x_2 - \cdots - \beta_p x_p\right\}^2 \qquad (3.3)$$

with respect to β_2, β_3, ..., β_p. This yields the equation

$$\underline{\sigma} = \Sigma_2 \underline{\beta} \, , \qquad (3.4)$$

where

$$\Sigma = \begin{bmatrix} \sigma_{11} & \underline{\sigma}' \\ \underline{\sigma} & \Sigma_2 \end{bmatrix} \begin{matrix} 1 \\ p\text{-}1 \end{matrix} \, , \quad \underline{\beta}' = [\beta_2, \, \cdots, \, \beta_p], \quad \underline{x}^* = \begin{bmatrix} x_2 \\ x_3 \\ \vdots \\ x_p \end{bmatrix} .$$
$$\begin{matrix} 1 & \ \ p\text{-}1 \end{matrix}$$
$$(3.5)$$

We assume Σ to be nonsingular, and so the solution of (3.4) is

$$\underline{\beta} = \Sigma_2^{-1} \, \underline{\sigma} \, , \qquad (3.6)$$

and the linear function that minimizes (3.3) is $\underline{\sigma}' \Sigma_2^{-1} \underline{x}^*$, which is called the mean-square regression line of x_1 on $x_2, x_3, \ldots, x_p$. The β's are called the partial regression coefficients. It should be noted here that this mean-square regression line and the regression line (3.1) will be the same if $E(x_1 | x_2, \, \ldots, \, x_p)$ is a linear function of x_2, x_3, ..., x_p. We often have to consider several mean-square regression lines simultaneously; it then becomes necessary to specify explicitly which are the independent variables, which are the dependent variables, and which regression coefficients are attached to which variables. If so, the above regression line can best be written as

$$\beta_{12 \cdot 34 \ldots p} x_2 + \beta_{13 \cdot 24 \ldots p} x_3 + \cdots + \beta_{1p \cdot 23 \ldots p\text{-}1} x_p \, ,$$

i.e., each regression coefficient β has two groups of subscripts, separated by a dot. The subscripts before the dot are known as the primary subscripts; they are only two, the first one of which denotes the dependent variable and the second denotes the independent variable to which it is attached. The secondary subscripts, those after the dot, simply indicate the remaining independent variables in the regression; their order is immaterial and they are even omitted, if no confusion is likely to arise. This dot notation is due to Yule [2].

For a given $\underline{x}^*$, the predicted value of x_1 is $\underline{\sigma}'\Sigma_2^{-1}\underline{x}^*$, and so the difference $x_1 - \underline{\sigma}'\Sigma_2^{-1}\underline{x}^*$ is called the residual of x_1 from its mean-square regression line on $x_2, \ldots, x_p$. This is denoted by $x_{1 \cdot 23 \ldots p}$ in an obvious dot notation. The variance of $x_{1 \cdot 23 \ldots p}$, denoted by $\sigma_{11 \cdot 23 \ldots p}$, is [remembering $E(\underline{x}) = 0$ and $E(\underline{x}^*\underline{x}^{*\prime}) = V(\underline{x}^*) = \Sigma_2$]

$$\begin{aligned} V(x_1 &- \underline{\sigma}'\Sigma_2^{-1}\underline{x}^*) \\ &= E(x_1 - \underline{\sigma}'\Sigma_2^{-1}\underline{x}^*)(x_1 - \underline{\sigma}'\Sigma_2^{-1}\underline{x}^*)' \\ &= V(x_1) - \underline{\sigma}'\Sigma_2^{-1}\,\mathrm{Cov}(\underline{x}^*, x_1) - \mathrm{Cov}(x_1, \underline{x}^*)\Sigma_2^{-1}\underline{\sigma} \\ &\quad + \underline{\sigma}'\Sigma_2^{-1}\,V(\underline{x}^*)\Sigma_2^{-1}\underline{\sigma} \\ &= \sigma_{11} - \underline{\sigma}'\Sigma_2^{-1}\underline{\sigma}, \quad \text{or,} \quad \sigma_{11} - \underline{\sigma}'\underline{\beta} \end{aligned} \tag{3.7}$$

where $\underline{\beta}$ is given by (3.6). It is an important property of the residual $x_{1 \cdot 23 \ldots p}$ that it is uncorrelated with any of the independent variables $x_2, x_3, \ldots, x_p$. This follows from

$$E\left[x_{1 \cdot 23 \ldots p} \underline{x}^{*\prime}\right] = E\left[(x_1 - \underline{\sigma}'\Sigma_2^{-1}\underline{x}^*)\underline{x}^{*\prime}\right]$$
$$= \underline{\sigma}' - \underline{\sigma}' = \underline{0} \ .$$

The mean-square regression line $\sigma'\Sigma_2^{-1}\underline{x}^*$ has another optimum property: among all linear functions of $\underline{x}^*$ it has the maximum correlation with x_1. This can be shown by writing down explicitly the correlation coefficient between x_1 and a linear function $\underline{\beta}'\underline{x}^*$ and maximizing it with respect to $\underline{\beta}$. This again leads to virtually the same solution (3.6), and the result follows. This maximum value of the correlation coefficient or, in other words, the correlation coefficient between x_1 and $\underline{\sigma}'\Sigma_2^{-1}\underline{x}^*$, is known as the multiple correlation coefficient between x_1 and $x_2, x_3, \ldots, x_p$. It is denoted by $\rho_{1(23 \ldots p)}$ and it can be expressed in terms of the elements of Σ as below:

$$V(\underline{\sigma}'\Sigma_2^{-1}\underline{x}^*) = \underline{\sigma}'\Sigma_2^{-1}\underline{\sigma} \ ,$$
$$\mathrm{Cov}(x_1, \underline{\sigma}'\Sigma_2^{-1}\underline{x}^*) = \underline{\sigma}'\Sigma_2^{-1}\sigma \ ,$$

and hence

$$\rho^2_{1(23 \ldots p)} = \left\{\mathrm{Cov}(x_1, \underline{\sigma}'\Sigma_2^{-1}x^*)\right\}^2 / \left\{V(x_1) \ V(\sigma'\Sigma_2^{-1}\underline{x}^*)\right\}$$
$$= \underline{\sigma}'\Sigma_2^{-1}\underline{\sigma}/\sigma_{11} \ . \qquad (3.8)$$

The multiple correlation coefficient $\rho_{1(23 \ldots p)}$ is defined as the positive value of the square root of the above expression, and hence $0 \leq \rho_{1(23 \ldots p)} \leq 1$ and not $-1 \leq \rho_{1(23 \ldots p)} \leq 1$, like other correlation coefficients.

Consider now the mean-square regression lines of x_1 on $x_3, x_4, \ldots, x_p$ and x_2 on $x_3, x_4, \ldots, x_p$ simultaneously. They are, from (3.6), $\underline{\sigma}'_{(i)}\Sigma_3^{-1}\underline{x}^{**}$, $(i = 1, 2)$, respectively, where

$$\underline{x}^{**} = \begin{bmatrix} x_3 \\ x_4 \\ \vdots \\ x_p \end{bmatrix} \quad \text{and} \quad \Sigma = \left[\begin{array}{cc|c} \sigma_{11} & \sigma_{12} & \underline{\sigma}'_{(1)} \\ \sigma_{21} & \sigma_{22} & \underline{\sigma}'_{(2)} \\ \hline \underline{\sigma}_{(1)} & \underline{\sigma}_{(2)} & \Sigma_3 \end{array}\right] \begin{matrix} 1 \\ 1 \\ p-2 \end{matrix} \; . \tag{3.9}$$
$$\qquad\qquad\qquad\qquad\qquad 1 \qquad 1 \qquad p-2$$

The residuals of x_1 and x_2 from these regression lines are

$$\begin{aligned} x_{1\cdot 34\ldots p} &= x_1 - \underline{\sigma}'_{(1)}\Sigma_3^{-1}\underline{x}^{**} , \\ x_{2\cdot 34\ldots p} &= x_2 - \underline{\sigma}'_{(2)}\Sigma_3^{-1}\underline{x}^{**} . \end{aligned} \tag{3.10}$$

The correlation coefficient between these residuals is known as the partial correlation coefficient between x_1 and x_2, when $x_3, x_4, \ldots, x_p$ are eliminated. It will be denoted by $\rho_{12\cdot 34\ldots p}$. From (3.9) and (3.10), after evaluating the covariance between $x_{1\cdot 34\ldots p}$ and $x_{2\cdot 34\ldots p}$ and using (3.7), we obtain

$$\rho_{12\cdot 34\ldots p} = \frac{\sigma_{12} - \underline{\sigma}'_{(1)}\Sigma_3^{-1}\underline{\sigma}_{(2)}}{\left(\sigma_{11} - \underline{\sigma}'_{(1)}\Sigma_3^{-1}\underline{\sigma}_{(1)}\right)^{1/2}\left(\sigma_{22} - \underline{\sigma}'_{(2)}\Sigma_3^{-1}\underline{\sigma}_{(2)}\right)^{1/2}} . \tag{3.11}$$

All these expressions can, alternatively, be expressed in different and sometimes simpler forms, if we use the elements σ^{ij} $(i, j = 1, 2, \ldots, p)$ of the matrix Σ^{-1}. For this, and some future applications, we need some results about matrices, which we give below.

Let the partitioned matrix

$$\left[\begin{array}{c|c} P & Q \\ \hline R & S \end{array}\right] \tag{3.12}$$

be the inverse of the non-singular matrix

$$\left[\begin{array}{c|c} A & B \\ \hline C & D \end{array}\right] \tag{3.13}$$

which is also partitioned in exactly the same manner. Then it is well known that the determinant

$$\left|\begin{array}{c|c} A & B \\ \hline C & D \end{array}\right| = |A|\ |D - CA^{-1}B| \quad \text{or} \quad |D|\ |A - BD^{-1}C| . \tag{3.14}$$

Multiplying (3.12) by (3.13) and equating to the identity matrix, we get four matrix equations in P, Q, R, S, from which it can be shown, by a little algebra, that $RP^{-1}Q = -D^{-1}CQ$ and $S = D^{-1} - D^{-1}CQ$. From these and by applying (3.14) to (3.12), it follows that

$$\frac{|P|}{\left|\begin{array}{c|c} P & Q \\ \hline R & S \end{array}\right|} = |D| . \tag{3.15}$$

By solving the four equations in P, Q, R, S, it can be seen that

$$\begin{aligned} P &= (A - BD^{-1}C)^{-1} = A^{-1} + A^{-1}B(D - CA^{-1}B)^{-1}CA^{-1} \\ Q &= -A^{-1}B(D - CA^{-1}B)^{-1} \\ R &= -D^{-1}C(A - BD^{-1}C)^{-1} \\ S &= (D - CA^{-1}B)^{-1} . \end{aligned} \tag{3.16}$$

By applying (3.14) to Σ, as given by (3.5), we obtain

$$\sigma_{11 \cdot 23 \ldots p} = \frac{|\Sigma|}{|\Sigma_2|} = \frac{1}{\sigma^{11}} \tag{3.17}$$

and

$$\rho^2_{1(23 \ldots p)} = 1 - \frac{|\Sigma|}{\sigma_{11}|\Sigma_2|} = 1 - \frac{1}{\sigma_{11}\sigma^{11}} \,. \tag{3.18}$$

Again, applying (3.14) to the cofactor of σ_{12} in Σ, given by (3.8), we get

$$-\sigma^{12} = |\Sigma_3|\left(\sigma_{12} - \underset{\sim}{\sigma}'_{(1)}\Sigma_3^{-1}\underset{\sim}{\sigma}_{(2)}\right)/|\Sigma| \,, \tag{3.19}$$

and similarly,

$$\sigma^{11} = |\Sigma_3|\left(\sigma_{22} - \underset{\sim}{\sigma}'_{(2)}\Sigma_3^{-1}\underset{\sim}{\sigma}_{(2)}\right)/|\Sigma| \,, \tag{3.20}$$

$$\sigma^{22} = |\Sigma_3|\left(\sigma_{11} - \underset{\sim}{\sigma}'_{(1)}\Sigma_3^{-1}\sigma_{(1)}\right)/|\Sigma| \,. \tag{3.21}$$

Using these results in (3.11), we get

$$\rho_{12 \cdot 34 \ldots p} = -\sigma^{12}/\sqrt{\sigma^{11}\sigma^{22}} \,. \tag{3.22}$$

From (3.7), (3.9) and (3.21), we have

$$\frac{\sigma_{11 \cdot 23 \ldots p}}{\sigma_{11 \cdot 34 \ldots p}} = \frac{1/\sigma^{11}}{\sigma^{22}|\Sigma|/|\Sigma_3|} \,. \tag{3.23}$$

But on using (3.15) for Σ^{-1}, we have

$$\begin{vmatrix} \sigma^{11} & \sigma^{12} \\ \sigma^{21} & \sigma^{22} \end{vmatrix} / |\Sigma^{-1}| = |\Sigma_3| \,. \tag{3.24}$$

Using (3.24) and (3.22) in (3.23), one obtains

$$\frac{\sigma_{11 \cdot 23 \ldots p}}{\sigma_{11 \cdot 34 \ldots p}} = 1 - \rho^2_{12 \cdot 34 \ldots p} \,.$$

A repeated application of this result yields,

$$1 - \rho^2_{1(23\ldots p)} = \frac{\sigma_{11\cdot 23\ldots p}}{\sigma_{11}}$$
$$= (1 - \rho^2_{12\cdot 34\ldots p})(1 - \rho^2_{13\cdot 45\ldots p}) \cdots$$
$$\cdots (1 - \rho^2_{1(p-1)\cdot p})(1 - \rho^2_{1p}) \ . \qquad (3.25)$$

In all this theory, we have assumed that the true variance-covariance matrix Σ of $\underline{x}$ is known. But in practical situations, we rarely know it, and only an unbiased estimate of it is available, from a sample of n observations from the distribution of $\underline{x}$. A matrix X of order p x n and elements x_{ir} ($i = 1, 2, \ldots, p$; $r = 1, 2, \ldots, n$) is said to be a sample of size n from the p-variate distribution of $\underline{x}$ if all the n columns of X are independently and identically distributed as $\underline{x}$. The quantities

$$\bar{x}_i = \sum_{r=1}^{n} x_{ir}/n, \quad s_{ii} = \sum_{r=1}^{n} (x_{ir} - \bar{x}_i)^2, \quad s_{ij} = \sum_{r=1}^{n} (x_{ir} - \bar{x}_i)(x_{jr} - \bar{x}_j) \qquad (3.26)$$

are known as the sample mean of x_i, the sample corrected sum of squares of x_i, and the sample corrected sum of products of x_i and x_j, respectively ($i, j = 1, 2, \ldots, p$). The vector

$$\underline{\bar{x}} = \frac{1}{n} X E_{n1} , \qquad (3.27)$$

where E_{ab} denotes an a x b matrix with all elements unity, is called the vector of sample means of $\underline{x}$, and the matrix

$$S = [s_{ij}] = X(I - \frac{1}{n} E_{nn})X' \qquad (3.28)$$

is called the corrected sum of squares and sum of products (abbreviated as s.s. and s.p. matrix) matrix of sample observations on $\underline{x}$. It can be shown that $[1/(n-1)]S$ is an unbiased estimate of Σ, and $\bar{\underline{x}}$ of the true means of $\underline{x}$. Hence all the previous formulas about regression, residual, partial, and multiple correlation can be rewritten, with σ_{ij} replaced by $[1/(n-1)]s_{ij}$ $(i, j = 1, \ldots, p)$ and x_i by $x_i - \bar{x}_i$ $(i = 1, \ldots, p)$, to yield the corresponding sample quantities. Thus, for example, the sample residual of x_1 with respect to $x_2, x_3, \ldots, x_p$ is

$$(x_1 - \bar{x}_1) - b_2(x_2 - \bar{x}_2) - \ldots - b_p(x_p - \bar{x}_p) , \quad (3.29)$$

where $\underline{b}$, the column vector of $b_2, b_3, \ldots, b_p$, the sample regression coefficients, is given by $S_2^{-1}\underline{s}$, and S_2, $\underline{s}$ are defined in relation to S in the same way as Σ_2 and $\underline{\sigma}$ are for Σ. The square of the sample multiple correlation coefficient between x_1 and $x_2, \ldots, x_p$ will be

$$1 - \frac{1}{s_{11}s^{11}} , \quad (3.30)$$

where s^{ij} are the elements of S^{-1}, and so on.

In the next section, we shall illustrate the concepts introduced in the foregoing sections, with the help of a particular distribution. Since, throughout this book, we shall be dealing mostly with the multivariate normal distribution, which is a continuous distribution, here, for a change, we shall choose the multinomial distribution, which is a discrete multivariate distribution.

4. THE MULTINOMIAL DISTRIBUTION

The probability function of a multinomial distribution is

$$u(n_1, n_2, \ldots, n_k) = \frac{n!}{\prod_{i=1}^{k} n_i!} \prod_{i=1}^{k} p_i^{n_i} , \qquad (4.1)$$

where $n_1, n_2, \ldots, n_k$ are the random variables, which take the values 0, 1, 2, ..., but are subject to the constraint $\sum_1^k n_i = n$, which is a fixed number. The p_i's are all in the range (0, 1) and $\sum_{i=1}^{k} p_i = 1$. The p_i's and n are thus the parameters of this distribution. The characteristic function of this distribution is

$$\phi(t_1, t_2, \ldots, t_k) = \Sigma\, e^{\sqrt{-1}(t_1 n_1 + \ldots + t_k n_k)}\, u(n_1, n_2, \ldots, n_k) , \qquad (4.2)$$

where the summation is over all the possible values of $n_1, n_2, \ldots, n_k$. The expression reduces to $\left(\sum_1^p p_i\, e^{\sqrt{-1}\, t_i}\right)^n$ on summation. By differentiating $\phi(t_1, t_2, \ldots, t_k)$ an appropriate number of times, we find

$$E(n_i) = np_i, \quad V(n_i) = np_i(1 - p_i), \quad Cov(n_i, n_j) = -np_i p_j \; (i \neq j)$$

$$\text{Correlation coefficient between } n_i, n_j = -\left\{p_i p_j / (1-p_i)(1-p_j)\right\}^{1/2} . \qquad (4.3)$$

The variance-covariance matrix of $n_1, n_2, \ldots, n_k$ is, therefore,

$$n \operatorname{diag}(p_1, p_2, \ldots, p_k) - n\, \underline{p}\, \underline{p}' , \qquad (4.4)$$

where $\underline{p}$ is the column vector of the p_i's. The determinant of this variance-covariance matrix is easily seen to be

$$n^k\, p_1 p_2 \ldots p_k \left| I - \mathrm{diag}\left(\frac{1}{p_1}, \frac{1}{p_2}, \ldots, \frac{1}{p_k}\right)\underline{p}\,\underline{p}' \right| ,$$

and on using the well-known formula

$$|I - LM| = |I - ML| \tag{4.5}$$

(subject to the condition that the product ML is possible), the determinant reduces to $n^k\, p_1 p_2 \ldots p_k \left(1 - \underline{p}'\, \mathrm{diag}(\frac{1}{p_1}, \ldots, \frac{1}{p_k})\underline{p}\right)$, which is zero, as $\sum_1^k p_i = 1$. The variance-covariance matrix is thus singular. The distribution itself is sometimes called singular in such a case. This is due to the fact that $n_1, n_2, \ldots, n_k$ are constrained by $\sum_1^k n_i = n$, and in reality, therefore, are only k-1 variates, as any one can be expressed in terms of the others. The marginal distribution of $n_1, n_2, \ldots, n_m$ $(m < k)$ can be obtained from (4.1) by summing with respect to $n_{m+1}, \ldots, n_k$ and yields

$$u_1(n_1, n_2, \ldots, n_m) = \sum_{n_{m+1}, \ldots, n_k} u(n_1, n_2, \ldots, n_k)$$

$$= \frac{n!}{n_1!\, n_2! \ldots n_m!\,(n-n_o)!} \left(\prod_{i=1}^{m} p_i^{n_i}\right)$$

$$\times\ (1 - p_1 - p_2 - \ldots - p_m)^{n-n_o} \cdot g\ , \tag{4.6}$$

where

$$n_o = n_1 + n_2 + \ldots + n_m \tag{4.7}$$

and

$$g = \sum_{n_{m+1}, \ldots, n_k} \frac{(n - n_o)!}{n_{m+1}!\, n_{m+2}! \ldots n_k!}\, p_{m+1}^{n_{m+1}} p_{m+2}^{n_{m+2}} \ldots p_k^{n_k} (1-p_1-\ldots-p_m)^{-(n-n_o)}$$

$$= (p_{m+1} + \ldots + p_k)^{n-n_o}/(1 - p_1 - \ldots - p_m)^{n-n_o}$$

$$= 1\ . \tag{4.8}$$

The marginal distribution of $n_1, n_2, \ldots, n_m$ is thus again a multinomial distribution. The frequency function of the conditional distribution of n_1, when $n_3, n_4, \ldots, n_m$ are fixed, is thus, on using (4.6) to (4.8),

$$u_2(n_1|n_3, n_4, \ldots, n_m) = \frac{\text{probability function of } (n_1, n_3, \ldots, n_m)}{\text{probability function of } (n_3, n_4, \ldots, n_m)}$$

$$= \frac{(n-n_3-n_4-\ldots-n_m)!p_1^{n_1}(1-p_1-p_3-\ldots-p_m)^{n-n_1-n_3-\ldots-n_m}}{n_1!(n-n_1-n_3-\ldots-n_m)!(1-p_3-\ldots-p_m)^{n-n_3-\ldots-n_m}} . \qquad (4.9)$$

This is also a multinomial distribution. On comparing (4.9) with (4.1) and using (4.3), we find

$$E(n_1|n_3, n_4, \ldots, n_m) = \frac{(n-n_3-n_4-\ldots-n_m)p_1}{(1-p_3-p_4-\ldots-p_m)} , \qquad (4.10)$$

showing that the regression of n_1 on $n_3, n_4, \ldots, n_m$ is linear and thus coincides with the mean-square regression line. The residual of n_1 from this regression line is

$$n_{1 \cdot 34 \ldots m} = n_1 - E(n_1|n_3, n_4, \ldots, n_m) . \qquad (4.11)$$

Similarly,

$$E(n_2|n_3, n_4, \ldots, n_m) = \frac{(n-n_3-n_4-\ldots-n_m)p_2}{(1-p_3-p_4-\ldots-p_m)} \qquad (4.12)$$

and

$$n_{2 \cdot 34 \ldots m} = n_2 - E(n_2|n_3, n_4, \ldots, n_m) . \qquad (4.13)$$

Using (4.3) to evaluate the correlation coefficient between (4.11) and (4.13), we find the partial correlation coefficient between n_1 and n_2, when $n_3, n_4, \ldots, n_m$ are fixed, to be

$$\rho_{12\cdot 34\ldots m} = -\left\{\frac{p_1 p_2}{(1-p_1-p_3-\ldots-p_m)(1-p_2-p_3-\ldots-p_m)}\right\}^{1/2} . \quad (4.14)$$

This could also have been derived by applying (3.22) directly to the variance-covariance matrix,

$$\Sigma = n\ \mathrm{diag}(p_1, p_2, \ldots, p_m) - n\ \underline{p}_o\ \underline{p}_o' , \quad (4.15)$$

of $n_1, n_2, \ldots, n_m$, where $\underline{p}_o$ is the column vector of $p_1, p_2, \ldots, p_m$. Observe that

$$\Sigma^{-1} = \frac{1}{n}\left(I - \mathrm{diag}(\frac{1}{p_1}, \ldots, \frac{1}{p_m})\underline{p}_o\underline{p}_o'\right)^{-1} \mathrm{diag}(\frac{1}{p_1}, \frac{1}{p_2}, \ldots, \frac{1}{p_m}) .$$

Now use the formula

$$(I + LM)^{-1} = I - L(I + ML)^{-1} M \quad (4.16)$$

to invert Σ. Taking L to be $\mathrm{diag}(\frac{1}{p_1}, \ldots, \frac{1}{p_m})\underline{p}_o$, we readily obtain

$$\Sigma^{-1} = \frac{1}{n}\ \mathrm{diag}(\frac{1}{p_1}, \ldots, \frac{1}{p_m}) + \frac{1}{n(1-p_1-p_2-\ldots-p_m)} E_{mm} . \quad (4.17)$$

This shows that $(i, j = 1, 2, \ldots, m, i \neq j)$

$$\sigma^{ii} = \frac{1}{np_i} + \frac{1}{n(1-p_1-p_2-\ldots-p_m)}$$

and

$$\sigma^{ij} = \frac{1}{n(1-p_1-p_2-\ldots-p_m)} .$$

Using these in (3.22), (4.14) follows. Similarly, using (3.18), the square of the multiple correlation coefficient of n_1 with $n_2, \ldots, n_m$ is seen to be

$$\rho^2_{1(23\ldots p)} = \frac{p_1(p_2 + p_3 + \ldots + p_m)}{(1 - p_1)(1 - p_2-\ldots-p_m)} . \quad (4.18)$$

REFERENCES

[1] Barnard, M. M. (1935). "The secular variations of skull characters in four series of Egyptian skulls", Ann. Eugen., 6 p. 352.

[2] Bartlett, M. S. (1948). "A note on the statistical estimation of demand and supply relations from time series", Econometrica, 16 p. 323.

[3] Bartlett, M. S. (1951). "The goodness of fit of a single hypothetical discriminant function in the case of several groups", Ann. Eugen., 16 p. 199.

[4] Cramer, H. (1946). *Mathematical Methods of Statistics.* Princeton University Press, Princeton.

[5] Fisher, R. A. (1946). *Statistical Methods for Research Workers.* 10th edition, Oliver and Boyd, Edinburgh.

[6] Rao, C. Radhakrishna (1965). *Linear Statistical Inference and its Applications.* John Wiley and Sons, New York.

[7] Yule, G. V. (1907). "On the theory of correlation for any number of variables treated by a new system of notation", Proc. Roy. Soc. London A, 79 p. 182.

CHAPTER 2

MULTIVARIATE NORMAL DISTRIBUTION

1. MULTIVARIATE NORMAL DISTRIBUTION

A (column) vector $\underline{x}$ of p components $x_1, x_2, \ldots, x_p$ is said to have a p-variate nonsingular normal distribution if its p.d.f. is

$$(2\pi)^{-p/2} \; |\Sigma|^{-1/2} \; \exp\left\{-(1/2)(\underline{x} - \underline{\mu})' \; \Sigma^{-1}(\underline{x} - \underline{\mu})\right\} \qquad (1.1)$$

$$-\infty < x_i < \infty, \; (i = 1, 2, \ldots, p) .$$

$\underline{\mu}$ and Σ are the parameters of this distribution; $\underline{\mu}$ is a column vector of elements μ_i $(i = 1, 2, \ldots, p)$ such that $-\infty < \mu_i < \infty$ and $\Sigma = [\sigma_{ij}]$ is a positive definite symmetric matrix of order p. The p.d.f. (1.1) will be denoted by $N_p(\underline{x}|\underline{\mu}|\Sigma)$ and the notation $\underline{x} \sim N_p(\underline{x}|\underline{\mu}|\Sigma)$ will be used to indicate that the variates $\underline{x}$ have a p-variate nonsingular normal distribution with parameters $\underline{\mu}$ and Σ. When Σ is a diagonal matrix, the p.d.f. (1.1) is the product of the p.d.f.'s of p univariate normal variates, showing that the x's are independently distributed in that case.

One can always determine a lower triangular (abbreviated as l.t. hereafter) nonsingular matrix:

$$C = \begin{bmatrix} c_{11} & 0 & 0 & \cdots & 0 \\ c_{21} & c_{22} & 0 & \cdots & 0 \\ c_{p1} & c_{p2} & c_{p3} & \cdots & c_{pp} \end{bmatrix}, \tag{1.2}$$

such that

$$\Sigma = CC'' . \tag{1.3}$$

If we, therefore, transform from the variates $\underline{x}$ of (1.1) to p new variates $\underline{y}' = [y_1, y_2, \ldots, y_p]$ by the transformation

$$\underline{y} = C^{-1}(\underline{x} - \underline{\mu}) , \tag{1.4}$$

the Jacobian of transformation, denoted by $J(\underline{x} \to \underline{y})$, will be (see [2]) $|C| = |\Sigma|^{1/2}$ and the p.d.f. of $\underline{y}$ will be

$$(2\pi)^{-p/2} \exp[-(1/2)\underline{y}'\underline{y}] , \ -\infty < y_i < \infty \tag{1.5}$$
$$(i = 1, 2, \ldots, p),$$

as $(\underline{x} - \underline{\mu})' \ \Sigma^{-1} \ (\underline{x} - \underline{\mu}) = \underline{y}'\underline{y}$. Thus $\underline{y} \sim N_p(\underline{y}|0|I_p)$, where I_p is the identity matrix of order p. The p.d.f. (1.5) is the product of the p.d.f.'s of p standard normal variables, and the range of variation of each y_i is $-\infty < y_i < \infty$. So the y_i's are independently distributed. The notation $y_i \sim NI(0, 1)$, $(i = 1, 2, \ldots, p)$ will be used to denote that the y_i's are independent standard normal variables. The characteristic function of $\underline{x}$ is (using 1.4)

$$\begin{aligned} \phi(\underline{t}) &= E(e^{\sqrt{-1}\ \underline{t}'\underline{x}}) \\ &= E(e^{\sqrt{-1}\ \underline{t}'(c\underline{y} + \underline{\mu})}) \\ &= e^{\sqrt{-1}\ \underline{t}'\underline{\mu}}\ E(e^{\sqrt{-1}\ \underline{u}'\underline{y}}) , \end{aligned}$$

where

$$\underset{-}{t}' = [t_1, t_2, \ldots, t_p] \quad \text{and} \quad \underset{-}{u}' = [u_1, u_2, \ldots, u_p] = \underset{-}{t}'C \,. \tag{1.6}$$

But as $y_i \sim NI(0, 1)$,

$$\begin{aligned} \phi &= e^{\sqrt{-1}\, \underset{-}{t}'\underset{-}{\mu}} \prod_{i=1}^{p} \left\{E(e^{\sqrt{-1}\, u_i y_i})\right\} \\ &= e^{\sqrt{-1}\, \underset{-}{t}'\underset{-}{\mu}} \prod_{i=1}^{p} \left\{\begin{array}{l}\text{characteristic function of} \\ y_i \text{ with } u_i \text{ as the argument}\end{array}\right\} \\ &= e^{\sqrt{-1}\, \underset{-}{t}'\underset{-}{\mu}} \prod_{i=1}^{p} \left(e^{-u_i^2/2}\right) \\ &= e^{\{\sqrt{-1}\, \underset{-}{t}'\underset{-}{\mu} - (1/2)\underset{-}{u}'\underset{-}{u}\}} \\ &= e^{\{\sqrt{-1}\, \underset{-}{t}'\underset{-}{\mu} - (1/2)t'\Sigma t\}} , \end{aligned} \tag{1.7}$$

on account of (1.6) and (1.3).

Since the y_i's are NI(0, 1), their variance-covariance matrix is I_p and hence from (1.4), noting that $\underset{-}{x} = C\underset{-}{y} + \underset{-}{\mu}$, we find

$$E(\underset{-}{x}) = \underset{-}{\mu} \quad \text{and} \quad V(\underset{-}{x}) = C\, I_p\, C' = \Sigma \,. \tag{1.8}$$

This could also have been obtained by differentiating the characteristic function (1.7) appropriately and setting $\underset{-}{t} = \underset{-}{0}$. This shows that the parameters $\underset{-}{\mu}$ and Σ occurring in the p.d.f. (1.1) are nothing but the first- and second-order moments of $\underset{-}{x}$, respectively. It should also be remembered in this connection that the exponent in the p.d.f. (1.1) is a quadratic form, the matrix of which is Σ^{-1}, while the exponent in the characteristic function (1.7) includes, among other terms, a quadratic form whose matrix is Σ itself.

Throughout this section, it is assumed that Σ is a positive-definite matrix. In fact, the p.d.f. (1.1) will not exist unless it is so, even though the characteristic function (1.7) can exist. However, we shall not consider the case of a singular Σ here. That will be deferred to Chapter 11. Since the characteristic function uniquely determines a distribution, we have the following theorem.

Theorem 1: If the characteristic function of the distribution of a vector variate $\underline{x}$ of p components is

$$\phi(\underline{t}) = \exp\left\{\sqrt{-1}\ \underline{t}'\underline{\mu} - (1/2)\underline{t}'\Sigma\underline{t}\right\} ,$$

where $\underline{t}' = [t_1, t_2, \ldots, t_p]$ is the vector of the arguments, $\underline{\mu}$ is a real p-compartment vector, and Σ is a symmetric positive definite matrix of order p, then $\underline{x} \sim N_p(\underline{x}|\underline{\mu}|\Sigma)$.

Let A be any m x p matrix of rank m. Then

$$A\ \Sigma\ A' = (AC)\ (AC)' ,$$

where C is given by (1.3). Hence, by well-known results about the rank of a matrix,

$$\text{rank } A\ \Sigma\ A' = \text{rank } AC = \text{rank } A = m ,$$

as C is nonsingular. $A\ \Sigma\ A'$ is thus a positive-definite matrix. Consider now the variates

$$\underline{z} = \begin{bmatrix} z_1 \\ z_2 \\ \vdots \\ z_m \end{bmatrix} = A\underline{x} . \tag{1.9}$$

Let us derive the joint distribution of these m linear combinations of $x_1, x_2, \ldots, x_p$. The characteristic function of $\underline{z}$, viz. $E(e^{\sqrt{-1}\,\underline{t}'A\underline{x}})$, where $\underline{t}' = [t_1, t_2, \ldots, t_m]$, is

$$\begin{aligned} E(e^{\sqrt{-1}\,\underline{t}'A\underline{x}}) &= E(e^{\sqrt{-1}\,\underline{u}'\underline{x}}) , \text{ where } \underline{u}' = \underline{t}'A \\ &= \exp\left\{\sqrt{-1}\,\underline{u}'\underline{\mu} - (1/2)\underline{u}'\Sigma\underline{u}\right\}, \text{ by (1.7)} \\ &= \exp\left\{\sqrt{-1}\,\underline{t}'(A\mu) - (1/2)\underline{t}'(A\,\Sigma\,A')\underline{t}\right\} . \end{aligned}$$

Hence, by Theorem 1, $\underline{z} \sim N_m(\underline{z}|A\mu|A\Sigma A')$. This is a very important result associated with a multivariate normal distribution and so we state it again as:

<u>Theorem 2</u>: If $\underline{x}$ has the p-variate nonsingular normal distribution $N_p(\underline{x}|\underline{\mu}|\Sigma)$, then any m linear combinations of $\underline{x}$, say $A\underline{x}$, have the m-variate nonsingular normal distribution $N_m(A\underline{x}|A\underline{\mu}|A\Sigma A')$, provided rank A = m, or in other words, the linear combinations are linearly independent.

As a particular case of this theorem, it follows that the marginal distribution of any x_i is a univariate normal distribution with mean μ_i and variance σ_{ii}, i.e., $x_i \sim N(\mu_i, \sigma_{ii})$.

2. A RANDOM SAMPLE FROM $N_p(\underline{x}|\underline{\mu}|\Sigma)$

Consider the p x n matrix

$$X = \begin{bmatrix} x_{11} & x_{12} & \cdots & x_{1n} \\ x_{21} & x_{22} & \cdots & x_{2n} \\ x_{p1} & x_{p2} & \cdots & x_{pn} \end{bmatrix} \tag{2.1}$$

whose columns are independently and identically distributed as a p-variate nonsingular normal distribution with parameters $\underline{\mu}$ and Σ. In other words, X is a random sample of size n from the $N_p(\underline{x}|\underline{\mu}|\Sigma)$ distribution. On account of the independence of the columns of X, the covariance matrix of any two columns is the null matrix and the variance-covariance matrix of any column is Σ. It can be easily seen from this that

$$V(X) = \Sigma \otimes I_n . \tag{2.2}$$

Also,

$$E(X) = \underline{\mu}\, E_{1n} . \tag{2.3}$$

The p.d.f. of X is the product of the p.d.f.'s of the n columns of X, on account of their independence, and can be written in a compact form as

$$(2\pi)^{-np/2} |\Sigma|^{-n/2} \exp\left\{-(1/2)\mathrm{tr}\, \Sigma^{-1}(X - \underline{\mu}\, E_{n1})(X - \underline{\mu}\, E_{n1})'\right\}, \tag{2.4}$$

where tr stands for the trace of a matrix. In obtaining this form from the p.d.f. (1.1) of any column of X, we have used the property

$$\text{tr } PQR = \text{tr } RPQ = \text{tr } QRP \tag{2.5}$$

of the trace of the product of matrices.

For verification of certain theoretical results, empirically, it is sometimes necessary to generate artifically a random sample from $N_p(\underline{x}|\underline{\mu}|\Sigma)$. For this refer to equations (1.4) and (1.3), namely,

$$\underline{y} = C^{-1}(\underline{x} - \underline{\mu}) , \quad CC' = \Sigma .$$

We have already proved that the y_i's are NI(0, 1). We assume that the reader is already familiar with the method of generating a random sample from a N(0, 1) population, with the help of a table of random numbers. In fact, ready-made samples from a N(0, 1) population are also available [10, 13], and computer programs for this also exist. From these, take a random sample of size np from a N(0, 1) population and arrange these observations in the form of a p x n matrix:

$$Y = [y_{ir}] \ (i = 1, 2, \ldots, p; \ r = 1, 2, \ldots, n) . \tag{2.6}$$

Convert this sample to a sample X from the $N_p(\underline{x}|\underline{\mu}|\Sigma)$ distribution by applying the reciprocal transformation of (1.4) to each column of Y:

$$X = CY + \underline{\mu} E_{1n} . \tag{2.7}$$

The determination of the l.t. matrix C is thus an essential step for this and this can be done by solving directly the equations $CC' = \Sigma$. Since C is l.t., the equations can easily be solved recursively, starting with c_{11}, then c_{21} and c_{22}, and so on. There is an indeterminacy about the signs of the diagonal terms c_{ii}, and any sign can be chosen arbitrarily; the convention is either to take all c_{ii} positive or choose them in such a way as to have all the elements in the first column of C positive.

3. ORTHOGONAL TRANSFORMATION

An orthogonal transformation is a very important tool in the distribution theory associated with normal distributions. The transformation

$$\underline{y} = P\underline{x} \ , \tag{3.1}$$

where P is a p x p orthogonal matrix, i.e., PP' = PP' = I and

$$\underline{y}' = [y_1, y_2, \ldots, y_p], \ \underline{x}' = [x_1, x_2, \ldots, x_p] \ , \tag{3.2}$$

is called an orthogonal transformation, and if x_i are NI(0, σ^2), the y_i are also NI(0, σ^2). This follows from Theorem 2. In this result, it is understood that the orthogonal matrix P is a matrix of constants and not random variables. In many cases, one uses an "incomplete" orthogonal matrix P_1 of order m x p (m < p), such that only $P_1 P_1' = I_m$ holds. But one can always complete this matrix P_1 by choosing p - m more rows suitably, so that the completed matrix

$$P = \left[\begin{array}{c} P_1 \\ \hline P_2 \end{array}\right]\begin{array}{l} m \\ p\text{-}m \end{array} \tag{3.3}$$

is orthogonal. The transformation

$$\left[\begin{array}{c} \underline{z} \\ \hline \underline{u} \end{array}\right] = P\underline{x} \tag{3.4}$$

then shows that $z_i (i = 1, 2, \ldots, m)$, the elements of $\underline{z}$, and $u_j (j = 1, 2, \ldots, p-m)$, the elements of $\underline{u}$, are all $NI(0, \sigma^2)$, and hence $\underline{z}$ is independently distributed of $\underline{u}'\underline{u}/\sigma^2$, which has a χ^2 distribution with (p-m) degrees of freedom (d.f.). But, as P is orthogonal,

$$\left[\begin{array}{c} \underline{z} \\ \hline \underline{u} \end{array}\right]' \left[\begin{array}{c} \underline{z} \\ \hline \underline{u} \end{array}\right] = x'P'Px = \underline{x}'\underline{x} ,$$

and hence $\underline{u}'\underline{u} = \underline{x}'\underline{x} - \underline{z}'\underline{z}$. We, therefore, have the following theorem:

Theorem 3: If $\underline{x} \sim N_p(\underline{x}|\underline{0}|\sigma^2 I_p)$ and if P_1 is an m x p matrix such that $P_1P_1' = I_m$, then $\underline{z} = P_1\underline{x} \sim N_m(\underline{z}|\underline{0}|\sigma^2 I_m)$ and is independent of $(\underline{x}'\underline{x} - \underline{z}'\underline{z})/\sigma^2$, which has a χ^2 distribution with p-m d.f.

Recently, much of the distribution theory associated with the multivariate normal distribution has been greatly simplified, and elegant, concise derivations have been introduced with the help of what is known as a random orthogonal transformation [12]. One employs an orthogonal or incomplete orthogonal matrix, with elements which are not constants but are random variables. So, we hold these elements fixed and apply this orthogonal transformation to the conditional distribution of some other variables, when these are fixed. Then, it so happens many times that the resulting distribution of the transformed variables does not involve the random elements of the orthogonal matrix at all, and hence the distribution is no longer a conditional distribution but an absolute one or unconditional one. Further more, it is independent of these random elements of the orthogonal matrix. We shall illustrate this with a simple example there, but more interesting and involved applications of this technique of random orthogonal transformations will occur again and again throughout this book.

Let (x_i, y_i) $(i = 1, 2, \ldots, n)$ be a random sample of size n from the bivariate normal distribution of x, y, which has zero means and variance-covariance matrix I_2. In other words x_i, y_i are all NI(0, 1). Consider the incomplete orthogonal matrix

$$P_1 = \begin{bmatrix} 1/\sqrt{n} \;, & 1/\sqrt{n} \;, & & 1/\sqrt{n} \\ (y_1 - \bar{y})/s, & (y_2 - \bar{y})/s, & \ldots, & (y_n - \bar{y})/s \end{bmatrix}, \tag{3.5}$$

where $\bar{y} = (1/n) \sum_1^n y_i$ and $s^2 = \sum_1^n (y_i - \bar{y})^2$. Since the x_i's are independently distributed of the y_j's, the conditional distribution of the x_i's, when P_1 is held fixed, is still that of NI(0, 1) variates and so, applying Theorem 3 to the transformed variables,

$$\begin{bmatrix} U_1 \\ U_2 \end{bmatrix} = P_1 \begin{bmatrix} x_1 \\ x_2 \\ \vdots \\ x_n \end{bmatrix}, \tag{3.6}$$

we find that the conditional distributions of U_1, U_2 and $v = \sum_1^n x_i^2 - U_1^2 - U_2^2$ are, respectively,

$$\frac{1}{\sqrt{2\pi}} e^{-U_i^2/2} dU_i, \quad -\infty < U_i < \infty \quad (i = 1, 2) , \tag{3.7}$$

and

$$\frac{1}{2^{(n-2)/2}\,\Gamma(n-2)/2} v^{[(n-2)/2]-1} e^{-v/2} dv, \quad 0 < v < \infty, \tag{3.8}$$

and the three are independent. But (3.7) and (3.8) do not involve the conditioning variates y_i $(i = 1, 2, \ldots, n)$ at all, and hence (3.7) and (3.8) represent the absolute (marginal) distributions of

$$u_1 = \frac{1}{\sqrt{n}} \sum_1^n x_i = \sqrt{n}\,\bar{x} , \tag{3.9}$$

$$u_2 = \sum_1^n (x_i - \bar{x})(y_i - \bar{y}) \Big/ \left\{\sum_1^n (y_i - \bar{y})^2\right\}^{1/2} , \tag{3.10}$$

$$v = \sum_1^n (x_i - \bar{x})^2 - U_2^2 = (1 - r^2) \sum_1^n (x_i - \bar{x})^2 , \tag{3.11}$$

where

$$r = \sum_1^n (x_i - \bar{x})(y_i - \bar{y})/s \left\{\sum_1^n (x_i - \bar{x})^2\right\}^{1/2} , \tag{3.12}$$

is the sample correlation coefficient between x and y. Further U_1, U_2, v and the y_i's are all independently distributed. Observe that v has a χ^2 distribution with n - 2 d.f. and s^2 has a χ^2 distribution with n - 1 d.f. From these results, it is now an easy exercise to show that r, s^2 and $\sum_1^n (x_i - \bar{x})^2$ are independently distributed, the distribution of r being

$$\frac{\Gamma(n-1)/2}{\pi\,\Gamma(n-2)/2}\,(1-r^2)^{(n-4)/2}\,dr,\ -1 \le r \le 1\ . \qquad (3.13)$$

s^2 and $\sum_1^n (x_i - \bar{x})^2$ are obviously χ^2 variates with n - 1 d.f.

One is not always as fortunate as in the example above, and the distribution of the new variates, after applying a random orthogonal transformation, sometimes involves the random elements of the orthogonal matrix. In that case, one will have to multiply the resulting conditional p.d.f. by the p.d.f. of the random elements of the orthogonal matrix entering the conditional p.d.f. and then integrate them out. Such a situation, for example, would have arisen if the bivariate normal distribution of x, y with which we started had the variance-covariance matrix

$$\begin{bmatrix} 1 & \rho \\ \rho & 1 \end{bmatrix}, \qquad (3.14)$$

with $\rho \neq 0$, and not $\rho = 0$ as we had.

4. CONDITIONAL DISTRIBUTION

We assume $\underline{x} \sim N_p(\underline{x}|\underline{\mu}|\Sigma)$. Let $\underline{x}' = [\underset{k}{\underline{x}'(1)}|\underset{p-k}{\underline{x}'(2)}]$. We now consider the conditional distribution of

$$\underline{x}'(2) = [x_{k+1}, x_{k+2}, \ldots, x_p] \qquad (4.1)$$

when

$$\underline{x}'(1) = [x_1, x_2, \ldots, x_k] \qquad (4.2)$$

are fixed. By Theorem 2, the p.d.f. of the marginal distribution of $\underline{x}(1)$ is $N_k(\underline{x}(1)|\underline{\mu}(1)|\Sigma_{11})$ where

$$\underline{\mu} = \left[\begin{array}{c} \underline{\mu}(1) \\ \hline \underline{\mu}(2) \end{array}\right]\begin{array}{l} k \\ p-k \end{array}, \quad \Sigma = \left[\begin{array}{c|c} \Sigma_{11} & \Sigma_{12} \\ \hline \Sigma_{21} & \Sigma_{22} \end{array}\right]\begin{array}{l} k \\ p-k \end{array} \tag{4.3}$$
$$\qquad\qquad\qquad\qquad k \qquad p-k$$

The p.d.f. of the conditional distribution of $\underline{x}(2)$, given $\underline{x}(1)$, is, therefore,

$$N_p(\underline{x}|\underline{\mu}|\Sigma)/N_k(\underline{x}(1)|\underline{\mu}(1)|\Sigma) \ . \tag{4.4}$$

In order to express this in a suitable form, we apply (1.3.16) to Σ as partitioned in (4.3). This yields

$$\Sigma^{-1} = \left[\begin{array}{c|c} \Sigma_{11\cdot 2}^{-1} & -\Sigma_{11}^{-1}\Sigma_{12}\Sigma_{22\cdot 1}^{-1} \\ \hline -\Sigma_{22}^{-1}\Sigma_{21}\Sigma_{11\cdot 2}^{-1} & \Sigma_{22\cdot 1}^{-1} \end{array}\right], \tag{4.5}$$

where

$$\Sigma_{11\cdot 2} = \Sigma_{11} - \Sigma_{12}\Sigma_{22}^{-1}\Sigma_{21} \ ,$$
$$\Sigma_{22\cdot 1} = \Sigma_{22} - \Sigma_{21}\Sigma_{11}^{-1}\Sigma_{12} \ . \tag{4.6}$$

Using (1.4.16), it is easy to show that

$$\Sigma_{11\cdot 2}^{-1} = (I - \Sigma_{11}^{-1}\ \Sigma_{12}\ \Sigma_{22}^{-1}\ \Sigma_{21})^{-1}\ \Sigma_{11}^{-1}$$
$$= \Sigma_{11}^{-1} + \Sigma_{11}^{-1}\ \Sigma_{12}\ \Sigma_{22\cdot 1}^{-1}\ \Sigma_{21}\ \Sigma_{11}^{-1} \ . \tag{4.7}$$

From (4.5), (4.6), and (4.7) we can write $(\underline{x} - \underline{\mu})'\ \Sigma^{-1}(\underline{x} - \underline{\mu})$ as

$$(\underline{x}(1) - \underline{\mu}(1))'\ \Sigma_{11}^{-1}(\underline{x}(1) - \underline{\mu}(1)) + \Big\{\underline{x}(2) - \underline{\mu}(2)$$
$$- \beta(\underline{x}(1) - \underline{\mu}(1))\Big\}'\Sigma_{22\cdot 1}^{-1}\ \Big\{\underline{x}(2) - \underline{\mu}(2)$$
$$- \beta(\underline{x}(1) - \underline{\mu}(1))\Big\} , \tag{4.8}$$

where

$$\beta = \Sigma_{21}\Sigma_{11}^{-1} \ . \tag{4.9}$$

Expression (4.4) therefore simplifies to

$$N_{p-k}\left(\underline{x}(2) \,|\, \underline{\mu}(2) + \beta(\underline{x}(1) - \underline{\mu}(1)) \,|\, \Sigma_{22\cdot 1}\right) , \qquad (4.10)$$

showing that the conditional distribution of $\underline{x}(2)$ when $\underline{x}(1)$ is fixed is a (p-k)-variate normal distribution, with mean vector

$$E(\underline{x}(2) \,|\, \underline{x}(1)) = \underline{\mu}(2) + \beta(\underline{x}(1) - \underline{\mu}(1)) \qquad (4.11)$$

and variance-covariance matrix

$$V(\underline{x}(2) \,|\, \underline{x}(1)) = \Sigma_{22\cdot 1} \cdot \qquad (4.12)$$

Equation (4.11) is called the regression of the vector $\underline{x}(2)$ on the vector $\underline{x}(1)$, and as each component of the vector on the right side of (4.11) is a linear function of the variates $\underline{x}(1)$, the regression is said to be linear. The matrix β is called the matrix of regression coefficients of $\underline{x}(2)$ on $\underline{x}(1)$. Had we considered the conditional distribution of $\underline{x}(1)$, when $\underline{x}(2)$ is fixed, we would have obtained for its p.d.f.

$$N_k\left(\underline{x}(1) \,|\, \underline{\mu}(1) + \Sigma_{12}\Sigma_{22}^{-1}(\underline{x}(2) - \underline{\mu}(2)) \,|\, \Sigma_{11\cdot 2}\right) , \qquad (4.13)$$

showing that the matrix of regression coefficients of $\underline{x}(1)$ on $\underline{x}(2)$ is $\Sigma_{12}\Sigma_{22}^{-1}$. The variates

$$\underline{x}_{2\cdot 1} = \underline{x}(2) - E(\underline{x}(2) \,|\, \underline{x}(1)) \qquad (4.14)$$

are the residuals of $\underline{x}(2)$ with respect to $\underline{x}(1)$, as the regression is linear and obviously the variance-covariance matrix of these residuals is

$$V(\underline{x}_{2\cdot 1}) = V(\underline{x}(2) \,|\, \underline{x}(1)) = \Sigma_{22\cdot 1} \cdot \qquad (4.15)$$

In particular, when $k = p-2$, the variances and covariances of the two residuals $x_{p-1\cdot 12\ldots p-2}$ and $x_{p\cdot 12\ldots p-2}$ will be given by the 2×2 matrix $\Sigma_{22\cdot 1}$, and so the partial correlation coefficient between x_{p-1} and x_p, when $x_1, \ldots, x_{p-2}$ are fixed, will be given by

$$\rho_{p,p-1\cdot 12\ldots p-2} = \sigma_{p,p-1\cdot 12\ldots p-2}/(\sigma_{p,p\cdot 12\ldots p-2}\,\sigma_{p-1,p-1\cdot 12\ldots p-2})^{1/2}, \tag{4.16}$$

where $\sigma_{ij\cdot 12\ldots p-2}$ $(i, j = p-1,p)$ are the elements of $\Sigma_{22\cdot 1}$.

Sample Quantities

Let X be the $p \times n$ matrix of sample observations when a sample of size n is drawn from the $N_p(\underline{x}|\underline{\mu}|\Sigma)$ distribution. Let it be partitioned as

$$X = \underset{n}{\begin{bmatrix} X[1] \\ \hline X[2] \end{bmatrix}}\begin{matrix} k \\ p-k \end{matrix}, \tag{4.17}$$

so that the corrected s.s. and s.p. matrix (as defined in 1.3.28) is

$$S = \underset{\quad k \qquad p-k}{\left[\begin{array}{c|c} s_{11} & s_{12} \\ \hline s_{21} & s_{22} \end{array}\right]}\begin{matrix} k \\ p-k \end{matrix} = X(I - (1/n)E_{nn})X', \tag{4.18}$$

where

$$s_{ij} = X[i]\ (I - (1/n)E_{nn})\ X[j] \quad i, j = 1, 2. \tag{4.19}$$

By analogy with (4.11), the sample regression of $\underline{x}(2)$ on $\underline{x}(1)$ is

$$\bar{\underline{x}}(2) + B(\underline{x}(1) - \bar{\underline{x}}(1)), \tag{4.20}$$

where

$$\bar{\underline{x}} = (1/n) X E_{n1} = \left[\begin{array}{c} \bar{\underline{x}}(1) \\ \hline \bar{\underline{x}}(2) \end{array}\right] \begin{array}{l} k \\ p-k \end{array} , \qquad (4.21)$$

$B' = S_{11}^{-1} S_{12}$, the matrix of sample regression coefficients in the regression of $\underline{x}(2)$ on $\underline{x}(1)$. (4.22)

The matrix of sample observations on the regression function (4.20) is

$$R = \bar{\underline{x}}(2) E_{1n} + B(X[1] - \bar{\underline{x}}(1) E_{1n}) , \qquad (4.23)$$

and hence the corrected s.s. and s.p. matrix of these observations is

$$\begin{aligned} R(I - (1/n) E_{nn})R' &= B S_{11} B' \\ &= S_{21} S_{11}^{-1} S_{12} \end{aligned} \qquad (4.24)$$

and is known as the s.s. and s.p. matrix due to the regression of $\underline{x}(2)$ on $\underline{x}(1)$, in a sample of size n. The sample residual of $\underline{x}(2)$ is, from (4.20),

$$\underline{x}(2) - \bar{\underline{x}}(2) - B(\underline{x}(1) - \bar{\underline{x}}(1)) ,$$

and the matrix of observations on these residuals is

$$X_{2\cdot 1} = X[2] - R , \qquad (4.25)$$

and the matrix of corrected s.s. and s.p. due to these is

$$\begin{aligned} X_{2\cdot 1}(I - (1/n) E_{nn}) X'_{2\cdot 1} &= S_{22} - S_{21} S_{11}^{-1} S_{12} \\ \text{or } &= S_{22} - B S_{11} B' \end{aligned} \qquad (4.26)$$

and will be denoted by $S_{22\cdot 1}$. Observe that (4.26) and (4.24) add up to the "total" corrected s.s. and s.p. matrix of $\underline{x}(2)$, namely, S_{22}. When the regression of $\underline{x}(1)$ on $\underline{x}(2)$ is considered, all the above quantities can be easily obtained by interchanging the subscripts 1 and 2. In particular, we will have the matrix of regression coefficients to be

$$S_{22}^{-1}S_{21} \tag{4.27}$$

and the residual s.s. and s.p. matrix to be

$$S_{11\cdot 2} = S_{11} - S_{12}S_{22}^{-1}S_{21} \ . \tag{4.28}$$

Using (1.3.14), it can be readily seen that

$$|S| = |S_{11}|\ |S_{22\cdot 1}| \quad \text{or} \quad |S_{22}|\ |S_{11\cdot 2}| \ . \tag{4.29}$$

All the sample quantities introduced here play an important part in the study of the relationship between two vectors and will occur often in subsequent chapters.

5. QUADRATIC FORMS IN NORMAL VARIABLES

A considerable literature is now available on quadratic forms in normal variables and the associated distribution theory. Quadratic forms play a very important role in statistical theory. A particular class of quadratic forms, namely, the ones with idempotent matrices, are still more important, because the various sums of squares in an analysis of variance table - whether it is a case of regression or of a fixed-effect model in experimental design - are of this type. It is not the object of this book to give an extensive account of this theory, as the reader is assumed to be familiar with least squares theory associated with linear models and its applications in Design of experiments, Multiple regression, and Analysis of variance. However, for developing the theory of multivariate analysis of variance later on, it is necessary to give a brief account of some of the results in this area. An exhaustive account of all distributional theory of quadratic forms in normal variables is given by Johnson and Kotz [5]. The reader may also refer to C. Radhakrishna Rao [9], Kendall and Stuart [6], Lancaster [8], or Graybill [3] for various aspects of this theory.

Let $\underline{x}'A\underline{x}$ be a quadratic form in the p NI(0, 1) variables x_i, forming the column vector $\underline{x}$. By transforming A to a diagonal form, using an orthogonal matrix, it can be seen that the quadratic form reduces to $\sum_1^m \lambda_i y_i^2$, where the λ's are the nonzero eigenvalues of A and the y_i are NI(0, 1). The characteristic function of $\underline{x}'A\underline{x}$ is then easily seen to be

$$\prod_{i=1}^{m} (1 - 2\sqrt{-1}\,\lambda_i t)^{-1/2} = |I - 2\sqrt{-1}\,At|^{-1/2} . \qquad (5.1)$$

From this it has been shown that a necessary and sufficient condition for $\underline{x}'A\underline{x}$ to be distributed as a χ^2 is that the matrix A be idempotent, i.e., $A^2 = A$. The nonzero eigenvalues of A, in that case will all be equal to unity, and the d.f. of the χ^2 distribution will be

$$m = \text{rank } A = \text{tr } A . \qquad (5.2)$$

Two quadratic forms $\underline{x}'A\underline{x}$ and $\underline{x}'B\underline{x}$, where $\underline{x}$ is as before, will be distributed independently if and only if $AB = 0$. This is known as Craig's [1] theorem, though Craig's original proof is incorrect. For a correct proof, the reader may refer to Lancaster [8]. The necessity of the condition is proved by equating the joint characteristic function to the product of the two separate characteristic functions. The sufficiency of the condition $AB = 0$ is proved by simultaneously diagonalizing A and B with the help of an orthogonal matrix and showing that no non-zero terms appear at the same position in each of the diagonal matrices. Let

$$\underline{x}'\underline{x} = \sum_{i=1}^{k} \underline{x}'A_i\underline{x} , \tag{5.3}$$

then for the quadratic forms $\underline{x}'A_i\underline{x}$ to be independently distributed as χ^2, any one of the following three equivalent conditions is necessary and sufficient:

$$\text{(I)} \quad A_i^2 = A_i; \quad i = 1, 2, \ldots, k$$

$$\text{(II)} \quad A_iA_j = 0, \; i \neq j; \quad i, j = 1, 2, \ldots, k$$

$$\text{(III)} \quad \sum_{i=1}^{k} \operatorname{rank} A_i = p . \tag{5.4}$$

This is known as James's theorem [4] and is a generalized version of what is popularly known as Cochran's theorem which states only the last condition above. The theorem also holds if the left-hand side of (5.3) is replaced by $\underline{x}'A\underline{x}$, where A is idempotent as, in that case, we can add $\underline{x}'(I - A)\underline{x}$ to both sides and use the fact that $I - A$ is also idempotent.

These results can be generalized to deal with quadratic forms $\underline{x}'A\underline{x}$, where the $\underline{x}_i$ are not NI(0, 1) but have the $N_p(\underline{x}|0|\Sigma)$ distribution. This is achieved by using the transformation (1.4) to express $\underline{x}'A\underline{x}$ as a quadratic form $\underline{y}'D\underline{y}$, where $\underline{y} \sim N_p(\underline{y}|\underline{0}|I)$, i.e., y_i are NI(0, 1), and then translating the conditions on D into those on A, by using the relation between A and D, which involves Σ. In certain cases, one has to deal with quadratic forms of the type $\underline{x}'A\underline{x}$, where the x_i's are NI(μ_i, 1). Thus if A is idempotent, $\underline{x}'A\underline{x}$ can be expressed as $\sum_1^m y_i^2$, where m is the rank of A and the y_i are NI, the variance of each y_i being unity. The characteristic function of $\underline{x}'A\underline{x}$ therefore turns out to be

$$\exp\left\{-(1/2)\lambda^2 + \frac{\lambda^2}{2(1-2\sqrt{-1}\,t)}\right\} \cdot (1-2\sqrt{-1}\,t)^{-m/2}\ , \qquad (5.5)$$

where

$$\lambda^2 = E(\underline{x}')A\,E(\underline{x})\ . \qquad (5.6)$$

By expanding $\exp\left\{\lambda^2/2(1-2\sqrt{-1}\,t)\right\}$ and applying the Inversion theorem, or by identifying $(1-2\sqrt{-1}\,t)^{-(m/2+r)}$ with the characteristic function of a χ^2 distribution with $m+2r$ d.f., we find the p.d.f. of $u = \underline{x}'A\underline{x}$ to be

$$\chi'^2_m(u|\lambda^2) = e^{-\lambda^2/2} \sum_{r=0}^{\infty} \frac{(\lambda^2)^r}{2^r\ r!}\ \chi^2_{m+2r}(u)\ , \qquad (5.7)$$

where

$$\chi^2_f(v) = \frac{1}{2^{f/2}\ \Gamma(f/2)}\ v^{(f/2)-1}\ e^{-v/2}\ . \qquad (5.8)$$

Equation (5.7) is known as the p.d.f. of a noncentral χ^2 with m d.f. and noncentrality parameter λ^2; (5.8) is, of course, the p.d.f. of a χ^2 distribution with f d.f. For another instructive derivation of the density of a noncentral χ^2, the reader may refer to Kerridge [7]. We shall use $u \sim \chi'^2_m(u|\lambda^2)$ to denote that u has a noncentral χ^2 distribution with m d.f. and noncentrality parameter λ^2. When $\lambda^2 = 0$, (5.7) reduces to the p.d.f. of a χ^2 distribution, and in that case, we shall denote it by $u \sim \chi^2_m(u)$. It should be observed from (5.6) that λ^2 is the same function of $E(\underline{x})$ as u is of $\underline{x}$. In certain cases, the following result is also useful in obtaining λ^2:

$$\begin{aligned} E(\underline{x}'A\underline{x}) &= E(\text{tr } \underline{x}'A\underline{x}) \\ &= E(\text{tr } \underline{x}\underline{x}'A) \\ &= \text{tr}\{E(\underline{x}\underline{x}')\cdot A\} \\ &= \text{tr}(I + E(\underline{x})\ E(\underline{x}'))\cdot A \\ &= m + \lambda^2 . \end{aligned} \tag{5.9}$$

Many of the results on quadratic forms have been recently extended to cover the case of a singular variance-covariance matrix of $\underline{x}$ also (see for example [11] and references in it).

In linear models, one comes across the distribution of quadratic forms in variables $x_1, x_2, \ldots, x_p$, with variance-covariance matrix $\sigma^2 I_p$ and means given by

$$E(\underline{x}) = E[x_1, x_2, \ldots, x_p]' = A\theta , \tag{5.10}$$

where A is a $p \times m$ matrix of known constants and $\underline{\theta}$ is an $m \times 1$ vector of unknown parameters. The rank of A is r and σ^2 is unknown. For testing the significance of some linear functions $B\underline{\theta}$, where B is a $k \times m$ matrix of rank k, the following quantities are useful:

$$SSE = \min_{\theta} (\underline{x} - A\underline{\theta})' (\underline{x} - A\underline{\theta}) , \tag{5.11}$$

$$SSH = \min_{\theta} [(\underline{x} - A\underline{\theta})' (\underline{x} - A\underline{\theta}) | \text{subject to } B\underline{\theta} = \underline{0}] - SSE . \tag{5.12}$$

In finding (5.12), it is further assumed that there exists a matrix D of order $k \times m$, such that $B = DA'A$. This is known as the condition of estimability of $B\underline{\theta}$, for if D exists, $E(DA'\underline{x}) = DA'A\underline{\theta} = B\underline{\theta}$, and $B\underline{\theta}$ is estimable. Both (5.11) and (5.12) can be expressed as $\underline{x}'Q_1\underline{x}$ and $\underline{x}'Q_2\underline{x}$, where Q_1, Q_2 are idempotent, and $Q_1Q_2 = 0$. From this, by application of James's theorem, it follows that

$$q_1 = SSE/\sigma^2 \sim \chi^2_{p-r}(q_1) . \tag{5.13}$$

However,

$$q_2 = SSH/\sigma^2 \sim \chi^2_k(q_2) \tag{5.14}$$

if and only if $B\underline{\theta} = 0$; q_2 is independently distributed of q_1. If $B\underline{\theta}$ is not null, q_2 has a noncentral χ^2 distribution but is still independent of q_1. All linear inference problems in analysis of variance, analysis of covariance, regression analysis, and design of experiments can be put into this framework, and so these results about q_1 and q_2 are extremely important. It may be noted here that SSE is called the s.s. due to error and is usually calculated as

$$\text{SSE} = \underline{x}'\underline{x} - \hat{\underline{\theta}}'A'\underline{x} , \qquad (5.15)$$

where $\hat{\underline{\theta}}$ is any solution of the equations

$$A'x = A'A\hat{\underline{\theta}} . \qquad (5.16)$$

Similarly SSH is known as the s.s. due to the hypothesis $B\underline{\theta} = \underline{0}$. $\text{SSE} + \text{SSH} = \min_{\theta} (\underline{x} - A\underline{\theta})'(\underline{x} - A\underline{\theta})$ subject to $B\underline{\theta} = 0$ is known as the conditional error s.s. SSH is thus the difference between the conditional and unconditional error sums of squares. The best unbiased linear estimate of $B\underline{\theta}$ is $B\hat{\underline{\theta}}$. It has been shown that SSH is also expressible as

$$(B\hat{\underline{\theta}})' \left\{1/\sigma^2 V(B\hat{\underline{\theta}})\right\}^{-1} (B\hat{\underline{\theta}}) . \qquad (5.17)$$

Of the three expressions for SSH [(5.12), (5.17), and $\underline{x}'Q_2\underline{x}$], $\underline{x}'Q_2\underline{x}$ is rarely used in applications. In other words, it is not usually necessary to spell out the matrix Q_2 explicitly, but it can be done from the other two expressions, if required. The same is true for Q_1 of SSE, given by (5.16).

An excellent account of all this "least squares theory", in its full generality is available in C. Radhakrishna Rao [9].

REFERENCES

[1] Craig, A. T. (1943). "Note on the independence of certain quadratic forms", Ann. Math. Statist., 14, p. 195.

[2] Deemer, Walter L., and Olkin, I. (1951). "The Jacobians of certain matrix transformations useful in multivariate analysis", Biometrika, 38, p. 345.

[3] Graybill, Franklin A. (1961). An Introduction to Linear Statistical Models. McGraw-Hill Book Co., Inc., New York.

[4] James, G. S. (1952). "Note on a theorem of Cochran", Proc. Camb. Philos. Soc. 48, p. 443.

[5] Johnson, N. L. and Kotz, S. Continuous Univariate Distributions-, 2 Houghton Mifflin Co., Boston.

[6] Kendall, M. G., and Stuart, A. (1968). The Advanced Theory of Statistics. Hafner Publishing Co., New York.

[7] Kerridge, D. (1965). "A probabilistic derivation of the non-central χ^2 distribution", Aust. J. Statist., 7, p. 37 (see also p. 114).

[8] Lancaster, H. O. (1969). The Chi-Squared Distribution. John Wiley and Sons, Inc., New York.

[9] Rao, C. Radhakrishna (1965). Linear Statistical Inference and Its Applications. John Wiley and Sons, Inc., New York.

[10] Sengupta, J. M. and Bhattacharya, N. (1958). "Tables of random normal deviates", Sankhyā, 20, p. 250.

[11] Styan, G. P. H. (1970). "Notes on the distribution of quadratic forms in singular normal variables", Biometrika, 57, p. 567.

[12] Wijsman, R. A. (1957). "Random orthogonal transformations", Ann. Math. Statist., 28, p. 415.

[13] Wold, H. (1948). Random Normal Deviates (Tracts for computers, XXV). Cambridge University Press, London.

CHAPTER 3

THE WISHART DISTRIBUTION

1. MAXIMUM LIKELIHOOD ESTIMATES OF THE MEAN VECTOR AND THE VARIANCE-COVARIANCE MATRIX

Let $X = [x_{ir}]$, $(i = 1, 2, \ldots, p;\ r = 1, 2, \ldots, n)$ be a $p \times n$ matrix representing a random sample of size n from the p-variate nonsingular normal distribution, whose p.d.f. is $N_p(\underline{x}|\underline{\mu}|\Sigma)$. We shall derive the maximum likelihood estimates of the parameters $\underline{\mu}$ and Σ in this section. From (2.2.4), the p.d.f. of X is

$$(2\pi)^{-np/2}\ |\Sigma|^{-n/2}\ \exp\left\{-\frac{1}{2}\ \mathrm{tr}\ \Sigma^{-1}(X - \underline{\mu}E_{1n})(X - \underline{\mu}E_{1n})'\right\}.$$

Hence the logarithm of the likelihood of the sample observations is

$$\begin{aligned}
\log_e L(\underline{\mu},\Sigma) &= -\frac{np}{2}\log_e(2\pi) - \frac{n}{2}\log_e|\Sigma| - \frac{1}{2}\mathrm{tr}\,\Sigma^{-1}(X-\underline{\mu}E_{1n})(X-\underline{\mu}E_{1n})' \\
&= -\frac{np}{2}\log_e(2\pi) - \frac{n}{2}\log_e|\Sigma| - \frac{1}{2}\mathrm{tr}\,\Sigma^{-1}(X-\bar{\underline{x}}E_{1n})(X-\bar{\underline{x}}E_{1n})' \\
&\quad - \frac{1}{2}\mathrm{tr}\,\Sigma^{-1}(\bar{\underline{x}}E_{1n} - \underline{\mu}E_{1n})(\bar{\underline{x}}E_{1n} - \underline{\mu}E_{1n})' \\
&= -\frac{np}{2}\log_e(2\pi) - \frac{n}{2}\log_e|\Sigma| - \frac{1}{2}\mathrm{tr}\,\Sigma^{-1}S \\
&\quad - \frac{n}{2}\mathrm{tr}\,\Sigma^{-1}(\bar{\underline{x}} - \underline{\mu})(\bar{\underline{x}} - \underline{\mu})' \qquad (1.1)
\end{aligned}$$

where $\bar{\underline{x}} = (n^{-1})X\,E_{n1}$ is the vector of the sample means and $S = X[I - (1/n)E_{nn}]X'$ is the matrix of corrected s.s. and s.p.

as defined in (1.3.28). To find the maximum likelihood estimates, we differentiate (1.1) with respect to $\underline{\mu}$ and Σ. For this we need the following results:

$$\frac{\partial \log |\Sigma|}{\partial \sigma_{ij}} = (2 - \delta_{ij})\sigma^{ij} \quad (i, j = 1, 2, \ldots, p) , \qquad (1.2)$$

where σ_{ij}, σ^{ij} are the elements in the ith row and j^{th} column of Σ and Σ^{-1}, respectively, and δ_{ij} is the Kronecker delta, which is equal to 1 if $i = j$ and 0 if $i \neq j$. Also, from

$$\Sigma \Sigma^{-1} = I_p ,$$

we obtain by differentiation

$$\Sigma \frac{\partial(\Sigma^{-1})}{\partial \sigma_{ij}} + \frac{\partial \Sigma}{\partial \sigma_{ij}} \cdot \Sigma^{-1} = 0 , \qquad (1.3)$$

where $\partial\Sigma/\partial\sigma_{ij}$ is the p x p matrix of partial derivatives of the elements of Σ, with respect to σ_{ij}. From (1.3), we obtain

$$\frac{\partial \Sigma^{-1}}{\partial \sigma_{ij}} = - \Sigma^{-1} \frac{\partial \Sigma}{\partial \sigma_{ij}} \Sigma^{-1} . \qquad (1.4)$$

The maximum likelihood equations are obtained by equating $\partial \log L/\partial\underline{\mu}$ and $\partial \log L/\partial\sigma_{ij}$ to zero. We write $\hat{\underline{\mu}}$ for $\underline{\mu}$ and $\hat{\Sigma}$ for Σ in these equations, to distinguish the estimates from the original parameters. From $\partial \log L/\partial\underline{\mu} = 0$, we get

$$\hat{\Sigma}^{-1}(\bar{\underline{x}} - \hat{\underline{\mu}}) = \underline{0} . \qquad (1.5)$$

From $\partial \log L/\partial\sigma_{ij} = 0$, we get, using (1.2) and (1.4),

$$- \frac{n}{2} (2 - \delta_{ij})\hat{\sigma}^{ij} + \frac{1}{2} \operatorname{tr} \left\{\hat{\Sigma}^{-1} \frac{\partial \hat{\Sigma}}{\partial \hat{\sigma}_{ij}} \hat{\Sigma}^{-1} S\right\}$$

$$+ \frac{n}{2} \operatorname{tr} \left\{ \frac{\partial \hat{\Sigma}^{-1}}{\partial \hat{\sigma}_{ij}} (\underline{\bar{x}} - \underline{\hat{\mu}})(\underline{\bar{x}} - \underline{\hat{\mu}})' \right\} = 0 \quad (i, j = 1, 2, \ldots, p) . \tag{1.6}$$

From (1.5), premultiplying by $\hat{\Sigma}$, we get

$$\underline{\hat{\mu}} = \underline{\bar{x}} . \tag{1.7}$$

Substituting this in (1.6), we obtain

$$\hat{\sigma}^{ij} = \frac{1}{n(2 - \delta_{ij})} \operatorname{tr} \left\{ \frac{\partial \hat{\Sigma}}{\partial \hat{\sigma}_{ij}} \hat{\Sigma}^{-1} S \hat{\Sigma}^{-1} \right\} \quad (i, j = 1, 2, \ldots, p) . \tag{1.8}$$

But $\partial\hat{\Sigma}/\partial\hat{\sigma}_{ij}$ is a p x p matrix, all the elements of which are zero, except those in the ith row and j^{th} column and j^{th} row and i^{th} column, which are unity; when i = j, only the ith diagonal element is unity and the others are all zero. Hence (1.8) reduces to

$$\hat{\sigma}^{ij} = \frac{1}{n(2 - \delta_{ij})} (2 - \delta_{ij}) \left\{ \begin{matrix} \text{the element in the } i^{th} \text{ row} \\ \text{and } j^{th} \text{ column of } \hat{\Sigma}^{-1} S \hat{\Sigma}^{-1} \end{matrix} \right\} . \quad (i, j = 1, 2, \ldots, p) . \tag{1.9}$$

Putting all these together, (1.9) can be also expressed as

$$\hat{\Sigma}^{-1} = \frac{1}{n} \hat{\Sigma}^{-1} S \hat{\Sigma}^{-1} . \tag{1.10}$$

Premultiplying by $\hat{\Sigma}$ and postmultiplying again by $\hat{\Sigma}$, we get

$$\hat{\Sigma} = \frac{1}{n} S . \tag{1.11}$$

Throughout this derivation, we have assumed that $\hat{\Sigma}$ is nonsingular. That this is so, with probability one, will be shown later in this chapter, when the p.d.f. of S is derived, it will also be shown that not $(1/n)S$, but $[1/(n-1)]S$ is unbiased for Σ.

We have thus seen that the maximum likelihood estimates of $\underline{\mu}$ and Σ are $\underline{\bar{x}}$ and $(1/n)S$ respectively. We shall now show that these estimates actually maximize the likelihood. From (1.1), we obtain [on substituting $\hat{\underline{\mu}} = \underline{\bar{x}}$ and $\hat{\Sigma} = (1/n)S$]

$$\begin{aligned}\log_e L(\hat{\underline{\mu}}, \hat{\Sigma}) &- \log L(\mu, \Sigma)\\ &= -\frac{n}{2}\log_e \left|\frac{1}{n}S\right| - \frac{np}{2} + \frac{n}{2}\log_e |\Sigma| + \frac{1}{2}\operatorname{tr}\Sigma^{-1}S\\ &\qquad + \frac{n}{2}\operatorname{tr}\Sigma^{-1}(\underline{\bar{x}} - \underline{\mu})(\underline{\bar{x}} - \underline{\mu})'\\ &\geq -\frac{n}{2}\log_e \frac{\left|\frac{1}{n}S\right|}{|\Sigma|} - \frac{np}{2} + \frac{1}{2}\operatorname{tr}\Sigma^{-1}S, \end{aligned} \tag{1.12}$$

as $\operatorname{tr}\Sigma^{-1}(\underline{\bar{x}} - \underline{\mu})(\underline{\bar{x}} - \underline{\mu})' = (\underline{\bar{x}} - \underline{\mu})'\Sigma^{-1}(\underline{\bar{x}} - \underline{\mu})$ is positive definite. Let $\theta_1, \theta_2, \ldots, \theta_p$ be the eigenvalues of the matrix $(1/n)\Sigma^{-1}S$. Then

$$\left|\frac{1}{n}\Sigma^{-1}S\right| = \prod_{i}^{p} \theta_i, \text{ and } \frac{1}{n}\operatorname{tr}\Sigma^{-1}S = \sum_{i=1}^{p} \theta_i .$$

Hence (1.12) reduces to

$$\begin{aligned}\log_e L(\hat{\underline{\mu}}, \hat{\Sigma}) - \log_e L(\underline{\mu}, \Sigma) &= \frac{n}{2}\left[-\log_e\left(\prod_{i}^{p}\theta_i\right) - p + \sum_{1}^{p}\theta_i\right]\\ &= \frac{n}{2}\sum_{i=1}^{p}(-\log_e \theta_i - 1 + \theta_i).\end{aligned} \tag{1.13}$$

For any non-negative α, $\alpha \leq e^{\alpha-1}$ and so by taking logs,

$$\alpha - \log_e^{\alpha} - 1 \geq 0 .$$

Using this in (1.13), we find that

$$\log_e L(\hat{\underline{\mu}}, \hat{\Sigma}) \geq \log_e L(\underline{\mu}, \Sigma) ,$$

for any $\underline{\mu}$ and Σ. This proves the required result. This proof is due to Watson [24].

2. THE WISHART DISTRIBUTION IN THE CANONICAL CASE

The Wishart matrix is defined as the $p \times p$ symmetric matrix of the s.s. and s.p. of the sample observations, from a p-variate nonsingular normal distribution. The sample observations could be measured either from the true means of the normal population or from the sample means. In this section, we shall consider the canonical case, i.e., we shall assume that the sample is from a p-variate normal population, for which $\underline{\mu} = \underline{0}$ and $\Sigma = I$. The Wishart matrix is then

$$[a_{ij}] = A = XX' , \tag{2.1}$$

where, as in Section 1, X is the $p \times n$ matrix of the sample observations in a sample of size n, and the s.s. and s.p. of the observations are about the true means, which are $\underline{0}$. In the next section, we shall consider the more general case when the variance-covariance matrix of the multivariate normal population is not I, but Σ, and $\underline{\mu} \neq \underline{0}$. Later, we shall consider the case when the s.s. and s.p. are about the sample means.

By distribution of A, we mean the joint distribution of the $p(p+1)/2$ distinct elements $a_{ij}(i, j = 1, 2, \ldots, p; i \leq j)$ of the symmetric matrix A, and we shall use dA to denote the product of the differentials of these distinct elements. Let $\underline{X}_i'$ denote the ith $(i = 1, 2, \ldots, p)$ row of the matrix X. We also assume n, the sample size, to be greater than p. Then, by using the Gram-Schmidt process of orthogonalization on the vectors $\underline{X}_1, \underline{X}_2, \ldots, \underline{X}_p$, we can obtain new vectors $\underline{Y}_1, \underline{Y}_2, \ldots, \underline{Y}_p$ which satisfy

$$\underline{Y}_i'\underline{Y}_j = \delta_{ij} \quad (i, j = 1, 2, \ldots, p) , \tag{2.2}$$

where δ_{ij} is the Kronecker delta. The relationship of the $\underline{Y}$'s and the $\underline{X}$'s is of the form

$$\underline{Y}_i = b^{i1}\underline{X}_1 + b^{i2}\underline{X}_2 + \ldots + b^{ii}\underline{X}_i \quad (i = 1,2,\ldots,p), \tag{2.3}$$

i.e., each $\underline{Y}_i$ is obtained by constructing a linear combination of $\underline{X}_1, \underline{X}_2, \ldots, \underline{X}_i$ only and the coefficients $b^{i1}, b^{i2}, \ldots, b^{ii}$ are so chosen that $\underline{Y}_i$ is orthogonal to $\underline{Y}_1, \underline{Y}_2, \ldots, \underline{Y}_{i-1}$ and is of unit length. Equation (2.3) can also be written in the compact form

$$Y = B^{-1}X , \tag{2.4}$$

where the $p \times n$ matrix Y is given by

$$Y = [\underline{Y}_1 | \underline{Y}_2 | \cdots | \underline{Y}_p]' , \tag{2.5}$$

and

$$B^{-1} = \begin{bmatrix} b^{11} & 0 & 0 & \ldots & 0 \\ b^{21} & b^{22} & 0 & \ldots & 0 \\ \hline b^{p1} & b^{p2} & b^{p3} & \ldots & b^{pp} \end{bmatrix}$$

is a l.t. matrix. The inverse of this matrix is also l.t. and, if we denote it by B and its nonzero elements by b_{ij}, we obtain, from (2.4), the reciprocal relation

$$X = BY . \tag{2.6}$$

The Wishart matrix, from (2.1), is therefore

$$A = BB' , \tag{2.7}$$

as

$$YY' = I_p \tag{2.8}$$

on account of (2.2). Equation (2.8) shows that Y is an incomplete orthogonal matrix. The matrix B of (2.7) provides what is known as the Bartlett decomposition of the Wishart matrix A. We shall first derive the distribution of B and, from that, derive the distribution of A, using (2.7) for the transformation. From (2.6), we obtain

$$\underset{-}{X}_i = b_{i1}\underset{-}{Y}_1 + b_{i2}\underset{-}{Y}_2 +\ldots+ b_{ii}\underset{-}{Y}_i \quad (i=1,2,\ldots,p) . \tag{2.9}$$

Premultiplying by $\underset{-}{Y}_j'$ and noting (2.2), we obtain

$$b_{ij} = \underset{-}{Y}_j'\underset{-}{X}_i ; \quad j = 1,2,\ldots,i, \ (i = 1,2,\ldots,p) . \tag{2.10}$$

Also, from (2.9),

$$a_{ii} = \underset{-}{X}_i'\underset{-}{X}_i = b_{i1}^2 + b_{i2}^2 +\ldots+ b_{ii}^2 \quad (i=1,2,\ldots,p) .$$

and, therefore,

$$b_{ii}^2 = \underset{-}{X}_i'\underset{-}{X}_i - \sum_{j=1}^{i-1} b_{ij}^2 \quad (i = 1, 2, \ldots, p) . \tag{2.11}$$

Since the p.d.f. of the population is $N_p(\underset{-}{x}|0|I)$, all the x_{ir} $(i = 1, 2, \ldots, p;\ r = 1, 2, \ldots, n)$ are NI(0, 1). We now use a random orthogonal transformation on these variates. The reader is already familiar with this technique, and the arguments associated with its application, from Section 3 of Chapter 2.

From (2.10), the transformation from $\underline{X}_i$ to the (i-1) variates $b_{i1}, b_{i2}, \ldots, b_{ii-1}$ is

$$\begin{bmatrix} b_{i1} \\ b_{i2} \\ \vdots \\ b_{i,i-1} \end{bmatrix} = \begin{bmatrix} \underline{Y}_1' \\ \hline \underline{Y}_2' \\ \hline \vdots \\ \hline \underline{Y}_{i-1}' \end{bmatrix} \underline{X}_i \tag{2.12}$$

This is an incomplete random orthogonal transformation. It is incomplete because (2.12) specifies only i-1 new variables, while $\underline{X}_i$ has n variates. It is orthogonal because the $\underline{Y}_i$'s satisfy (2.2). It is random because $\underline{Y}_1, \underline{Y}_2, \ldots, \underline{Y}_{i-1}$ are functions of $\underline{X}_1, \underline{X}_2, \ldots, \underline{X}_{i-1}$ which are random variates. We, therefore, hold $\underline{X}_1, \underline{X}_2, \ldots, \underline{X}_{i-1}$ fixed. This does not affect the distribution of $\underline{X}_i$. Hence, by theorem 3 of Chapter 2,

$$b_{i1}, b_{i2}, \ldots, b_{i,i-1} \text{ are } NI(0, 1); \tag{2.13}$$

$$\underline{X}_i'\underline{X}_i - b_{i1}^2 - \ldots - b_{ii-1}^2 = b_{ii}^2 \quad \text{(see 2.11)} \tag{2.14}$$

has the χ^2 distribution with n-(i-1) d.f. All these variables are independently distributed. This is, however, their conditional distribution, when $\underline{X}_1, \underline{X}_2, \ldots, \underline{X}_{i-1}$ are fixed. But $\underline{X}_1, \ldots, \underline{X}_{i-1}$ do not appear in the distribution of the variates $b_{i1}, b_{i2}, \ldots, b_{ii}$, given by (2.13) and (2.14). Hence (2.13) and (2.14) represent the unconditional distribution of $b_{i1}, b_{i2}, \ldots, b_{ii}$ and further these are independently distributed of $\underline{X}_1, \underline{X}_2, \ldots, \underline{X}_{i-1}$ and hence of $b_{rs}(r, s = 1, 2, \ldots, i-1; r \geq s)$

and $\underline{Y}_1, \ldots, \underline{Y}_{i-1}$, which are functions of $\underline{X}_1, \ldots, \underline{X}_{i-1}$ only. Using this result for each i, starting from $i = p$ down to $i = 1$, we obtain the following theorem.

Theorem 1: If the $p \times n$ matrix $X = [x_{ir}]$ represents a random sample of size n ($n > p$) from the $N_p(\underline{x}|\underline{0}|I_p)$ population, and if $XX' = BB'$, where $B = [b_{ij}]$ ($i = 1, 2, \ldots, p$; $j = 1, 2, \ldots, i$) is a lower triangular matrix [with $b_{ii} > 0$ ($i = 1, 2, \ldots, p$)] then the variates b_{ij} ($i, j = 1, 2, \ldots, p$; $i > j$) are NI(0,1); the variates b_{ii}^2 ($i = 1, 2, \ldots, p$) are independently distributed as χ^2 variates with n-(i-1) d.f. and are also independent of the b_{ij}'s ($i > j$).

The distribution of the matrix B is, therefore,

$$\prod_{i=1}^{p} \prod_{j=1}^{i-1} \left\{\frac{1}{\sqrt{2\pi}} e^{-(1/2)b_{ij}^2} d\, b_{ij}\right\} \prod_{k=1}^{p} \left\{\chi^2_{n-(k-1)}(b_{kk}^2)\, d(b_{kk}^2)\right\}.$$

$$-\infty < b_{ij} < \infty \quad \text{for} \quad i > j$$

$$0 < b_{ii} < \infty . \tag{2.15}$$

Transform now from B to A by (2.7). The Jacobian of transformation [4] is

$$J(B \to A) = \frac{1}{J(A \to B)} = 2^p \prod_{i=1}^{p} (b_{ii})^{i-p-1} . \tag{2.16}$$

Observing, from (2.7), that

$$|A| = \prod_{i=1}^{p} b_{ii}^2 \tag{2.17}$$

and

$$\text{tr } A = \text{tr } BB' = \sum_{i=1}^{b} \sum_{j=1}^{i} b_{ij}^2 , \tag{2.18}$$

the p.d.f. of A comes out as

$$W_p(A|n|I) = \begin{cases} K(p,n)\,|A|^{\frac{n-p-1}{2}} e^{-(1/2)\text{tr } A}, & A > 0 \\ 0, & \text{otherwise} , \end{cases} \tag{2.19}$$

where

$$1/K(p,n) = 2^{np/2} \pi^{p(p-1)/4} \prod_{i=1}^{p} \left\{\Gamma\left(\frac{n+1-i}{2}\right)\right\} , \tag{2.20}$$

and $A > 0$ means A is positive-definite. $W_p(A|n|I)$ stands for this Wishart density of A, given by (2.19). The parameter n is called the d.f. of the Wishart distribution. The matrix I in $W_p(A|n|I)$ is to remind us that the variance-covariance matrix of the p-variate normal distribution, from which X was drawn, is I. As before, we shall use $A \sim W_p(A|n|I)$ to denote that the $p \times p$ symmetric matrix A has the Wishart distribution given by (2.19).

By retracing the steps by which we derived the distribution of A from that of B, we can easily show that, if A has the $W_p(A|n|I)$ distribution, and if $A = BB'$, where B is l.t. with positive diagonal elements, then the distribution of B is given by (2.15). This is known as the Bartlett decomposition theorem.

3. THE WISHART DISTRIBUTION IN THE GENERAL CASE

Let us now consider the case when the matrix X of the previous section is a random sample of size n from the $N_p(\underline{x}|\underline{\mu}|\Sigma)$ population. We shall derive the distribution of the $p \times p$

symmetric matrix:

$$D = (X - \underline{\mu}\, E_{1n})\, (X - \underline{\mu}\, E_{1n})' \ . \tag{3.1}$$

This is the matrix of the s.s. and s.p. of the sample observations measured from the population means. Consider the transformation

$$\underline{y} = C^{-1}(\underline{x} - \underline{\mu}), \text{ where } \Sigma = CC', \text{ c being l.t.} \tag{3.2}$$

Then, from (1.1.5), $\underline{y} \sim N_p(\underline{y}|\underline{0}|I_p)$ and

$$Y = C^{-1}(X - \underline{\mu}\, E_{1n}) \tag{3.3}$$

is a random sample of size n from this distribution of $\underline{y}$. Hence, $A = YY'$ has the $W_p(A|n|I_p)$ distribution, as proved in the previous section. But, from (3.3) and (3.1),

$$A = C^{-1}\, D(C^{-1})' \ . \tag{3.4}$$

Using this relation, transform from A to D. The Jacobian of transformation is [4]

$$J(A \to D) = |C^{-1}|^{p+1} = |\Sigma|^{-(p+1)/2} \ . \tag{3.5}$$

Also observe that

$$|A| = |C^{-1}|\ |D|\ |C^{-1}| = |D|/|\Sigma| \tag{3.6}$$

and

$$\operatorname{tr} A = \operatorname{tr} C^{-1}\, D(C^{-1})' = \operatorname{tr}\, (CC')^{-1}D = \operatorname{tr} \Sigma^{-1}D \ . \tag{3.7}$$

Using (3.5), (3.6), and (3.7) to transform from A to D, the distribution of D comes out as

$$W_p(d|n|\Sigma)dD = \begin{cases} \dfrac{K(p,\ n)}{|\Sigma|^{n/2}}\, |D|^{(n-p-1)/2}\, e^{-(1/2)\operatorname{tr}\Sigma^{-1}D} dD & D > 0 \\ 0, \text{ otherwise.} & \end{cases} \tag{3.8}$$

This is known as the Wishart distribution with n d.f., and Σ is its associated matrix.

It is worth noting that, when $p = 1$, D and Σ are scalar quantities and $W_p(D|n|\Sigma)$ reduces to the p.d.f. of Σ times a chi-squared variate with n d.f. The Wishart distribution is thus the multivariate generalization of a χ^2 distribution. In fact, it plays the same role as that of a χ^2 distribution in multivariate regression analysis, multivariate analysis of variance and covariance, and in general multivariate linear models. The Wishart distribution in the particular case $p = 2$ was first derived by Fisher [6] in 1915, and for a general p, Wishart [25] obtained it in 1928 using n-dimensional geometry. Several proofs of its derivation have since appeared. Wishart himself listed a number of derivations and commented on them in [26]. Ingham [8] obtained the characteristic function of D and then inverted it. Mahalanobis, Bose, and Roy [17] used the method of rectangular coordinates. Olkin and Roy [21] later simplified this very much. Hsu [7] used the method of induction. Madow [16] derived it from the distribution of the correlation matrix. Elfving [5] used matrix methods to derive the Bartlett decomposition theorem first and then composed the Wishart distribution. The proof given here in this section is by the author [14] and is similar but uses the random orthogonal transformation as a tool. Sverdrup [23] uses the straightforward integration method to obtain the distribution of D from that of X.

Narain [19] uses the regression approach. Ogawa [20] uses a similar but more elegant method to derive the Wishart distribution as well as the Bartlett decomposition. Mauldon [18] evaluates the Jacobian of the transformation $X = BY$ (of 2.6), from X to B and Y, and from that derives the distribution of $D = XX' = BB'$. This is very similar to Olkin and Roy's method. Rasch [22] obtained a functional equation first and used it to obtain the Wishart distribution. Khatri [13] also gives a much simpler and direct derivation of the Wishart distribution by transforming from X to D and other variables, step by step, and then integrating out these latter variables.

All these derivations have an inherent similarity that is brought forth by the Bartlett decomposition theorem [1]. When we are concerned with $x_1, x_2, \ldots, x_p$ simultaneously, we can consider x_1 first, then the regression of x_2 on x_1, then of x_3 on x_1 and x_2, and so on. The regression coefficients are independently distributed of the residual sum of squares and are normally distributed, while the residual s.s. has a $\chi^2\sigma^2$ distribution, but its d.f. are n-m, where m is the number of the independent variables in the multiple regression. [This result is proved in its most general form in the next chapter.] We have then, from all these regressions (of x_i on $x_1, x_2, \ldots, x_{i-1}$; $i = 2, 3, \ldots, p$) the various regression coefficients and residual sums of squares. The b_{ij}'s $(i > j)$ of Section 2 are related to these regression coefficients and the b_{ii}^2's to the residual sums of squares, and that is why their d.f. go on decreasing from n to n-(p-1), as more and more independent variables are involved in the successive regressions. [Incidentally this process justifies the name d.f. also.] Bartlett [1] was the first to state this decomposition explicitly. Although his theorem can now be proved more easily and elegantly [15] his discussion in [1] is extremely interesting and provides added insight into the nature of the problem.

4. DISTRIBUTIONS OF THE MAXIMUM LIKELIHOOD ESTIMATES OF $\underline{\mu}$ AND Σ

With X representing a random sample of size n, from $N_p(\underline{x}|\underline{\mu}|\Sigma)$, as in the previous section, we already know that

$$\bar{\underline{x}} = \frac{1}{n} X E_{n1} \quad \text{and} \quad \frac{1}{n} S = \frac{1}{n} X(I - \frac{1}{n} E_{nn})X'$$

are the m.l.e.'s of $\underline{\mu}$ and Σ, respectively. We shall show, in this section, that they are independently distributed; $\sqrt{n}\,\bar{\underline{x}}$ has the $N_p(\sqrt{n}\,\bar{\underline{x}}/\sqrt{n}\,\underline{\mu}|\Sigma)$ distribution and S has the $W_p(S|n-1|\Sigma)$ distribution. For this we make the transformation

$$Y = XP \tag{4.1}$$

in the distribution (see 2.2.4)

$$\frac{1}{(2\pi)^{np/2}|\Sigma|^{n/2}} \exp^{[-(1/2)\mathrm{tr}\,\Sigma^{-1}(X - \underline{\mu}\,E_{1n})(X - \underline{\mu}\,E_{1n})']} dX$$

$$-\infty < X < \infty$$

of X. The matrix P is an n x n orthogonal matrix, such that its last column is $(1/\sqrt{n})E_{n1}$. The Jacobian of the transformation

[4] is

$$J(X \to Y) = |P|^p = 1 , \tag{4.2}$$

as P is orthogonal. Since the last column of P is $n^{-1/2} E_{n1}$, Y can be partitioned as

$$Y = [Z \vdots \sqrt{n}\, \underline{\bar{x}}] , \tag{4.3}$$

where Z is of order p x (n - 1). Also,

$$E_{1n}P = [0, 0, \ldots, 0, \sqrt{n}] , \tag{4.4}$$

as the first (n - 1) columns of P are orthogonal to the last one, which is proportional to E_{n1}. Hence

$$\begin{aligned}(X - \underline{\mu}\, E_{1n})(X - \underline{\mu}\, E_{1n})' &= (X - \underline{\mu}\, E_{1n})PP'\,(X - \underline{\mu}\, E_{1n})' \\ &= (Y - \underline{\mu}\, E_{1n}\, P)\,(Y - \underline{\mu}\, E_{1n}P)' \\ &= [Z|\sqrt{n}(\underline{\bar{x}} - \underline{\mu})]\,[Z|\sqrt{n}(\underline{\bar{x}} - \underline{\mu})]' \\ &= ZZ' + n(\underline{\bar{x}} - \underline{\mu})\,(\underline{\bar{x}} - \underline{\mu})' .\end{aligned} \tag{4.5}$$

On making the transformation (4.1), the distribution of Y, i.e., of Z and $\sqrt{n}\, \underline{\bar{x}}$ (see 4.3) becomes

$$\frac{1}{(2\pi)^{np/2}|\Sigma|^{n/2}} \exp^{[-(1/2)\mathrm{tr}\, \Sigma^{-1}\{ZZ' + n(\underline{\bar{x}} - \underline{\mu})(\underline{\bar{x}} - \underline{\mu})'\}]}\, d\, Z d(\sqrt{n}\, \underline{\bar{x}}) .$$

$$-\infty < z < \infty, \; -\infty < \sqrt{n}\, \underline{\bar{x}} < \infty \tag{4.6}$$

This shows that Z and $\sqrt{n}\, \underline{\bar{x}}$ are independently distributed, since (4.6) can be factorized into two parts, one containing only Z and the other containing only $\underline{\bar{x}}$. It also shows that

$$\sqrt{n}\, \underline{\bar{x}} \sim N_p(\sqrt{n}\, \underline{\bar{x}}|\sqrt{n}\, \underline{\mu}|\Sigma) , \tag{4.7}$$

and on comparing with (2.2.4), the factor containing Z only in (4.6) shows that Z is a random sample of size $n - 1$ (remember that Z has only $n - 1$ columns) from a p-variate normal population with zero means and variance-covariance matrix Σ. Hence from Section 3,

$$ZZ' \sim W_p(ZZ' | n - 1 | \Sigma) \ . \tag{4.8}$$

But as P is orthogonal, from (4.3),

$$\begin{aligned} ZZ' &= YY' - n \bar{\underline{x}} \bar{\underline{x}}' \\ &= XPP'X' - n \bar{\underline{x}} \bar{\underline{x}}' \\ &= X(I - \frac{1}{n} E_{nn})X' \\ &= S \text{ or } n \hat{\Sigma} \ . \end{aligned} \tag{4.9}$$

Hence the distribution of the m.l.e. of Σ is that of $(1/n)S$ and S has the $W_p(S | n - 1 | \Sigma)$ distribution, due to (4.8).

Note that S is the matrix of the s.s. and s.p. of sample observations, measured from the sample means and have $n - 1$ d.f. and not n, like the matrix D of (3.1), where the sample observations are measured from the true means $\underline{\mu}$. As $\bar{\underline{x}}$ and Z are independently distributed, $\bar{\underline{x}}$ and $S = ZZ'$ are also independently distributed. This is analogous to the univariate result of the independence of the sample mean and sample variance.

We observe that the p.d.f. of S is $W_p(S | n - 1 | \Sigma)$ is $S > 0$ and is zero otherwise and from

$$\int_{S>0} W_p(S | n - 1 | \Sigma) d\, S = 1 \tag{4.10}$$

conclude that S and hence $\hat{\Sigma}$ is positive-definite with probability 1, justifying the assumption made in Section 1.

Incidentally,

$$n(\bar{\underline{x}} - \underline{\mu})' \Sigma^{-1}(\bar{\underline{x}} - \underline{\mu}) = \underline{u}'\underline{u} , \tag{4.11}$$

where

$$\underline{u} = \sqrt{n}\, C^{-1} (\bar{\underline{x}} - \underline{\mu}), \quad \Sigma = CC' . \tag{4.12}$$

By Theorem 2 of Chapter 2, $\underline{u}$ has the $N_p(\underline{u}|\underline{0}|I_p)$ distribution and so

$$n(\bar{\underline{x}} - \underline{\mu})' \Sigma^{-1}(\bar{\underline{x}} - \underline{\mu}) \tag{4.13}$$

has the χ^2 distribution with p d.f. This provides a test of significance for the hypothesis

$$H_o: \quad \underline{\mu} = \underline{\mu}_o \text{ (specified)} ,$$

when a sample of size n is available from a p-variate normal population with unknown mean $\underline{\mu}$ but known variance-covariance matrix Σ.

5. MISCELLANEOUS RESULTS ASSOCIATED WITH THE WISHART DISTRIBUTION

Let the $p \times n$ matrix X represent a random sample of size n from the $N_p(\underline{x}|\underline{\mu}|\Sigma)$ population. Then HX is a random sample from $N_m(H\,\underline{x}|H\,\underline{\mu}|H\,\Sigma\,H')$, where H is any $m \times p$ matrix of constant elements and is of rank m. Then $(HX - H\underline{\mu}E_{1n})\,(HX - H\underline{\mu}E_{1n})' = HDH'$, where D is given by (3.1), and has the $W_m(HDH'|n|H\Sigma H')$ distribution, from the result of Section 3. This leads to:

Lemma 1: If a symmetric positive-definite matrix D has the $W_p(D|n|\Sigma)$ distribution, the matrix HDH' has the $W_m(HDH'|n|H\Sigma H')$ distribution, where H is any m x p matrix of constant elements and is of rank m.

In particular, if we take H to be

$$\begin{bmatrix} I_m & | & 0 \end{bmatrix}$$

we find that D_m has the $W_m(D_m|n|\Sigma_m)$ distribution, where D_m and Σ_m are the matrices formed by the first m rows and columns of D and Σ, respectively.

Let H consist of a single row vector $\underline{h}'$. We can even remove the restriction that the elements of $\underline{h}$ be constants and instead assume that $\underline{h}$ is independently distributed of D. Then, for fixed $\underline{h}$, the conditional distribution of $u = \underline{h}'D\underline{h}/\underline{h}'\Sigma\underline{h}$, by Lemma 1, is the χ^2 distribution with n d.f. But this distribution does not involve $\underline{h}$, and hence it is also the unconditional distribution of u and further it is independent of $\underline{h}$. We, therefore, have:

Lemma 2: If D has the $W_p(D|n|\Sigma)$ distribution, and $\underline{h}$, a p x 1 vector is independently distributed of D, $u = \underline{h}'D\underline{h}/\underline{h}'\Sigma\underline{h}$ has the $\chi_n^2(u)$ distribution and is independent of $\underline{h}$.

Consider now

$$A = C^{-1} D(C^{-1})' \quad \text{or} \quad D = CAC' , \qquad (5.1)$$

where $\Sigma = CC'$ and C is l.t. Then by Lemma 1, A has the

$W_p(A|n|I_p)$ distribution, in the canonical form. Let

$$A = BB' , \tag{5.2}$$

$$T = CB \quad \text{or} \quad B = C^{-1}T , \tag{5.3}$$

where B is l.t. T is also l.t. as C is l.t. Let b_{ij} be the nonzero elements of B. Observe that

$$D = TT' . \tag{5.4}$$

Hence

$$|D| = |T|^2 = |C|^2 \; |B|^2 = |\Sigma| \prod_{i=1}^{p} b_{ii}^2 . \tag{5.5}$$

From the last paragraph of Section 2, we know that the b_{ii}^2 are independently distributed as χ^2 with $n - (i - 1)$ d.f. Hence:

<u>Lemma 3</u>: If D has the $W_p(D|n|\Sigma)$ distribution, then $|D|/|\Sigma|$ is distributed as the product of p independent χ^2 variates with d.f. n, n-1, ..., $n - (p - 1)$, respectively.

It also follows from the preceding discussion that the hth moment of the distribution of $|D|$ is

$$E\{|D|^h\} = |\Sigma|^h \prod_{i=1}^{p} E(b_{ii}^2)^h$$

$$= 2^{hp}|\Sigma|^h \prod_{i=1}^{p} \left\{\frac{\Gamma\left(\frac{n+1-i}{2} + h\right)}{\Gamma\left(\frac{n+1-i}{2}\right)}\right\} . \tag{5.6}$$

Let, in general, M_i denote the matrix obtained from a matrix M, by retaining only its first i rows and columns. then, as B, C, T are all l.t.,

$$D_i = T_iT_i', \quad A_i = B_iB_i', \quad \Sigma_i = C_iC_i',$$

$$C_iB_i = T_i, \quad D_i = C_iA_iC_i' , \tag{5.7}$$

and therefore, $|A_i|/|A_{i-1}| = |B_iB_i'|/|B_{i-1}B_{i-1}'| = b_{ii}^2$ has the χ^2 distribution with $n-(i-1)$ d.f. In particular

$$b_{pp}^2 = \frac{|A_p|}{|A_{p-1}|} = \frac{|C_pA_pC_p'|}{|C_{p-1}A_{p-1}C_{p-1}'|} \cdot \frac{|C_{p-1}C_{p-1}'|}{|C_pC_p'|}$$

$$= \frac{|D_p|}{|D_{p-1}|} \cdot \frac{|\Sigma_{p-1}|}{|\Sigma_p|}$$

$$= \frac{\sigma^{pp}}{d^{pp}}, \tag{5.8}$$

where $D^{-1} = [d^{ij}]$ and $\Sigma^{-1} = [\sigma^{ij}]$. Hence:

<u>Lemma 4</u>: If D has the $W_p(D|n|\Sigma)$ distribution, then σ^{pp}/d^{pp} has the χ^2 distribution with $n-(p-1)$ d.f., where d^{pp}, σ^{pp} are the last elements of D^{-1} and Σ^{-1}, respectively.

Now apply Lemma 4 to the matrix $HDH'' = D^*$, where H is an orthogonal matrix of order p, such that its last row is $\underline{h}'$. By Lemma 1, $D^* \sim W_p(D^*|n|\Sigma^*)$, where $\Sigma^* = H\Sigma H'$. Hence σ^{*pp}/d^{*pp} has the χ^2 distribution with $n-(p-1)$ d.f., where $D^{*-1} = [d^{*ij}]$, $\Sigma^{*-1} = [\sigma^{*ij}]$. But as H is orthogonal, $D^{*-1} = HD^{-1}H'$ and $\Sigma^{*-1} = H\Sigma^{-1}H'$. Hence

$$\frac{\sigma^{*pp}}{d^{*pp}} = \frac{\underline{h}'\Sigma^{-1}\underline{h}}{\underline{h}'D^{-1}\underline{h}}.$$

If $\underline{h}$ is not a vector of constant elements but is distributed independently of D, we can make an argument similar to the one in Lemma 2 and obtain:

Lemma 5: If D has the $W_p(D|n|\Sigma)$ distribution and $\underline{h}$, a $p \times 1$ vector, is distributed independently of D, $\underline{h}'\Sigma^{-1}\underline{h}/\underline{h}'D^{-1}\underline{h}$ has the χ^2 distribution with $n-(p-1)$ d.f. and is independent of $\underline{h}$.

If the $p \times n$ matrix X is a random sample of size n from $N_p(\underline{x}|\underline{0}|\Sigma)$, $D = XX'$ has the $W_p(D|n|\Sigma)$ distribution. Theoretically, at least, this distribution of D can be obtained from that of the pn variables comprising X, by transforming from X to the $p(p+1)/2$ variates in D and $n - p(p+1)/2$ suitable other variates and then integrating out these latter variables. Let $\underline{u}$ denote the vector of these last $n - p(p+1)/2$ variates and let $J(D, \underline{u})$ denote the Jacobian of transformation from X to D and $\underline{u}$. The distribution of X is given by (2.2.4) if we set $\underline{\mu} = 0$. Transforming from X to D and $\underline{u}$, we obtain the joint distribution of D and $\underline{u}$ as

$$\frac{1}{(2\pi)^{np/2}|\Sigma|^{n/2}} e^{-(1/2)\mathrm{tr}\,\Sigma^{-1}D} J(D, \underline{u})\, dD\, d\underline{u} .$$

On integrating out $\underline{u}$, we must obtain $W_p(D|n|\Sigma)\, dD$, the distribution of D. So we obtain

$$\int_{\underline{u}} J(D, \underline{u})\, d\underline{u} = (2\pi)^{np/2} K(p, n) |D|^{\frac{n-p-1}{2}} , \quad \text{if} \quad n > p , \tag{5.9}$$

where the integration is over the appropriate range of the variables $\underline{u}$ and K(p, n) is given by (2.20). This result enables one to obtain the distribution of a matrix $D = XX'$, when X, a $p \times n$ matrix, has <u>any</u> distribution (not necessarily normal), such that its p.d.f. is a function of XX' alone, say $\psi(XX')$. We can then transform from X to D and $\underline{u}$. The joint distribution of D and $\underline{u}$ is $\psi(D)\, J(D, u)\, dD d\underline{u}$. Using (5.9), the distribution of D is

$$(2\pi)^{np/2}\, K(p, n)\, |D|^{\frac{n-p-1}{2}}\, \psi(D)\, dD\,. \tag{5.10}$$

This result is sometimes stated in the following form:

Lemma 6:

$$\int_{XX'=D} \psi(XX')\, dX = (2\pi)^{np/2}\, K(p, n)\, |D|^{\frac{n-p-1}{2}}\, \psi(D)\, dD\,,$$

where X is a $p \times n$ matrix, $n > p$ and $D > 0$.

The left-hand side in Lemma 6 represents the integral of $\psi(XX')$ with respect to X over the surface $XX' = D$ and is an abbreviation to indicate that a transformation is made from X to D and $\underline{u}$ and $\underline{u}$ is integrated out.

Integrating the p.d.f. $W_p(D|n|\Sigma)$ over the range $D > 0$ (i.e., all possible values of d_{ij}'s, such that $D > 0$), we obtain unity, as the total probability is 1. This leads to:

Lemma 7: (The Wishart-Bartlett lemma) [27]

$$\int_{D>0} |D|^{\frac{n-p-1}{2}}\, e^{-(1/2)\mathrm{tr}\, \Sigma^{-1} D}\, dD = \frac{|\Sigma|^{n/2}}{K(p,n)}\,,$$

where D and Σ are $p \times p$ symmetric positive-definite matrices and $n > p$. $K(p, n)$ is given by (2.20).

We now prove certain results about the $p \times p$ symmetric matrix A which has the $W_p(A|n|I_p)$ distribution in the canonical form.

Lemma 8: The distribution of A is invariant for a transformation of the type $A^* = HAH'$, where H is any $p \times p$ orthogonal matrix, the elements of which are either constants or are random variables, distributed independently of A; in the latter case HAH' is distributed independently of H.

First we consider the case when the elements of H are constants. The Jacobian of transformation from A to $A^* = HAH'$ is, by [4], $|H|^{p+1}$, which is 1 as H is orthogonal. Also, $|A^*| = |A|$ and $\text{tr } A^* = \text{tr } H'HA = \text{tr } A$. Hence the distribution of A^* is $W_p(A^*|n|I)\, dA^*$, by Lemma 1, and does not involve H. If H is a random orthogonal matrix, distributed independently of A, we can first consider the conditional distribution of A^*, when H is fixed, observe that it does not involve H, and prove Lemma 8 above.

Since A has the Wishart distribution in the canonical form, the elements a_{ij} of A can be looked upon as $a_{ij} = \sum_{r=1}^{n} x_{ir}x_{jr}$ where $x_{ir}(i = 1, 2, \ldots, p;\ r = 1, 2, \ldots, n)$ are all NI(0, 1). Hence, by symmetry it follows that the $a_{ii}(i = 1, 2, \ldots, p)$ are all identically distributed and the a_{ij} $(i, j = 1, 2, \ldots, p;\ i \neq j)$ are all identically distributed. Hence $E(A^k)$ must be of the form

$$\begin{bmatrix} \alpha & \beta & \beta \ldots \beta \\ \beta & \alpha & \beta \ldots \beta \\ \hdashline \beta & \beta & \beta \ldots \alpha \end{bmatrix} .$$

In other words,

$$E(A^k) = (\alpha - \beta)I_p + \beta E_{pp} . \tag{5.11}$$

But by Lemma 8, A and $A^* = HAH'$ have the same distribution. Hence,

$$E(A^k) = E(A^{*k}) = E(HAH' \cdot HAH' \ldots \cdot HAH')$$
$$= E(HA^kH') ,$$

as $H'H = I$, H being orthogonal. By (5.11), therefore,

$$(\alpha - \beta)I_p + \beta E_{pp} = H\left\{(\alpha - \beta)I_p + \beta E_{pp}\right\}H' . \tag{5.12}$$

Since Lemma 8 is true for any orthogonal matrix of constant elements, we choose H to be the orthogonal matrix that makes $H\left\{(\alpha - \beta)I + \beta E_{pp}\right\}H'$ a diagonal matrix. This, when used in (5.12), shows that β must be zero or that $E(A^k) = \alpha I_p$. Obviously α must depend on p, k and n. In an exactly similar way, we can show that $E(A^{-k})$ must be of the same form and hence we have:

<u>Lemma 9</u>: If A has the $W_p(A|n|I)$ distribution, then

(i) $E(A^k) = c(k, n, p)I_p$,

(ii) $E(A^{-k}) = d(k, n, p)I_p$,

where c(k, n, p) and d(k, n, p) are constants depending on k, n, p.

We shall now find c(1, n, p) and c(2, n, p). For this it is sufficient to find the expected value of any diagonal elements of A and A^2. Consider the Bartlett decomposition A = BB', where B is l.t., and b_{ij} (i, j = 1, 2, ..., p; j < i) are its nonzero elements. Then

$$c(1,n,p) = E(a_{11}) = E(b_{11}^2) = n \tag{5.13}$$

as $b_{11}^2 \sim \chi_n^2(b_{11}^2)$, by (2.15). The element in the first row and column of A^2 is

$$a_{11}^2 + a_{12}^2 + \ldots + a_{1p}^2 = b_{11}^4 + b_{11}^2(b_{12}^2 + b_{13}^2 + \ldots + b_{1p}^2) .$$

Hence, as the b_{ij} are all independent, $b_{11}^2 \sim \chi_n^2(b_{11}^2)$ and $b_{ij} \sim N(0, 1)$, $(i > j)$

$$c(2, n, p) = E(b_{11}^4) + E(b_{11}^2)\, E(b_{12}^2 + \ldots + b_{1p}^2)$$

$$= (n^2 + 2n) + n(p-1) = n(n + p + 1) . \tag{5.14}$$

Similarly, if $A^{-1} = [a^{ij}]$, from Lemma 9,

$$d(1, n, p) = E(a^{pp}) .$$

But by Lemma 4, $1/a^{pp}$ has the χ^2 distribution with n - (p - 1) d.f., and so

$$d(1, n, p) = \frac{1}{n-p-1} , \quad \text{if } n - p - 1 > 0 . \tag{5.15}$$

We shall derive d(2, n, p) later, as it requires additional results. Lemma 9 is due to Das Gupta [3].

If D has the $W_p(D|n|\Sigma)$ distribution and if we set D = CAC', where $\Sigma = CC'$, we find

$$\begin{aligned} E(D) &= E(CAC') \\ &= C\,E(A)C' \\ &= n\,\Sigma\,, \end{aligned} \tag{5.16}$$

as $A \sim W_p(A|n|I)$ by Lemma 1 and $E(A) = n\,I_p$ from Lemma 9. Also,

$$\begin{aligned} E(D^{-1}) &= E(CAC')^{-1} \\ &= (C')^{-1}\,E(A^{-1})C^{-1} \\ &= \frac{1}{n-p-1}\,\Sigma^{-1}, \quad \text{if } n - p - 1 > 0, \end{aligned} \tag{5.17}$$

as $E(A^{-1}) = \frac{1}{n-p-1}\,I_p$. Further

$$\begin{aligned} E(D^{-1}\,\Sigma\,D^{-1}) &= E(C'^{-1}\,A^{-2}\,C^{-1}) \\ &= d(2,\,n,\,p)\Sigma^{-1}\,. \end{aligned} \tag{5.18}$$

We thus have

<u>Lemma 10</u>: If D has the $W_p(D|n|\Sigma)$ distribution, then

(i) $E(D) = n\,\Sigma$,

(ii) $E(D^{-1}\,\Sigma\,D^{-1}) = d(2,\,n,\,p)\,\Sigma^{-1}$,

(iii) $E(D^{-1}) = \frac{1}{n-p-1}\,\Sigma^{-1}$, if $n - p - 1 > 0$.

If S is the matrix of the corrected s.s. and s.p. of sample observations in a sample of size n from $N_p(\underline{x}|\underline{\mu}|\Sigma)$, the maximum likelihood estimate of Σ is $(1/n)S$, but S has the $W_p(S|n-1|\Sigma)$ distribution, as proved in Section 4. By Lemma 10,

$$E(\frac{1}{n}\,S) = \frac{n-1}{n}\,S$$

and is thus biased, while $[1/(n-1)]S$ is unbiased for Σ.

6. THE CHARACTERISTIC FUNCTION OF THE WISHART DISTRIBUTION

Let X represent the p x n matrix of sample observations in a random sample of size n from the $N_p(\underline{x}|\underline{0}|\Sigma)$ population. The Wishart matrix is then $D = [d_{ij}] = XX'$, and its characteristic function i.e. the characteristic function of the $p(p+1)/2$ variates $d_{11}, d_{12}, \ldots, d_{1p}, d_{22}, d_{23}, \ldots, d_{2p}, \ldots, d_{pp}$ is, by definition

$$\begin{aligned}\phi &\equiv \phi(t_{11}, t_{12}, \ldots, t_{1p}, t_{22}, \ldots, t_{2p}, \ldots, t_{pp}) \\ &= E\left\{\exp[\sqrt{-1}(t_{11}d_{11} + t_{12}d_{12} + \ldots + t_{pp}d_{pp})]\right\} \\ &= E\left\{\exp[(1/2)\sqrt{-1} \sum_{i=1}^{p} \sum_{j=1}^{p} (1 + \delta_{ij})t_{ij}d_{ij}]\right\} \\ &= E\left\{\exp[(1/2)\sqrt{-1} \operatorname{tr} DK]\right\}, \end{aligned} \quad (6.1)$$

where

$$t_{ij} = t_{ji} \quad \text{for } i > j, \; k_{ij} = (1 + \delta_{ij})t_{ij}, \; K = [k_{ij}]$$

and δ_{ij} is the Kronecker delta. Using (2.2.4) with $\underline{\mu} = 0$ for the distribution of X, the characteristic function ϕ is

$$\phi = \int_{-\infty}^{\infty} \exp[(1/2)\sqrt{-1} \operatorname{tr} DK] \, .$$

$$\cdot \frac{1}{(2\pi)^{np/2}|\Sigma|^{n/2}} \exp[-(1/2) \operatorname{tr} \Sigma^{-1}XX']dX$$

$$= \int_{-\infty}^{\infty} \frac{1}{(2\pi)^{np/2}} \exp[-(1/2) \operatorname{tr} \{\Sigma^{-1} - \sqrt{-1}\, K\}XX']dX \, , \quad (6.2)$$

where a single integral sign is used to denote the np-fold integral. The above integral can be evaluated by using the following result:

$$\int_{-\infty}^{\infty} \exp[-(1/2)\operatorname{tr} G^{-1} YY']\, dY = (2\pi)^{mp/2}|G|^{m/2}, \quad (6.3)$$

where Y is an $p \times m$ matrix and G is any positive definite matrix. [This result can be established by observing that Y can be regarded as a random sample of size n from a p-variate normal population with zero means and variance-covariance matrix G. Then use the fact that the integral of the p.d.f. of Y over the entire range of Y is unity.] Therefore,

$$\phi = \frac{|\Sigma^{-1} - \sqrt{-1}\, K|^{-n/2}}{|\Sigma|^{n/2}}$$

$$= \frac{|\Sigma^{-1}|^{n/2}}{|\Sigma^{-1} - \sqrt{-1}\, K|^{n/2}} = \frac{1}{|I - \sqrt{-1}\, K\, \Sigma|^{n/2}}. \quad (6.4)$$

Since (6.3) holds, even when $m < p$, (6.4) also is true, even if $n < p$. But the p.d.f. of D, viz., $W_p(D|n|\Sigma)$, exists only when $n > p$, as we have seen. When $n < p$, the distribution of D does not have any p.d.f., however its characteristic function is given by (6.4). In such a case we shall call the distribution of D a pseudo-Wishart distribution. Thus, a matrix D of order p will be said to have a pseudo-Wishart distribution of n d.f. and associated matrix Σ if it can be expressed as $D = XX'$, where X is of order $p \times n$ with $n \le p$ and the n columns of X are independently and identically distributed as $N_p(\underline{x}|\underline{0}|\Sigma)$. Even if no p.d.f. exists, for such a pseudo-Wishart distribution, any probability statement about any elements of D can be translated into a suitable equivalent statement about elements of X, whose p.d.f. is known. The moments and product moments of the d_{ij}'s, can be obtained from the characteristic function by the usual technique (for the Wishart as well as the pseudo-Wishart distribution). Thus,

$$E(d_{ij}d_{kl}) = n^2\sigma_{ij}\sigma_{kl} + n\,\sigma_{ik}\sigma_{jl} + n\,\sigma_{il}\sigma_{jk}\ . \qquad (6.5)$$

where $\Sigma = [\sigma_{ij}]$, for i, j, k, l = 1, 2, ..., p. Hence

$$\mathrm{Cov}(d_{ij},\ d_{kl}) = n(\sigma_{ik}\sigma_{jl} + \sigma_{il}\sigma_{jk})\ , \qquad (6.6)$$

as $E(d_{ij}) = n\ \sigma_{ij}$.

The Wishart distribution has an additive property, similar to that of the χ^2 distribution. If $D_1, D_2, \ldots, D_k$ are k p x p matrices, having the independent Wishart distributions $W_p(D_r|n_r|\Sigma)$, (r = 1, 2, ..., k), then the matrix $D = D_1 + D_2 + \ldots + D_k$ has the $W_p(D|n|\Sigma)$ distribution, with $n = n_1 + n_2 + \ldots + n_k$ as the d.f. This can be proved by observing that the characteristic function of D is the product of the characteristic functions of the individual D_r's and then identifying the characteristic function of D with that of a Wishart distribution of n d.f. and associated matrix Σ. It should be noted that this result is true even if the D_r's have pseudo-Wishart distributions. The distribution of D, however, will be Wishart or pseudo-Wishart according as $n > p$ or $n \leq p$.

Throughout this discussion, we have assumed that Σ is nonsingular. If Σ is singular, the underlying normal distribution itself is singular and that introduces a further complication in the Wishart or pseudo-Wishart distribution, but we shall be considering only nonsingular normal distributions.

7. DISTRIBUTION OF XAX'

Let the p x n matrix X have the distribution

$$\frac{1}{(2\pi)^{np/2}|\Sigma|^{n/2}} e^{-(1/2)\text{tr}\,\Sigma^{-1}XX'}\,dX\ , \tag{7.1}$$

i.e., columns of X are independently distributed as $N_p(\underline{x}|\underline{0}|\Sigma)$. Let A be any n x n idempotent matrix, of rank m. Then there exists an orthogonal matrix P of order n, such that

$$P'AP = \text{diag}(\underbrace{1, 1, \ldots, 1}_{m},\ 0, 0, \ldots, 0)\ . \tag{7.2}$$

Hence

$$\begin{aligned} XAX' &= XPP'APP'X' = XP\ \text{diag}(1,1,\ldots,1,\ 0,0,\ldots,0)P'X' \\ &= YY'\ , \end{aligned} \tag{7.3}$$

where

$$XP = [\underset{m}{Y}|\underset{n-m}{Z}]\,p\ . \tag{7.4}$$

By using the derivation employed in Section 4, we can see that Y can be regarded as a random sample of size n from a p-variate normal population with zero means and variance-covariance matrix Σ. Hence, from (7.3) and Section 3, XAX' will have the $W_p(XAX'|m|\Sigma)$ distribution if $m > p$ and pseudo-Wishart if $m \leq p$.

Conversely, if XAX' has the $W_p(XAX'|m|\Sigma)$ distribution, and X has the distribution (7.1), from Lemma 2, taking $\underline{h}' = [1, 0, \ldots, 0]$, we find that $\underline{x}'A\underline{x}/\sigma_{11}$ has the χ^2 distribution with m d.f., where $\underline{x}'A\underline{x}$ is the first element of XAX' and $\underline{x}'$ is the first row of X. Therefore, as $\underline{x} \sim N_n(\underline{x}|\underline{0}|\sigma_{11}I_n)$, from Section 5 of Chapter 2 it follows that A must be an idempotent matrix of rank m.

It is also obvious that James's theorem on quadratic forms, stated in Section 5 of Chapter 2, is also applicable to XA_iX' (i = 1, 2, ..., k) if

$$XX' = \sum_{i=1}^{k} XA_iX' ,$$

and the distribution of X is as given by (7.1), provided we replace the word "χ^2 distribution", by the word "Wishart distribution" (or pseudo-Wishart, if the d.f. are less than p), with Σ as its associated matrix.

In univariate analysis of variance, the different sums of squares are quadratic forms in normal variables. In multivariate analysis of variance, their role is taken by s.s. and s.p. matrices, which are of the form XAX', described in the preceding paragraph, and are therefore important.

8. THE NONCENTRAL WISHART DISTRIBUTION

If the distribution of X is not (7.1) but

$$\frac{1}{(2\pi)^{np/2}|\Sigma|^{n/2}} \exp^{[-(1/2)\mathrm{tr}\,\Sigma^{-1}(X-M)(X-M)']} dX \quad (8.1)$$

i.e., if the columns of X are independently normally distributed, not about zero means, but about non-null means given by the columns of M, the distribution of D = XX' will not be $W_p(D|n|\Sigma)$. The distribution of D in such a case is called the noncentral Wishart distribution. Its p.d.f. (when $n > p$) in its most general form was first derived by A. T. James [9, 10, 12] (see also Constantine [2]). It involves the use of zonal polynomials [11]. They are certain symmetric polynomials in the eigenvalues of a matrix. Bessel functions and hypergeometric functions with matrix arguments are also needed. All this is beyond the scope of this book. We, however, add here that the p.d.f. of D involves the eigenvalues of the matrix $\Sigma^{-1}MM'$. Hence $\Sigma^{-1}MM'$ is called the noncentrality matrix of the noncentral Wishart distribution. The rank of $\Sigma^{-1}MM'$, which is the same as the rank of MM' or M, is therefore used to specify the noncentral Wishart distribution. Thus if the rank of M is only 1, the noncentral Wishart distribution is said to belong to the linear case. If it is 2, it is the 'planar' case, and so on.

In particular if $\Sigma = I$ and

$$M = \begin{bmatrix} \mu_1 & \mu_2 & \cdots & \mu_n \\ 0 & 0 & \cdots & 0 \\ \hline 0 & 0 & \cdots & 0 \end{bmatrix} \tag{8.2}$$

i.e., the means of x_{1r} $(r = 1, 2, \ldots, n)$, the elements in the first row only of X are non-null, and if we consider the Bartlett decomposition $D = BB'$, where B is l.t. (with b_{ij} as its nonzero elements), it can be easily seen, in exactly the same way as in Section 2, that the distribution of B remains as it was in the case of a central Wishart distribution [as given by (2.15)], except that b_{11}^2 will now have a noncentral χ^2 distribution with n d.f. and noncentrality parameter $\lambda^2 = \mu_1^2 + \mu_2^2 + \dots + \mu_n^2$. This is so because $b_{11}^2 = x_{11}^2 + \dots + x_{1n}^2$ and x_{1r} are $NI(\mu_r, 1)$. The distribution of D, therefore, turns out to be

$$W_p(D|n|I) = e^{-\lambda^2/2} \sum_{\alpha=0}^{\infty} \frac{(\lambda^2/2)^\alpha}{\alpha!} \frac{\Gamma(n/2)}{\Gamma(n/2+\alpha)} \left(\frac{d_{11}}{2}\right)^\alpha \cdot dD. \tag{8.3}$$

This is the canonical form of the noncentral Wishart distribution in the linear case. Observe that λ^2 is the only nonzero eigenvalue of MM'. In this case, the elements of the l.t. matrix B are still independently distributed. However, when we go beyond the linear case, this independence also breaks down and the derivation becomes complicated [15]. If the matrix M is not of the form (8.2), but still of rank 1, we can reduce the problem, by suitable transformations, to one in which M is of the form (8.2) and then use (8.3).

If the distribution of X is (8.1), and A is an idempotent matrix of order n and rank m, we can show, by using (7.3) and (7.4), that XAX' will have a noncentral Wishart distribution with m d.f., associated matrix Σ, and noncentrality matrix $\Sigma^{-1}MAM'$. However, if we need E(XAX'), we observe that $E(X - M) = 0$ and so $(X - M) A (X - M)' = U$ has the $W_p(U|m|\Sigma)$ distribution, and hence by Lemma 10, $E(U) = m \Sigma$, from which it follows that

$$E(XAX') = m \Sigma + MAM' . \tag{8.4}$$

Since the characteristic function of U is the same whether $m \leq p$ or $m > p$, the result $E(U) = m \Sigma$ is true for pseudo-Wishart distributions also, and therefore (8.4) holds even when $m \leq p$ and XAX' has a noncentral pseudo-Wishart distribution.

REFERENCES

[1] Bartlett, M. S. "On the theory of statistical regression", Proc. Roy. Soc. Edinb., 53, p. 260.

[2] Constantine, A. G. (1963). "Some non-central distribution problems in multivariate analysis", Ann. Math. Statist., 34, p. 1270.

[3] Das Gupta, S. (1968). "Some aspects of discrimination function coefficients", Sankhyā, A 30, p. 387.

[4] Deemer, Walter L. and Olkin, I. (1951). "The jacobians of certain matrix transformations useful in multivariate analysis", Biometrika, 38, p. 345.

[5] Elfving, G. (1947). "A simple method of deducing certain distributions connected with multivariate sampling", Skand. Aktuarietidskr; 30, p. 56.

[6] Fisher, R. A. (1915). "Frequency distribution of the values of the correlation coefficient in samples from an indefinitely large population", Biometrika, 10, p. 507.

[7] Hsu, P. L. (1939), "A new proof of the joint product moment distribution", Proc. Camb. Philos. Soc., 35, p. 336.

[8] Ingham, A. E. (1933). "An integral that occurs in statistics", Proc. Camb. Philos. Soc., 29, p. 271.

[9] James, A. T. (1954). "Normal multivariate analysis and the orthogonal group", Ann. Math. Statist., 25, p. 40.

[10] James, A. T. (1955). "The non-central Wishart distribution", Proc. Roy. Soc. A, 229, p. 364.

[11] James, A. T. (1961). "Zonal polynomials of the real positive definite symmetric matrices", Ann. of Math., 74, p. 456.

[12] James, A. T. (1964). "Distributions of matrix variates and latent roots derived from normal samples", Ann. Math. Statist., 35, p. 475.

[13] Khatri, C. G. (1963). "Wishart distribution" (Queries and answers section), J. Ind. Stat. Ass., 1, p. 30.

[14] Kshirsagar, A. M. (1959). "Bartlett decomposition and Wishart distribution", Ann. Math. Statist., 30, p. 239.

[15] Kshirsagar, A. M. (1963). "Effect of non-centrality on the Bartlett decomposition of a Wishart matrix", Ann. Inst. Stat. Math., 14, p. 217.

[16] Madow, W. G. (1938). "Contributions to the theory of multivariate statistical analysis", Trans. Amer. Math. Soc., 44, p. 454.

[17] Mahalanobis, P. C., Bose, R. C. and Roy, S. N. (1937). "Normalization of statistical variates and the use of rectangular coordinates in the theory of sampling distributions", Sankhyā, 3, p. 1.

[18] Mauldon, J. G. (1955). "Pivotal quantities for Wishart's and related distributions and a paradox in fiducial theory", J. Roy. Stat. Soc., B, 17, p. 79.

[19] Narain, R. D. (1948). "A new approach to the sampling distributions of the multivariate normal theory", J. Ind. Soc. Agric. Stat., 1, p. 59.

[20] Ogawa, J. (1953). "On the sampling distributions of classical statistics in multivariate analysis", Osaka Math. J., 5, p. 13.

[21] Olkin, I. and Roy, S. N. (1954). "On multivariate distribution theory", Ann. Math. Statist., 25, p. 329.

[22] Rasch, G. (1948). "A functional equation for Wishart's distribution", Ann. Math. Statist., 19, p. 262.

[23] Sverdrup, E. (1947). "Derivation of the Wishart distribution of the second order sample moments by straightforward integration of a multiple integral", Skand. Aktuarietidskr., 30, p. 151.

[24] Watson, G. W. (1964). "A note on maximum likelihood", Sankhyā, 26A, p.

[25] Wishart, J. (1928). "The generalized product moment distribution in samples from a normal multivariate population", Biometrika, 20A, p. 32.

[26] Wishart, J. (1948). "Proofs of the distribution law of the second order moment statistics", Biometrika, 35, p. 55.

[27] Wishart, John and Bartlett, M. S. (1932). "The distribution of second order moment statistics in a normal system", Proc. Camb. Phil. Soc., 28, p. 455.

CHAPTER 4

DISTRIBUTIONS ASSOCIATED WITH REGRESSION

1. DISTRIBUTION OF THE CORRELATION COEFFICIENT

In Section 3 of Chapter 2, we obtained the distribution of the correlation coefficient r in a sample of size n from a bivariate normal population in which the true (i.e., population) correlation coefficient ρ was null. Here we shall obtain the distribution of r when ρ is non-null.

Let $S = [s_{ij}]$ (i, j = 1, 2) be the matrix of the corrected sum of squares and products in a sample of size n from the bivariate normal population $N_2(\underline{x}|\underline{\mu}|\Sigma)$. The population correlation coefficient is $\rho = \sigma_{12}/(\sigma_{11}\ \sigma_{22})^{1/2}$, where $\Sigma = [\sigma_{ij}]$ and the sample correlation coefficient is $r = s_{12}/(s_{11}\ s_{22})^{1/2}$. We assume that $n > 3$. From Section 4 of Chapter 3, S has the $W_2(S|\Sigma|n-1)$ distribution. Let C and T be l.t. matrices of order 2 such that $\Sigma = CC'$ and $S = TT'$. Then from Lemma 1, Section 5 of Chapter 3, $A = C^{-1}\,S\,C'^{-1}$ has the $W_2(A|n-1|I)$ distribution, which is in the canonical form, and $A = BB'$, where $B = C^{-1}T$ is l.t. as C and T are both l.t. Therefore, the nonzero elements of B, viz., b_{11}, b_{22}, b_{21}, have, by the Bartlett decomposition theorem (Section 2, Chapter 3), a chi-distribution with n-1 d.f., a chi-distribution with n-2 d.f., and a N(0, 1) distribution, respectively, and they are all independent. By equating the elements of $S = TT'$, we find

$$r^* = \frac{r}{\sqrt{1 - r^2}} = \frac{s_{12}}{\sqrt{s_{11}\, s_{22} - s_{12}^2}} = \frac{t_{21}}{t_{22}}, \tag{1.1}$$

and from $T = CB$, we find

$$t_{21} = c_{21}\, b_{11} + c_{22}\, b_{21}, \quad t_{22} = c_{22}\, b_{22}, \tag{1.2}$$

where t_{ij}, c_{ij} $(i, j = 1, 2;\ i > j)$ are the nonzero elements of the l.t. matrices T and C, respectively. Hence, from (1.1),

$$r^* = \frac{b_{21} + \rho^*\, b_{11}}{b_{22}}, \tag{1.3}$$

where

$$\rho^* = \frac{\rho}{(1 - \rho^2)^{1/2}} = \frac{c_{21}}{c_{11}}, \tag{1.4}$$

as $\Sigma = CC'$.

As b_{21} is N(0, 1), we find from (1.3) that the conditional distribution of r^* when b_{11} and b_{22} are fixed is normal with mean $\rho^*\, b_{11}/b_{22}$ and variance $1/b_{22}^2$. The joint distribution of r^*, b_{11}^2, b_{22}^2 is, therefore,

$$f(r^*|b_{11},b_{22})\chi^2_{n-1}(b_{11}^2)\chi^2_{n-2}(b_{22}^2)\,dr^*db_{11}^2\,db_{22}^2, \tag{1.5}$$

where $f(r^*|b_{11}, b_{22})$ denotes the p.d.f. of the conditional distribution of r^*, mentioned above. Transforming from r^* to r, b_{11}^2 to $v = b_{11}/(1 - \rho^2)^{1/2}$, and b_{22}^2 to $w = b_{22}/(1 - \rho^2)^{1/2}$, and using

$$2^{n-3}\,\Gamma\!\left(\frac{n-1}{2}\right)\Gamma\!\left(\frac{n-2}{2}\right) = \sqrt{\pi}\,\Gamma(n-2), \tag{1.6}$$

we find the joint distribution of r, v, w to be

$$\frac{(1-\rho^2)^{(n-1)/2}(1-r^2)^{(n-4)/2}}{\Gamma(n-4)\cdot\pi}(vw)^{n-2}\exp\{-(1/2)(r^2-2\rho rvw+w^2)\}drdvdw,$$
$$-1 \le r \le 1,$$
$$0 < v < \infty,$$
$$0 < w < \infty. \qquad (1.7)$$

We now expand exp(prvw) in (1.7) and then integrate out v and w by term-by-term integration. This yields the non-null distribution of r (i.e., the distribution of r, when $\rho \neq 0$), in the form

$$p_n(r)dr = \frac{2^{n-3}(1-\rho^2)^{(n-1)/2}(1-r^2)^{(n-4)/2}dr}{\pi\,\Gamma(n-2)}\sum_{j=0}^{\infty}\Gamma\left(\frac{n-1+j}{2}\right)\Gamma\left(\frac{n-1+j}{2}\right)\frac{(2\rho r)^j}{j!}$$
$$-1 \le r \le 1. \qquad (1.8)$$

An alternative expression for $p_n(r)$ can be obtained if we use

$$\int_0^\infty\int_0^\infty \exp\{-(1/2)(v^2 - 2prvw + w^2)\}\,dv\,dw = \frac{\cos^{-1}(-pr)}{(1-\rho^2r^2)^{1/2}}, \qquad (1.9)$$

which can be established by transforming from v, w to the polar coordinates a, ϕ by using $v = a\sin\phi$, $w = a\cos\phi$. Differentiating both sides of (1.9) (n-2) times with respect to ρr, we get

$$\int_0^\infty\int_0^\infty (vw)^{n-2}\exp\{-(1/2)(v^2 - 2prvw + w^2)\}\,dv\,dw$$
$$= \frac{\partial^{(n-2)}}{\partial(\rho r)^{n-2}}\left\{\frac{\cos^{-1}(-\rho r)}{(1-\rho^2r^2)^{1/2}}\right\}. \qquad (1.10)$$

Using this to integrate out v and w from (1.7), pn(r) can be expressed as

$$pn(r) = \frac{(1-\rho^2)^{(n-1)/2}(1-r^2)^{(n-4)/2}dr}{\pi\,\Gamma(n-2)}\frac{\partial^{(n-2)}}{\partial(\rho r)^{n-2}}\left\{\frac{\cos^{-1}(-\rho r)}{(1-\rho^2r^2)^{1/2}}\right\}. \qquad (1.11)$$

Fisher [7] was the first to obtain the distribution of r; however he had used n-dimensional geometry as a tool to derive the distribution.

By putting $\rho = 0$ in (1.8), or observing from (1.3) that $r^* = b_{21}/b_{22}$ when $\rho = 0$, one can easily see that the null distribution of r is as given by (2.3.13), which we derived earlier by a different method.

The distribution of r is very complicated. David [4] has prepared tables which give $p_n(r)$ and the cumulative distribution function (c.d.f.) $F_n(r_0) = \text{Prob}(r \leq r_0)$, for various values of n. The moments of the distribution of r have been obtained by Ghosh [11] and are in terms of the hypergeometric function. For large n, and as a useful approximation,

$$E(r) \doteq \rho - \frac{1}{2n}\rho(1-\rho^2) \tag{1.12}$$

and

$$V(r) \doteq \frac{1}{n}(1-\rho^2)^2 . \tag{1.13}$$

Hotelling [17] has given an interesting method of derivation of $p_n(r)$ when $\rho \neq 0$, from that of $p_n(r)$ when $\rho = 0$. He has also given a large number of useful results associated with r. Harley [15, 16] has proved by induction that

$$E(\sin^{-1} r) = \sin^{-1} \rho . \tag{1.14}$$

This is known as Harley's theorem. $\text{Sin}^{-1} r$ is the only function of r that is unbiased for the same function of ρ. A short geometrical proof of Harley's theorem is given by Daniels and

Kendall [3], while an analytical proof based on the characteristic function of $\sin^{-1} r$ can be found in Khatri [19]. He gives higher moments of $\sin^{-1} r$ also.

Since the distribution of r is complicated, it is not of much practical use, but one can consider a transformed variable g(r) and employ its Taylor expansion about $r = \rho$, viz.,

$$g(r) = g(\rho) + (r - \rho) \left(\frac{\partial g}{\partial r}\right)_{r=\rho} + \dots .$$

By neglecting higher order terms in this expansion, the approximate mean and variance of g(r) for large n are $g(\rho)$ and

$$V(r) \left[\left(\frac{\partial g}{\partial r}\right)_{r=\rho}\right]^2 . \tag{1.15}$$

We use (1.13) and choose g(r) so that the above variance of g(r) equals a constant, independent of ρ. This shows that g(r) must be of the form

$$g(r) = \text{constant} \int \frac{dr}{1 - r^2} = \tanh^{-1} r = (1/2)\log_e \frac{1+r}{1-r} . \tag{1.16}$$

This transformed variable is denoted by z and the transformation from r to z is known as Fisher's z transformation. The corresponding function of ρ is $\tanh^{-1}\rho$ or $(1/2)\log_e (1 + \rho)/(1 - \rho)$ and is denoted by ξ. Fisher [6] and later Gayen [10] and Nabeya [21] have obtained the moments of z and from that conclude that the distribution of z is much closer to normality than the original distribution of r. For moderately large n,

$$E(z) = \xi, \quad V(z) = 1/(n-3) \tag{1.17}$$

approximately, and thus $z \sim N(\xi, 1/(n-3))$. This is a useful approximation, providing tests of significance and confidence intervals in problems associated with inference on ρ. A slight improvement over (1.17) is

$$E(z) = \xi + \frac{\rho}{2(n-1)} .$$

A useful and important feature of z is that its variance is independent of ρ. This "stabilization" of variance incidentally also brought the distribution of z closer to normality. This happens even in the case of other transformations. Fisher's z transformation thus opened a whole new field of transformations. Normality and homogeneity of variance are two basic assumptions underlying the useful technique of analysis of variance in statistics, and it is indeed a fortunate thing that both are achieved simultaneously by transformation. A useful account of transformations used in statistics can be found in Bartlett [2].

Recently Ruben [23] has suggested another approximation to the distribution of r, and to obtain it he uses the following result due to Fisher:

> If U^2 has a χ^2 distribution with f d.f., the distribution of $2^{1/2}\{U - (f - 1/2)^{1/2}\}$ is approximately N(0, 1) for large f. (1.18)

In other words, a chi-variable with f d.f. can be replaced approximately by a normal variable with mean $(f - 1/2)^{1/2}$ and variance 1/2, for large f. Now, in (1.3) b_{11} and b_{22} are chi-variables with $n-1$ and $n-2$ d.f. and hence, for $r_o^* = r_o/(1-r_o^2)^{1/2}$,

$$\begin{aligned}\text{Prob}(r \le r_o) &= \text{Prob}(r^* \le r_o^*) \\ &= \text{Prob}(b_{21} + \rho^* b_{11} - r_o^* b_{22} \le 0) \\ &= \text{Prob}(L \le 0), \qquad (1.19)\end{aligned}$$

where

$$L = b_{21} + \rho^* b_{11} - r_o^* b_{22} \,. \qquad (1.20)$$

Using (1.18) for the chi-variables b_{11} and b_{22}, the approximate distribution of L is normal with mean

$$\alpha = \rho^*\left(\frac{2n-3}{2}\right)^{1/2} - r_o^*\left(\frac{2n-5}{2}\right)^{1/2} \qquad (1.21)$$

and variance

$$\beta^2 = 1 + (1/2)\rho^{*2} + (1/2)r_o^{*2} \,, \qquad (1.22)$$

as L is a linear combination of b_{21}, b_{11}, b_{22}, which are all normally distributed, either exactly or approximately for large n. Hence, from (1.19),

$$\begin{aligned}\text{Prob}(r \le r_o) &= \text{Prob}\left(\frac{L-\alpha}{\beta} \le -\frac{\alpha}{\beta}\right) \\ &\doteq \Phi\left(-\frac{\alpha}{\beta}\right), \qquad (1.23)\end{aligned}$$

where

$$\Phi(y) = \int_{-\infty}^{y} \frac{1}{\sqrt{2\pi}} e^{-y^2/2}\, dy \,, \qquad (1.24)$$

is the c.d.f. of a N(0, 1) variable. Thus

$$\text{Prob}(r \le r_o) \doteq \Phi(h_n(r_o, \rho)) \,, \qquad (1.25)$$

where

$$h_n(r, \rho) = \frac{\left(\frac{2n-5}{2}\right)^{1/2} r^* - \left(\frac{2n-3}{2}\right)^{1/2} \rho^*}{(1 + (1/2) r^{*2} + (1/2) \rho^{*2})^{1/2}} . \tag{1.26}$$

In other words,

$$\begin{aligned} \text{Prob}(r \le r_o) &= \text{Prob}\{h_n(r, \rho) \le h_n(r_o, \rho)\} \\ &\doteq \Phi\ \{h_n(r_o, \rho)\} , \end{aligned} \tag{1.27}$$

showing that $h_n(r, \rho)$ is approximately a N(0, 1) variable, for large n. Ruben's investigations have shown that this approximation is much superior to Fisher's z. Fisher's z and Ruben's $h_n(r, \rho)$ are useful in the following problems:

(a) To test the hypothesis

$$H_o: \quad \rho = \rho_o \text{ (specified)},$$

either calculate

$$\begin{aligned} U &= \sqrt{n-3}\ (\tanh^{-1} r - \tanh^{-1} \rho_o) \\ &= \sqrt{n-3}\ (z - \xi_o), \text{ say}, \end{aligned}$$

or calculate $h_n(r, \rho_o)$ from a sample of size n from a bivariate normal population and reject the hypothesis H_o if $|U| > a_\alpha$, or alternatively if $|h_n(r, \rho_o)| > a_\alpha$, where a_α is a constant such that $\Phi(a_\alpha) = 1 - \alpha/2$. The level of significance of these tests is then α. These tests are based on the approximate normal distributions of z or $h_n(r, \rho)$. If, however, ρ_o is zero, we have an exact test based on the t-distribution of $r^*\sqrt{n-2}$, with n-2 d.f. This follows from (1.3), after putting $\rho = 0$.

(b) A $100(1 - \alpha)\%$ confidence interval (approximate) for ρ, the population correlation coefficient, will be provided by

$$\tanh\left\{\tanh^{-1} r \pm a_\alpha \frac{1}{\sqrt{n-3}}\right\}.$$

This follows from

$$\text{Prob}\left\{\frac{|z - \xi|}{1/\sqrt{n-3}} \le a_\alpha\right\} = 1 - \alpha .$$

Alternatively, using Ruben's approximation, we have

$$\text{Prob}\left\{|h_n(r, \rho)| \le a_\alpha\right\} \doteq 1 - \alpha .$$

On solving the equation $h_n^2(r, \rho) = a_\alpha^2$ for ρ^*, we obtain (c_1, c_2) as a $100(1-\alpha)\%$ confidence interval (approximate) for ρ^*, where

$$C_1, C_2 = (2n-3 - a_\alpha^2)^{-1}\Big[(2n-5)^{1/2}(2n-3)^{1/2} r^*$$
$$\pm \left\{(4n-6-a_\alpha^2) + (4n-8-2a_\alpha^2)r^{*2}\right\}^{1/2} a_\alpha\Big].$$

A confidence interval for ρ will then be

$$\frac{C_1}{(1 + C_1^2)^{1/2}}, \quad \frac{C_2}{(1 + C_2^2)^{1/2}},$$

as $\rho^* = \rho/(1 - \rho^2)^{1/2}$.

(c) If we have two samples of sizes n_1 and n_2, respectively, from two independent bivariate normal populations with correlation coefficients ρ_1 and ρ_2, the hypothesis $\rho_1 = \rho_2$ can be tested by using the statistic

$$U = \frac{z_1 - z_2}{\left(\dfrac{1}{n_1-3} + \dfrac{1}{n_2-3}\right)^{1/2}},$$

where $z_i = \tanh^{-1} r_i$ $(i = 1, 2)$ and r_i is the sample correlation coefficient for the ith population. If the null hypothesis

$\rho_1 = \rho_2$ is true, $E(U) = 0$ and U, being a linear combination of the approximate normal variates z_i, has a N(0, 1) distribution for large n_1, n_2. The hypothesis will therefore be rejected, at a level of significance α, if $|U| > a_\alpha$. Ruben's approximation cannot be used in this situation, as the common value of ρ_1, ρ_2 is not specified, and is essential for calculating $h_n(r, \rho)$.

(d) If we have k samples of sizes $n_1, n_2, \ldots, n_k$ from k independent bivariate normal populations, with correlation coefficients $\rho_1, \rho_2, \ldots, \rho_k$, one can test the hypothesis of homogeneity of the correlation coefficients, i.e., the hypothesis

$$H_o: \rho_1 = \rho_2 = \cdots = \rho_k ,$$

by using the statistic

$$U = \sum_{i=1}^{k} (n_i - 3)z_i^2 - \left\{\sum_1^k (n_i - 3)z_i\right\}^2 \Big/ \sum_1^k (n_i - 3) , \quad (1.28)$$

where $z_i = \tanh^{-1} r_i$ (i = 1, 2, ..., k) and r_i is the sample correlation coefficient calculated from the ith sample. Let $\omega_i = (n_i - 3)^{1/2} z_i$ (i = 1, 2, ..., k). Then, under the null hypothesis, the w_i are all normal independent variables with unit variance and means $(n_i - 3)^{1/2} \rho$, where ρ is the common unspecified value of the ρ_i's. Consider now an orthogonal transformation from $\omega_1, \omega_2, \ldots, \omega_k$ to $\eta_1, \eta_2, \ldots, \eta_k$ in which

$$\eta_1 = \frac{(n_1-3)^{1/2}\omega_1 + \cdots + (n_k-3)^{1/2}\omega_k}{\{(n_1-3) + (n_2-3) + \cdots + (n_k-3)\}^{1/2}} .$$

Then it can be readily seen that $E(\eta_1) = \left\{\sum_1^k (n_i-3)\right\}^{1/2} \rho \neq 0$, but, on account of orthogonality of the other variables $\eta_2, \ldots, \eta_k$ to η_1, $E(\eta_i) = 0$ for $i = 2, 3, \ldots, k$, while $V(\eta_i) = 1$. Hence $\eta_2^2 + \ldots + \eta_k^2 = \sum_1^k \omega_i^2 - \eta_1^2 = u$ (given by 1.28) has a χ^2 distribution with $k-1$ d.f. The hypothesis of homogeneity will therefore be rejected, at a level of significance α, if u exceeds u_o given by

$$\int_o^{u_o} \chi_{k-1}^2(u)\, du = 1 - \alpha .$$

If the ρ_i are all equal and an estimate is desired of this common value, we have k independent estimates z_i of $\tan h^{-1} \rho$ with variances $1/(n_i-3)$. We can combine them linearly to obtain a better estimate. If $z = c_1 z_1 + c_2 z_2 + \ldots + c_k z_k$ is an arbitrary linear combination of the z_i's, we can choose the c's to have $E(z) = \tan h^{-1} \rho$ and $V(z)$ is minimum. In other words, we minimize $\sum_1^k c_i^2/(n_i-3)$ subject to $\sum_1^k c_i = 1$. This yields $c_i = (n_i-3)/(N-3k)$ where $N = n_1 + \ldots + n_k$. The better estimator of $\tanh^{-1} \rho$ is thus $z = \sum_1^k (n_i-3)z_i/(N-3k)$ and that of ρ is tanh z. The variance of z is easily seen to be $1/(N-3k)$.

Ruben's $h_n(r, \rho)$ cannot be used to test this hypothesis of homogeneity unless the common value of the ρ's is also specified. But if it is specified to be, say, ρ_o, the $h_n(r_i, \rho_o)$ $(i = 1, 2, \ldots, k)$ are approximately NI(0, 1) and so

$$\sum_{i=1}^k \left\{h_n(r_i, \rho_o)\right\}^2$$

will have a χ^2 distribution with k d.f. This will give a χ^2 test for the hypothesis.

2. DISTRIBUTION OF THE MULTIPLE CORRELATION COEFFICIENT

In this section, we shall derive the distribution of R, the multiple correlation coefficient of x_p with $x_1, x_2, \ldots, x_{p-1}$ in a sample of size n from a p-variate normal population $N_p(\underline{x}|\underline{\mu}|\Sigma)$, where $\underline{x}$ is the column vector of $x_1, x_2, \ldots, x_p$. Let S be the matrix of the corrected s.s. and s.p. of the sample observations. From (1.3.18) and (1.3.30) the population and sample multiple correlation coefficients γ and R are given by, respectively,

$$\gamma^2 = 1 - \frac{|\Sigma|}{|\Sigma_{p-1}|\sigma_{pp}}, \quad R^2 = 1 - \frac{|S|}{|S_{p-1}|s_{pp}}, \tag{2.1}$$

where

$$\Sigma = [\sigma_{ij}] = \begin{bmatrix} \Sigma_{p-1} & \underline{\sigma} \\ \underline{\sigma}' & \sigma_{pp} \end{bmatrix} \begin{matrix} p-1 \\ 1 \end{matrix} \tag{2.2}$$
$$\qquad\qquad\qquad p-1 \quad\; 1$$

and

$$S = [s_{ij}] = \begin{bmatrix} S_{p-1} & \underline{s} \\ \underline{s}' & s_{pp} \end{bmatrix} \begin{matrix} p-1 \\ 1 \end{matrix} \; . \tag{2.3}$$
$$\qquad\qquad\qquad p-1 \quad\; 1$$

Consider a p x p matrix H, partitioned as

$$H = \begin{bmatrix} H_{p-1} & \underline{0} \\ \underline{0}' & h_{pp} \end{bmatrix} \begin{matrix} p-1 \\ 1 \end{matrix} \;, \tag{2.4}$$
$$\qquad\qquad p-1 \quad\; 1$$

where

$$H_{p-1} = \begin{bmatrix} \underline{\sigma}' \, \Sigma_{p-1}^{-1}/\gamma\sqrt{\sigma_{pp}} \\ \underline{h}_2'/(\underline{h}_2' \, \Sigma_{p-1} \, \underline{h}_2)^{1/2} \\ \cdots \\ \underline{h}_{p-1}'/(\underline{h}_{p-1}'\Sigma_{p-1}\underline{h}_{p-1})^{1/2} \end{bmatrix}, \quad h_{pp} = \frac{1}{\sigma_{pp}^{1/2}} \tag{2.5}$$

and $\underline{h}_2, \underline{h}_3, \ldots, \underline{h}_{p-1}$ are p-2 vectors orthogonal to $\underline{\sigma}$, i.e.,

$$\underline{h}_i' \, \underline{\sigma} = 0 \quad (i = 2, 3, \ldots, p-1) \tag{2.6}$$

(since $\underline{\sigma}$ is a (p-1)-component vector, we can always find p-2 vectors which are linearly independent and orthogonal to $\underline{\sigma}$) and satisfying $\underline{h}_i' \, \Sigma_{p-1} \, \underline{h}_j = 0$ for $i \neq j$. It is then easy to verify that

$$\Sigma^* = [\sigma_{ij}^*] = H \Sigma H' = \left[\begin{array}{c|c} I_{p-1} & \begin{matrix}\gamma \\ 0 \\ \vdots \\ 0\end{matrix} \\ \hline \gamma, 0, \ldots, 0 & 1 \end{array}\right] = CC' \tag{2.7}$$

say, where C is l.t. and is defined by

$$C = \left[\begin{array}{c|c} I_{p-1} & \underline{0} \\ \hline \gamma, 0, \ldots, 0 & \sqrt{1 - \gamma^2} \end{array}\right] . \tag{2.8}$$

From Section 4 of Chapter 3, we know that S has the $W_p(s|n-1|\Sigma)$ distribution. Consider

$$S^* = [s_{ij}^*] = HSH'; \tag{2.9}$$

$$S^* = TT', \quad \text{where T is l.t.}; \tag{2.10}$$

$$\begin{aligned} A &= C^{-1} S^* C'^{-1} \\ &= C^{-1} TT' C'^{-1} \\ &= BB' , \end{aligned} \tag{2.11}$$

where

$$B = [b_{ij}] = C^{-1} T . \tag{2.12}$$

As C and T are l.t., B is also. By lemma 1 of Section 5, Chapter 3, S* has the $W_p(S^*|n-1|\Sigma^*)$ distribution, and a further application of the same result to S* shows that A has the $W_p(A|n-1|I)$ distribution, which is in the canonical form. B, therefore, provides the Bartlett decomposition of the Wishart matrix, and by the Bartlett decomposition theorem (Section 2, Chapter 3), the b_{ij}'s are NI(0, 1) for $i > j$ and the b_{ii}^2 are χ^2 variables with $n-1-(i-1)$ d.f. All the b_{ij}'s and b_{ii}'s are independently distributed. We shall now express R^2 in terms of these b_{ij}'s. Because of (2.4),

$$S^*_{p-1} = H_{p-1} S_{p-1} H'_{p-1}, \quad s^*_{pp} = h^2_{pp} s_{pp}, \tag{2.13}$$

where S^*_{p-1} is the matrix of the first p-1 rows and columns of S* and s^*_{pp} is the last element of S*. Hence, as $|H| = h_{pp}|H_{p-1}|$,

$$R^2 = 1 - \frac{|S|}{|S_{p-1}|s_{pp}} = 1 - \frac{|HSH'|}{|H_{p-1} S_{p-1} H'_{p-1}|h^2_{pp} s_{pp}}$$
$$= 1 - \frac{|S^*|}{|S^*_{p-1}|s^*_{pp}}. \tag{2.14}$$

But from (2.10),

$$|S^*| = t^2_{11} t^2_{22} \cdots t^2_{pp},$$
$$|S^*_{p-1}| = t^2_{11} t^2_{22} \cdots t^2_{p-1,\,p-1},$$
$$s^*_{pp} = \sum_{j=1}^{p} t^2_{pj},$$

where $t_{ij} (i > j)$ are the elements of the l.t. matrix T. Hence

$$R^2 = \frac{t^2_{p1} + t^2_{p2} + \cdots + t^2_{p,\,p-1}}{t^2_{p1} + t^2_{p2} + \cdots + t^2_{pp}}. \tag{2.15}$$

But from (2.12), T = CB, and so by equating elements in the last row of T and CB, we have

$$t_{p1} = \gamma\, b_{11} + (1 - \gamma^2)^{1/2}\, b_{p1} ,$$

$$t_{pj} = (1 - \gamma^2)^{1/2}\, b_{pj} \quad (j = 2, 3, \ldots, p) . \qquad (2.16)$$

Substituting these in (2.15), we obtain

$$\frac{R^2}{1 - R^2} = \frac{\left(b_{p1} + \gamma\, b_{11}/(1-\gamma^2)^{1/2}\right)^2 + \left(b_{p2}^2 + b_{p3}^2 + \ldots + b_{p,p-1}^2\right)}{b_{pp}^2} . \qquad (2.17)$$

Since $b_{p1}, b_{p2}, \ldots, b_{p,p-1}$ are NI(0, 1), the conditional distribution of

$$u = \left\{b_{p1} + \gamma\, b_{11}/(1 - \gamma^2)^{1/2}\right\}^2 + \sum_{j=2}^{p-1} b_{pj}^2 \qquad (2.18)$$

when b_{11} is fixed is noncentral χ^2 with (p-1) d.f. and noncentrality parameter $\lambda^2 = \gamma^2\, b_{11}^2/(1 - \gamma^2)$. The p.d.f. of u, given b_{11}^2 is thus

$$\chi'^{2}_{p-1}(u|\lambda^2) , \qquad (2.19)$$

the function χ'^{2}_{f} being already defined in (2.5.7). The p.d.f. of b_{pp}^2 is $\chi^2_{n-p}(b_{pp}^2)$. The joint distribution of u and b_{pp}^2, given b_{11}^2, is thus

$$\chi'^{2}_{p-1}(u|\lambda^2)\, \chi^2_{n-p}(b_{pp}^2)\, du\, db_{pp}^2 , \quad 0 < u < \infty, \quad 0 < b_{pp}^2 < \infty . \qquad (2.20)$$

From (2.17), $R^2 = u/(u + b_{pp}^2)$. So in (2.20) transform from u and b_{pp}^2 to R^2 and $v = u + b_{pp}^2$ and integrate out v. This yields, the conditional distribution of R^2, when b_{11}^2 is fixed, as

$$\phi(R^2|b_{11}^2)\,dR^2 = e^{-(1/2)\lambda^2} \sum_{\alpha=0}^{\infty} \frac{(\lambda^2/2)^\alpha}{\alpha!} \frac{(R^2)^{(p-1)/2+\alpha-1}(1-R^2)^{[(n-p)/2]-1}}{B\left(\frac{p-1}{2}+\alpha,\ \frac{n-p}{2}\right)}\, dR^2$$

$$0 \leq R^2 \leq 1\ . \quad (2.21)$$

where

$$B(\ell,\ m) = \frac{\Gamma(\ell)\ \Gamma(m)}{\Gamma(\ell+m)}\ . \quad (2.22)$$

Equation (2.21) can alternatively be expressed as

$$\phi(R^2|b_{11}^2)\,dR^2 = e^{-\lambda^2/2}\ B\left(R^2\Big|\frac{p-1}{2},\ \frac{n-p}{2}\right)$$

$${}_1F_1\left(\frac{n-1}{2},\ \frac{p-1}{2},\ \lambda^2R^2/2\right) dR^2$$

$$0 \leq R^2 \leq 1 \quad (2.23)$$

where

$${}_1F_1(\ell,\ m,\ z) = \frac{\Gamma(m)}{\Gamma(\ell)} \sum_{\alpha=0}^{\infty} \frac{\Gamma(\ell+\alpha)}{\Gamma(m+\alpha)} \frac{z^\alpha}{\alpha!} \quad (2.24)$$

is the hypergeometric series and $B(z|f_1,\ f_2)$ is the p.d.f. of a beta distribution, i.e.,

$$B(z|f_1,\ f_2) = \frac{1}{B(f_1,\ f_2)}\ z^{f_1-1}\ (1-z)^{f_2-1}\ ,\quad 0\leq z\leq 1\ . \quad (2.25)$$

If the unconditional distribution of R^2 is desired, we use the fact that b_{11}^2 has the χ^2 distribution with (n-1) d.f., and therefore the joint distribution of R^2 and b_{11}^2 is

$$\phi(R^2|b_{11}^2)\ \chi_{n-1}^2(b_{11}^2)\,dR^2\ db_{11}^2\ . \quad (2.26)$$

Substitute $\lambda^2 = \gamma^2\ b_{11}^2/(1-\gamma^2)$ and then integrate out b_{11}^2 over the range 0 to ∞. The distribution of R^2 is then given by

$$(1 - \gamma^2)^{(n-1)/2} \, B\left(R^2 \mid \frac{p-1}{2}, \frac{n-p}{2}\right) \, {}_2F_1\left(\frac{n-1}{2}, \frac{n-1}{2}, \frac{p-1}{2}, \gamma^2 R^2\right) d\,R^2 \,,$$
$$0 \le R^2 \le 1 \,, \qquad (2.27)$$

where

$$_2F_1(\ell, m, q, z) = \frac{\Gamma(q)}{\Gamma(\ell)\,\Gamma(m)} \sum_{\alpha=0}^{\infty} \frac{\Gamma(\ell + \alpha)\,\Gamma(m + \alpha)}{\Gamma(q + \alpha)} \, \frac{z^\alpha}{\alpha!} \,. \qquad (2.28)$$

Observe that, when $\gamma = 0$, the distribution of R^2 simplifies to

$$B\left(R^2 \mid \frac{p-1}{2}, \frac{n-p}{2}\right) d\,R^2 \,, \quad 0 \le R^2 \le 1 \,. \qquad (2.29)$$

This is known as the null distribution of R^2, while (2.27) is
the non-null distribution. From (2.29) or directly from (2.17),
one can see that

$$\frac{n-p}{p-1} \frac{R^2}{1-R^2} \qquad (2.30)$$

has an F-distribution with p-1, n-p d.f., if $\gamma^2 = 0$. This,
therefore, provides a test for the hypothesis $\gamma = 0$.

Perhaps it will be instructive here to examine the matrix
H used in transforming from S to S* and Σ to Σ*. If we consider
a new set of variables $\underline{y} = [y_1, y_2, \ldots, y_p]'$, obtained by the
transformation $\underline{y} = H\underline{x}$ on the original multinormal variables $\underline{x}$,
we will find that y_p is proportional to x_p and y_1 is proportional
to $\underline{\sigma}' \Sigma_{p-1}^{-1} [x_1, x_2, \ldots, x_{p-1}]' = \underline{\beta}'[x_1, \ldots, x_{p-1}]'$, which is
nothing but the population mean-square regression of x_p on x_1,
$\ldots, x_{p-1}$ and $\underline{\beta}$ is the vector of the regression coefficients
(see 1.3.6). From the definition of the multiple correlation
coefficient, γ is the ordinary correlation coefficient between

y_p and y_1, i.e., $\rho_{p1} = \gamma$, if we use ρ_{ij} to denote the correlation coefficient between y_i and y_j. Equation (2.14) and a similar result for Σ^* show that R^2 and γ^2 are invariant for the transformation $\underline{y} = H\underline{x}$, and thus γ is the multiple correlation coefficient between y_p and $y_1, \ldots, y_{p-1}$ also. So using (1.3.25) we get

$$1 - \gamma^2 = (1 - \rho_{p1}^2)(1 - \rho_{p2\cdot 1}^2) \ldots (1 - \rho_{p,p-1\cdot 12\ldots p-2}^2) . \quad (2.31)$$

But $\rho_{p1} = \gamma$; therefore $\rho_{p2\cdot 1}$ (and all other ρ's in the above formula except ρ_{p1}) must be zero. But by (1.3.11), it can be shown that

$$\rho_{p2\cdot 1} = \frac{\rho_{p2} - \rho_{p1}\,\rho_{21}}{(1 - \rho_{p1}^2)^{1/2}\,(1 - \rho_{21}^2)^{1/2}} . \quad (2.32)$$

In this formula $\rho_{p2\cdot 1}$ is zero, as shown earlier. Our choice of $\underline{h}_2, \underline{h}_3, \ldots, \underline{h}_{p-1}$ on H is such that $y_2, y_3, \ldots, y_{p-1}$ are uncorrelated with y_1, and so $\rho_{21} = 0$. Equation (2.32), therefore, shows that $\rho_{p2} = 0$. Since the subscripts 2, 3, ..., p-1 in (2.31) can be permuted, a similar argument will show that $\rho_{p3}, \ldots, \rho_{p,p-1}$ are all zero, i.e., $y_2, \ldots, y_{p-1}$ are uncorrelated with y_p. This is indirectly reflected in the zeros of the last column of Σ^*. Thus the entire linear relationship between x_p and $x_1, \ldots, x_{p-1}$ is taken over by the relationship between y_p and y_1 and $y_2, \ldots, y_{p-1}$ which are uncorrelated with y_1 and have no association whatsoever with y_p.

Also note that the b_{11}^2 occurring in (2.17) is nothing but s_{11}^*, which is the corrected s.s. of observations on y_1 and is thus

$$\underline{\sigma}' \Sigma_{p-1}^{-1} S_{p-1} \Sigma_{p-1}^{-1} \underline{\sigma}/\gamma^2 \sigma_{pp} = \underline{\beta}' S_{p-1} \underline{\beta}/\sigma_{pp\cdot 12, \ldots, p-1}$$

where $\sigma_{pp\cdot 12, \ldots, p-1} = \gamma^2 \sigma_{pp}$ is the residual variance of x_p, when $x_1, \ldots, x_{p-1}$ are eliminated (see 1.3.17).

The distribution of R^2 was first obtained by Fisher [8] using n-dimensional geometry. Since it is complicated, the following approximations have been suggested, though none is as elegant as Fisher's z-transformation in the case of γ.

(a) $\dfrac{n-p}{f} \dfrac{R^{*2}}{g} \doteq F_{f,n-p}$

where

$$R^{*2} = R^2/(1 - R^2);$$

$F_{f,n-p}$ is a random variable whose distribution is the F distribution with f, n-p d.f.,

$$g = \frac{n\,\gamma^{*2}\,(\gamma^{*2} + 2) + p-1}{n\,\gamma^{*2} + p-1},$$

$$f = \frac{(n\,\gamma^{*2} + p-1)^2}{n\,\gamma^{*2}(\gamma^{*2} + 2) + p-1},$$

$$\gamma^{*2} = \frac{\gamma^2}{1-\gamma^2}.$$

This result is due to Gurland [12, 13].

(b) $(p-1)^{-1}\,(n-p)\,(1-\gamma^2)\,\omega(\gamma^2)\,R^{*2}$

has an approximate noncentral F distribution with p-1, n-p d.f. and noncentrality parameter $(1/2)\gamma^2\,\omega(\gamma^2)\,(n-1)^{1/2}\,(n-p)^{1/2}$, where

$$\omega(\gamma^2) = [p-1 + \{n-p + (n-1)^{1/2}(n-p)^{1/2}\}\gamma^2]\;[p-1 + (n-p)(2-\gamma^2)\gamma^2]^{-1}.$$

[By noncentral F with f_1, f_2 d.f. and noncentrality parameter λ^2, we understand a variable which is distributed as f_2u/f_1v, where u has the $\chi'^2_{f_1}(u|\lambda^2)$ distribution and v has an independent $\chi^2_{f_2}(v)$ distribution.] This approximation is due to Khatri [18].

(c) Khatri has suggested a second approximation also. It states that

$$(n-p)\ (1-\gamma^2)\ \{(n-p)\ \gamma^2 + (p-1)\}^{-1}$$

has an F distribution with

$$[(n-p)\ \gamma^2 + k]^2\ [(n-p)\ \gamma^2\ (2-\gamma^2) + (p-1)]^{-1}$$

and (n-p) d.f. According to Khatri, these approximations are good for $n-p \geq 100$, and the latter approximation should be preferred if γ^2 is large.

Using the representation (2.17) (for which see Wijsman [24]), Hodgson [14] has also suggested a couple of approximations.

Kramer [20] has given a table of the upper 5 percent points of the distribution of R. This is being extended to provide upper and lower 5 and 1 percent points. This table is being included in the "Tables For Multivariate Analysis", edited by E. S. Pearson [22].

The moments of the distribution of R have been given by Banerjee [1].

3. DISTRIBUTION OF THE MATRIX OF CORRELATION COEFFICIENTS

As defined before, let S be the matrix of the corrected s.s. and s.p. of the observations, in a sample of size n from the $N_p(\underline{x}|\underline{\mu}|\Sigma)$ population. Then $\rho_{ij} = \sigma_{ij}/(\sigma_{ii}\ \sigma_{jj})^{1/2}$ is the population correlation coefficient between x_i and x_j, and $r_{ij} = s_{ij}/(s_{ii}\ s_{jj})^{1/2}$ is the sample correlation coefficient between x_i and x_j, where $\Sigma = [\sigma_{ij}]$ and $S = [s_{ij}]$. The matrix $P = [\rho_{ij}]$, where the diagonal elements ρ_{ii} are all 1, is called the population correlation matrix, and $R = [r_{ij}]$, with $r_{ij} = 1$, is the sample correlation matrix. We shall consider the distribution of the matrix R, i.e., of the $p(p-1)/2$ variables r_{ij} (i, j = 1, 2, ..., p; i > j). Observe that

$$\Sigma = \mathrm{diag}\left(\sigma_{11}^{1/2}, \sigma_{22}^{1/2}, \ldots, \sigma_{pp}^{1/2}\right) P\, \mathrm{diag}\left(\sigma_{11}^{1/2}, \sigma_{22}^{1/2}, \ldots, \sigma_{pp}^{1/2}\right)$$

and (3.1)

$$S = \mathrm{diag}\left(s_{11}^{1/2}, s_{22}^{1/2}, \ldots, s_{pp}^{1/2}\right) R\, \mathrm{diag}\left(s_{11}^{1/2}, s_{22}^{1/2}, \ldots, s_{pp}^{1/2}\right). \tag{3.2}$$

Hence

$$|\Sigma| = |P| \cdot \prod_{i=1}^{p} \sigma_{ii} \quad \text{and} \quad |S| = |R| \cdot \prod_{i=1}^{p} s_{ii}. \tag{3.3}$$

Now, in the $W_p(S|n-1|\Sigma)$ distribution of S, transform from the $p(p-1)/2$ variables s_{ij} (i, j = 1, 2, ..., p; i > j) to the $p(p-1)/2$ distinct variables r_{ij} (i > j) of R, by the transformation

$$r_{ij} = s_{ij}/(s_{ii}\ s_{jj})^{1/2}$$

and keep the remaining p variables s_{ii} (i = 1, 2, ..., p) untransformed. Since $J(s_{ij} \to r_{ij})$ is $(s_{ii}\ s_{jj})^{1/2}$, the Jacobian of transformation from S to R and s_{ii} (i = 1, 2, ..., p) is $\prod_{i=1}^{p} s_{ii}^{(p-1)/2}$. By making the transformation, the joint distribution of R and s_{ii} (i = 1, 2, ..., p) turns out to be

$$\frac{K(p, n-1)}{|\Sigma|^{(n-1)/2}} |R|^{(n-p-2)/2} \prod_{i=1}^{p} s_{ii}^{(n-3)/2} \exp\{-(1/2)\mathrm{tr}\, DP^{-1} DR\} \times dR \prod_{1}^{p} ds_{ii}$$

$$R > 0, \; s_{ii} > 0 \quad (i = 1, 2, \ldots, p) \tag{3.4}$$

where

$$D = \mathrm{diag}\left(\frac{s_{11}^{1/2}}{\sigma_{11}^{1/2}}, \frac{s_{22}^{1/2}}{\sigma_{11}^{1/2}}, \frac{s_{pp}^{1/2}}{\sigma_{pp}^{1/2}}\right), \tag{3.5}$$

and K(p, n-1) is given by (3.2.20). Now transform from s_{ii} (i = 1, 2, ..., p) to u_i (i = 1, 2, ..., p) by

$$u_i = \left(\frac{s_{ii}}{\sigma_{ii}}\right)^{1/2} \sqrt{\rho^{ii}}, \tag{3.6}$$

where ρ^{ij} (i, j = 1, 2, ..., p) are the elements of P^{-1}. Also let

$$\gamma_{ij} = -\rho^{ij} r_{ij}/(\rho^{ii} \rho^{jj})^{1/2} = -r_{ij} \rho_{ij}^{*}, \tag{3.7}$$

where

$$\rho_{ij}^{*} = \rho^{ij}/(\rho^{ii} \rho^{jj})^{1/2}. \tag{3.8}$$

The joint distribution of $\underline{u} = [u_1, u_2, \ldots, u_p]'$ and R, therefore, turns out to be

$$\frac{|R|^{(n-p-2)/2} \exp((1/2)\underline{u}'\Gamma\underline{u}) \prod_{1}^{p} u_i^{n-2} \, dR \, d\underline{u}}{\pi^{p(p-1)/4} \, 2^{p(n-3)/2} \prod_{i=1}^{p} \Gamma\left(\frac{n-i}{2}\right) |P|^{(n-1)/2} \prod_{i=1}^{p} (\rho^{ii})^{(n-1)/2}}, \tag{3.9}$$

where

$$\Gamma = \begin{bmatrix} 1 & -\gamma_{12} & -\gamma_{13} & \cdots & -\gamma_{1p} \\ -\gamma_{21} & 1 & -\gamma_{23} & \cdots & -\gamma_{2p} \\ \vdots & & & & \\ -\gamma_{p1} & -\gamma_{p2} & -\gamma_{p3} & \cdots & 1 \end{bmatrix}. \tag{3.10}$$

The distribution of R can be obtained from (3.9) if we can integrate out the u_i's. Observe that each u_i varies from 0 to ∞, as u_i is equal to $s_{ii}^{1/2}$ times a positive constant and $0 < s_{ii} < \infty$. We therefore need the following p-fold integral:

$$\int_0^\infty \exp(-(1/2)\underline{u}'\,\Gamma\,\underline{u}) \prod_{i=1}^{p} u_i^{n-2}\, d\underline{u} . \tag{3.11}$$

Since the integral depends only on n and Γ, we shall denote it by $F_{n-2}(\Gamma)$. The distribution of R is therefore,

$$\frac{|P|^{-(n-1)/2} \prod_{i=1}^{p} (\rho^{ii})^{-(n-1)/2} |R|^{(n-p-2)/2} F_{n-2}(\Gamma)\, dR}{\pi^{p(p-1)/4}\, 2^{p(n-3)/2} \prod_{i=1}^{p} \Gamma\left(\frac{n-i}{2}\right)} ,$$

$$R > 0 \tag{3.12}$$

First of all, notice that, in the particular case, when all ρ_{ij}'s $(i > j)$ are zero, i.e., $P = I$, all the γ_{ij}'s are zero and $|\rho_{ij}^*| = 1$. $F_{n-2}(\Gamma)$ then reduces to

$$2^{p(n-3)/2} \left\{\Gamma\left(\frac{n-1}{2}\right)\right\}^p$$

and the distribution of R in this null case is

$$A(p, n-1)\ |R|^{(n-p-2)/2}\ dR, \quad R > 0 , \tag{3.13}$$

where

$$A(p, n-1) = \frac{\left\{\Gamma\left(\frac{n-1}{2}\right)\right\}^p}{\pi^{p(p-1)/4} \prod_{i=1}^{p} \Gamma\left(\frac{n-i}{2}\right)} . \tag{3.14}$$

Also, in this case R and $\underline{u}$, and hence R and s_{ii} $(i = 1, \ldots, p)$, are independently distributed.

An explicit expression for $F_{n-2}(\Gamma)$ in the general case has not as yet been obtained. Fisher [9] has studied some properties of this function. He has given a simple geometric interpretation to $F_0(\Gamma)$. He has given, for $p = 3$, an explicit expression for $F_1(\Gamma)$.

We shall now demonstrate Wilks's technique for obtaining moments of $|R|$ (which is sometimes known as the square of the scatter coefficient of the sample) in the particular case $P = I$. Since (3.13) is a p.d.f.,

$$\int_{R>0} A(p, n-1)\ |R|^{(n-p-2)/2}\ dR = 1 \tag{3.15}$$

is an identify in n. So we change n to $n + 2h$ in this identity and obtain

$$\int_{R>0} |R|^h\ |R|^{(n-p-2)/2}\ dR = \frac{1}{A(p,\ n-1+2h)}\ . \tag{3.16}$$

Now multiply both sides by $A(p, n-1)$. The left-hand side is then $E(|R|^h)$, given by

$$E(|R|^h) = \frac{A(p,\ n-1)}{A(p,\ n-1+2h)}\ . \tag{3.17}$$

The mean and variance of $|R|$ can be found from this easily.

4. DISTRIBUTIONS OF THE MATRIX OF REGRESSION COEFFICIENTS AND THE RESIDUAL s.s. AND s.p. MATRIX

Let S be the matrix of the corrected s.s. and s.p. of observations in a sample of size n from the p-variate normal population $N_p(\underline{x}|\underline{\mu}|\Sigma)$. Let $\underline{x}$, $\underline{\mu}$, Σ, S be partitioned as

$$\underline{x} = \begin{bmatrix} \underline{x}(1) \\ \hline \underline{x}(2) \end{bmatrix} \begin{matrix} k \\ p-k \end{matrix}, \qquad \underline{\mu} = \begin{bmatrix} \underline{\mu}(1) \\ \hline \underline{\mu}(2) \end{bmatrix}$$

$$\Sigma = \begin{bmatrix} \Sigma_{11} & \Sigma_{12} \\ \Sigma_{21} & \Sigma_{22} \end{bmatrix} \begin{matrix} k \\ p-k \end{matrix}, \qquad S = \begin{bmatrix} S_{11} & S_{12} \\ S_{21} & S_{22} \end{bmatrix} \begin{matrix} k \\ p-k \end{matrix} \tag{4.1}$$
$$\qquad k \quad p-k \qquad\qquad\qquad k \quad p-k$$

and let

$$\beta = \Sigma_{21} \Sigma_{11}^{-1}, \qquad B = S_{21} S_{11}^{-1},$$

$$\Sigma_{22\cdot1} = \Sigma_{22} - \Sigma_{21} \Sigma_{11}^{-1} \Sigma_{12}, \qquad S_{22\cdot1} = S_{22} - S_{21} S_{11}^{-1} S_{12} . \tag{4.2}$$

From Section 4 of Chapter 2, we know that β is the $(p-k) \times k$ matrix of the population regression coefficients in the regression of the vector $\underline{x}(2)$ on the vector $\underline{x}(1)$. The corresponding matrix of sample regression coefficients is B, and $\Sigma_{22\cdot1}$ and $S_{22\cdot1}$ are, respectively, the population and sample residual s.s. and s.p. matrices. We shall now derive the distributions of B and $S_{22\cdot1}$ from that of S, which is $W_p(S|n-1|\Sigma)\, dS$. From (2.4.5) and a similar partitioning of S^{-1}, we can readily see that

$$\begin{aligned} \operatorname{tr} \Sigma^{-1} S &= \operatorname{tr} \Sigma_{22\cdot1}^{-1} S_{22} - \operatorname{tr} \Sigma_{22\cdot1}^{-1} \beta S_{12} - \operatorname{tr} \beta' \Sigma_{22\cdot1}^{-1} S_{21} \\ &\quad + \operatorname{tr} \Sigma_{11}^{-1} S_{11} + \operatorname{tr} \beta' \Sigma_{22\cdot1}^{-1} \beta S_{11} \\ &= \operatorname{tr} \Sigma_{22\cdot1}^{-1} S_{22\cdot1} + \operatorname{tr} \Sigma_{11}^{-1} S_{11} \\ &\quad + \operatorname{tr} (B - \beta)' \Sigma_{22\cdot1}^{-1} (B - \beta) S_{11} , \end{aligned} \tag{4.3}$$

as $S_{22} = S_{22\cdot1} + S_{21} S_{11}^{-1} S_{12} = S_{22\cdot1} + B S_{11} B'$.

The distribution of S is the same as the distribution of S_{11}, S_{21}, and S_{22}, and by (4.3) it can be written as

$$\frac{K(p,\ n-1)}{|\Sigma_{11}|^{(n-1)/2}\ |\Sigma_{22\cdot 1}|^{(n-1)/2}}\ |S_{11}|^{(n-p-2)/2}\ |S_{22\cdot 1}|^{(n-p-2)/2}$$

$$\times \exp\left\{-(1/2)\ \mathrm{tr}\ \Sigma_{22\cdot 1}^{-1}\ S_{22\cdot 1}\right\}$$

$$\times \exp\left\{-(1/2)\ \mathrm{tr}\ \Sigma_{11}^{-1}\ S_{11}\right\}$$

$$\times \exp\left\{-(1/2)\ \mathrm{tr}\ (B-\beta)'\Sigma_{22\cdot 1}^{-1}(B-\Sigma)S_{11}\right\}$$

$$dS_{11}\ dS_{21}\ dS_{22}\ , \qquad (4.4)$$

as, by (2.4.29),

$$|S| = |S_{11}|\ |S_{22\cdot 1}| \quad \text{and} \quad |\Sigma| = |\Sigma_{11}|\ |\Sigma_{22\cdot 1}|\ .$$

Using expression (4.4), transform from S_{22} to $S_{22\cdot 1}$ and S_{21} to B, by employing relations (4.2). Obviously $J(S_{22} \to S_{22\cdot 1}) = 1$ and on using Deemer and Olkin's [5] results, $J(S_{21} \to B) = |S_{11}|^{(p-k)}$ The joint distribution of S_{11}, $S_{22\cdot 1}$, and B, therefore, becomes

$$W_k(S_{11}|n-1|\Sigma_{11}) \quad W_{p-k}(S_{22\cdot 1}|n-1-k|\Sigma_{22\cdot 1})$$

$$\times \frac{\exp\left\{-(1/2)\mathrm{tr}(B-\beta)'\Sigma_{22\cdot 1}^{-1}(B-\beta)S_{11}\right\}dS_{11}dS_{22\cdot 1}dB}{|S_{11}|^{-(p-k)/2}\ |\Sigma_{22\cdot 1}|^{k/2}\ (2\pi)^{[k(p-k)]/2}}\ . \qquad (4.5)$$

Observe that the joint density of $S_{22\cdot 1}$, S_{11} and B splits into two parts, one containing only $S_{22\cdot 1}$ and the other containing S_{11} and B. This, therefore, shows that $S_{22\cdot 1}$ is distributed as $W_{p-k}(S_{22\cdot 1}|n-1-k|\Sigma_{22\cdot 1})$ independently of S_{11} and B. We have therefore the following theorem:

<u>Theorem 1</u>: If the $p \times p$ matrix S has the $W_p(S|n-1|\Sigma)$ distribution and if S and Σ are partitioned as in (4.1), $S_{22\cdot 1}$

has the $W_{p-k}(S_{22\cdot 1}|n-1-k|\Sigma_{22\cdot 1})$ distribution, where $S_{22\cdot 1}$ and $\Sigma_{22\cdot 1}$ are defined by (4.2). Further this distribution is independent of the distribution of S_{11} and $B = S_{21}\, S_{11}^{-1}$.

By Lemma 1 of Section 5, Chapter 3, S_{22} has the Wishart distribution with n-1 d.f., while now se see that $S_{22\cdot 1}$ is Wishart with n-1-k d.f. The d.f. are reduced by k because, while S_{22} is the matrix of the s.s. and s.p. of observations on $\underline{x}(2)$, measured from their sample means only, $S_{22\cdot 1}$ is the matrix of the s.s. and s.p. of observations on $\underline{x}(2)$, measured from their sample regression on $x_1, x_2, \ldots, x_k$ (see 2.4.26), i.e., of

$$\underline{x}(2) - \text{sample mean of } \underline{x}(2) - B(\underline{x}(1) - \text{sample mean of } \underline{x}(1)) ,$$

which is the sample residual of $\underline{x}(2)$. The d.f. reduced are thus equal to the number of variables eliminated by regression, or in other words, each "eliminated" variable removes one d.f. In fact, this is the idea underlying the use of the term "degrees of freedom". Also, the associated matrix in the Wishart distribution of S_{22} is Σ_{22} but that of $S_{22\cdot 1}$ is $\Sigma_{22\cdot 1}$, i.e., the variance-covariance matrix of $\underline{x}(2)$ is replaced by $\Sigma_{22\cdot 1}$, the variance-covariance matrix of the population residuals

$$\underline{x}(2) - \underline{\mu}(2) - \boldsymbol{\beta}(\underline{x}(1) - \underline{\mu}(1)) .$$

Apart from these two changes, viz., n-1 to n-1-k and Σ_{22} to $\Sigma_{22\cdot 1}$, the distribution remains the same.

We now turn to the distribution of the matrix of regression coefficients. The joint distribution of S_{11} and B is, as already found out in (4.5),

$$W_k(S_{11}|n-1|\Sigma_{11}) \frac{\exp\left\{-(1/2)\mathrm{tr}\,(B-\beta)'\,\Sigma_{22\cdot1}^{-1}(B-\beta)S_{11}\right\}\,dS_{11}\,dB}{|S_{11}|^{-(p-k)/2}\,|\Sigma_{22\cdot1}|^{k/2}\,(2\pi)^{k(p-k)/2}} . \tag{4.6}$$

But by Lemma 1 of Section 5, Chapter 3, S_{11} has the $W_k(S_{11}|n-1|\Sigma)$ distribution, and so the conditional distribution of B, when S_{11} is fixed, is

$$|S_{11}|^{(p-k)/2}\,|\Sigma_{22\cdot1}|^{-k/2}\,(2\pi)^{-k(p-k)/2}$$
$$\times \exp\left\{-(1/2)\mathrm{tr}\,\Sigma_{22\cdot1}^{-1}(B-\beta)S_{11}\,(B-\beta)'\right\}dB . \tag{4.7}$$

That this is a multivariate normal distribution can be seen as follows. Let

$$S_{11} = MM' , \tag{4.8}$$

where M may be taken to be lower triangular, but this is not essential. Transform now from B to the $(p-k)\times k$ matrix U by the relation

$$U = BM \quad \text{or} \quad B = UM^{-1} . \tag{4.9}$$

The Jacobian of this transformation is, from [5],

$$J(B \to U) = |M^{-1}|^{p-k} = |S_{11}|^{-(p-k)/2} .$$

The distribution of U is, therefore,

$$\frac{1}{(2\pi)^{k(p-k)/2}\,|\Sigma_{22\cdot1}|^{k/2}} \exp\left\{-(1/2)\mathrm{tr}\,\Sigma_{22\cdot1}^{-1}(U-\beta M)(U-\beta M)'\right\}dU . \tag{4.10}$$

Compare this with (2.2.4). The comparison shows that every column of U has a (p-k)-variate normal distribution, with mean given by the corresponding column of βM and variance-covariance matrix $\Sigma_{22\cdot 1}$; all the k columns of U are independently distributed. This is, of course, the conditional distribution of U when S_{11} is fixed. The mean and variance-covariance matrix of U (see 2.2.2), when S_{11}, i.e., M, is fixed, are

$$E(U|M) = \beta M ; \quad V(U|M) = \Sigma_{22\cdot 1} \otimes I_k . \tag{4.11}$$

From (4.9), therefore,

$$E(B|S_{11}) = E(UM^{-1}|M) = \beta MM^{-1} = \beta \tag{4.12}$$

and

$$\begin{aligned} V(B|S_{11}) &= V(UM^{-1}|M) \\ &= \Sigma_{22\cdot 1} \otimes (M'^{-1} I_k M^{-1}) \\ &= \Sigma_{22\cdot 1} \otimes (MM')^{-1} \\ &= \Sigma_{22\cdot 1} \otimes S_{11}^{-1} . \end{aligned} \tag{4.13}$$

on account of (1.2.7). This is a nonsingular matrix, and the elements of B are linear functions of the elements of U, which has a multivariate normal distribution. Hence, by the use of Theorem 2 of Chapter 2, it follows that the conditional distribution of B, when S_{11} is fixed, is multivariate normal, with mean β and variance-covariance matrix $\Sigma_{22\cdot 1} \otimes S_{11}^{-1}$. We, therefore, state:

Theorem 2: If the $p \times p$ matrix S has the $W_p(S|n-1|\Sigma)$ distribution and if S, Σ are partitioned as in (4.1), the conditional distribution of $B = S_{21} S_{11}^{-1}$, when S_{11} is fixed, is $N_{(p-k)k}(B|\beta|\Sigma_{22\cdot 1} \otimes S_{11}^{-1})$. The p.d.f. is given by(4.7).

Theorems 1 and 2 concerning the regression and residual s.s. and s.p. matrices, in the regression of one vector on another, are extremely important in the theory of multivariate analysis, as all important applications of multivariate analysis of variance and tests associated with it depend on these, implicitly or explicitly. We therefore devote one section to a particular case, viz., the regression of x_p only on the remaining variables $x_1, x_2, \ldots, x_{p-1}$.

5. MULTIPLE REGRESSION OF x_p ON $x_1, x_2, \ldots, x_{p-1}$

This corresponds to the particular case $k = p-1$, of the previous section. We employ a slightly different notation and partition S and Σ as

$$S = \left[\begin{array}{c|c} S_{p-1} & \underline{s} \\ \hline \underline{s}' & s_{pp} \end{array}\right]\begin{array}{l} p-1 \\ 1 \end{array}, \quad \Sigma = \left[\begin{array}{c|c} \Sigma_{p-1} & \underline{\sigma} \\ \hline \underline{\sigma}' & \sigma_{pp} \end{array}\right]\begin{array}{l} p-1 \\ 1 \end{array} . \tag{5.1}$$
$$\qquad p-1 \quad\ 1 \qquad\qquad\qquad p-1 \quad\ 1$$

The population regression of x_p on $x_1, x_2, \ldots, x_{p-1}$ is

$$E(x_p|x_1, \ldots, x_{p-1}) = \alpha + \beta_1 x_1 + \ldots + \beta_{p-1} x_{p-1} , \tag{5.2}$$

where

$$\alpha = \mu_p - \beta_1\mu_1 - \cdots - \beta_{p-1}\mu_{p-1},$$

$$\underline{\mu}' = [\mu_1, \mu_2, \ldots, \mu_p],$$

$$\underline{\beta}' = [\beta_1, \beta_2, \ldots, \beta_{p-1}] = \underline{\sigma}' \Sigma_{p-1}^{-1}.$$

The population residual variance of x_p, i.e., the conditional variance of x_p when $x_1, x_2, \ldots, x_{p-1}$ are fixed, is

$$V(x_p | x_1, \ldots, x_{p-1}) = \sigma_{pp} - \underline{\sigma}' \Sigma_{p-1}^{-1} \underline{\sigma} = \sigma^{*2}. \quad (5.3)$$

These results follow either from putting $k = 1$ in the previous section or from using results of Section 4 of Chapter 2. The sample regression coefficients are given by the vector

$$\underline{b}' = \underline{s}' S_{p-1}^{-1}, \quad (5.4)$$

and the residual s.s. is

$$s_{pp} - s' S_{p-1}^{-1} \underline{s} = E. \quad (5.5)$$

Then, by Theorem 1 of the previous section, as $k = 1$,

E/σ^{*2} has a χ^2 distribution with n-p d.f. and is independently distributed of $\underline{b}$. (5.6)

Further, by Theorem 2 of the previous section, the conditional distribution of $\underline{b}$ when S_{p-1} is fixed is

$$N_{p-1}(\underline{b} | \underline{\beta} | \sigma^{*2} S_{p-1}^{-1})\, d\underline{b}. \quad (5.7)$$

Transforming from $\underline{b}$ to $\underline{u}$ by $\underline{u} = (1/\sigma^*)M'\underline{b}$, where $S_{p-1} = MM'$, we can readily see that $\underline{u}$ has the $N_{p-1}(\underline{u} | (1/\sigma^*)M'\boldsymbol{\beta} | I_{p-1})$ distribution and hence $\underline{u}'\underline{u} = (1/\sigma^{*2})b' S_{p-1} \underline{b}$ has a noncentral χ^2

distribution with (p-1) d.f. and noncentrality parameter $(1/\sigma^{*2})$ $\underline{\beta}' S_{p-1} \underline{\beta}$.

In particular, when $\underline{\beta} = 0$, $(1/\sigma^{*2})\underline{b}' S_{p-1} \underline{b}$ has a χ^2 distribution with p-1 d.f. and, as $\underline{b}$, S_{p-1} and E are independently distributed,

$$\frac{n-p}{p-1} \frac{\underline{b}' S_{p-1} b}{E} \tag{5.8}$$

has an $F_{p-1,n-p}$ distribution. This F-test can be used to test the hypothesis $\underline{\beta} = \underline{0}$. The quantity $\underline{b}' S_{p-1} \underline{b}$ in the numerator of (5.8) is called the regression s.s. in the regression of x_p on $x_1, x_2, \ldots, x_{p-1}$ and, due to (5.4), can be written as $\underline{s}' S_{p-1}^{-1} \underline{s}$ also. From Chapter 1, in particular from (1.3.7) or (1.3.30), it can be readily identified that $\underline{b}' S_{p-1} \underline{b}/E$ is nothing but $R^2/(1-R^2)$, where R is the multiple correlation coefficient of x_p with $x_1, x_2, \ldots, x_{p-1}$. Expression (5.8) is thus the same result as (2.30), as $\underline{\beta} = 0$ implies x_p is independently distributed of $x_1, x_2, \ldots, x_{p-1}$ and, in that case, γ, the population multiple correlation coefficient between x_p and $x_1, \ldots, x_{p-1}$, is zero.

When $\underline{\beta} \neq 0$, on using the noncentral χ^2 distribution of $\underline{b}' S_{p-1} \underline{b}$, one can derive the non-null distribution of R^2 in almost the same way as we did in Section 2 of this chapter. It will only be necessary to note that

$$\gamma^2 = \underline{\beta}' \Sigma_{p-1} \underline{\beta}/\sigma_{pp}$$

or that

$$\frac{\gamma^2}{1-\gamma^2} = \frac{\underline{\sigma}'\Sigma_{p-1}^{-1}\underline{\sigma}}{\sigma^{*2}} .$$

6. DISTRIBUTION OF THE PARTIAL CORRELATION COEFFICIENT

Let r' be the partial correlation coefficient in a sample of size n, between x_{p-1} and x_p, when the effect of $x_1, x_2, \ldots, x_{p-2}$ is eliminated. Let ρ' be the corresponding population partial correlation coefficient. Let Σ, the variance-covariance matrix of $\underline{x} = [x_1, x_2, \ldots, x_p]'$, having the $N_p(\underline{x}|\underline{\mu}|\Sigma)$ distribution, be partitioned as

$$\Sigma = \left[\begin{array}{c|c} \Sigma_{11} & \Sigma_{12} \\ \hline \Sigma_{21} & \Sigma_{22} \end{array}\right] \begin{array}{l} p-2 \\ 2 \end{array}$$
$$\qquad\quad p-2 \quad\; 2$$

and S, the matrix of the corrected s.s. and s.p. of the sample observations, be partitioned similarly. By definition, ρ' is the ordinary correlation coefficient between the residuals $x_{p-1\cdot a}$ and $x_{p\cdot a}$, where a denotes the group of subscripts 1, 2, ..., p-2. The variance-covariance matrix of these two residuals is by (2.4.12), $\Sigma_{22\cdot 1} = \Sigma_{22} - \Sigma_{21}\Sigma_{11}^{-1}\Sigma_{12}$. If we denote the elements of this matrix by $\sigma_{ij\cdot a}$ (i, j = p-1, p), the partial correlation coefficient ρ' is

$$\rho' = \frac{\sigma_{p-1,p\cdot a}}{(\sigma_{p-1,p-1\cdot a}\,\sigma_{pp\cdot a})^{1/2}} . \tag{6.1}$$

Similarly, the sample partial correlation coefficient r' is given by

$$r' = \frac{s_{p-1,p\cdot a}}{(s_{p-1,p-1\cdot a}\,s_{pp\cdot a})^{1/2}} , \tag{6.2}$$

where $s_{ij\cdot a}$ (i, j = p-1, p) are the elements of $S_{22\cdot 1} = S_{22} - S_{21}S_{11}^{-1}S_{12}$.

In Section 4, we noticed that the distribution of $S_{22\cdot 1}$ is the same as that of S_{22} with the only change being that n-1 is reduced to n-1-k and Σ_{22} is replaced by $\Sigma_{22\cdot 1}$. Here k, the number of variables eliminated from x_{p-1} or x_p by regression, is (p-2). Now, r, the ordinary correlation coefficient between x_{p-1} and x_p, is

$$\frac{s_{p-1,p}}{(s_{p-1,p-1}\ s_{pp})^{1/2}}$$

and is the same function of S_{22} as r' is of $S_{22\cdot 1}$. Also ρ, the true correlation coefficient between x_p and x_{p-1}, is the same function of Σ_{22} as ρ' is of $\Sigma_{22\cdot 1}$. Hence it follows that the distribution of r' is the same as that of r, provided we change n to n-(p-2) and ρ to ρ' in the distribution of r, which is given by (1.11). Fisher's z-transformation, Ruben's transformation, and all tests associated with r and ρ thus hold, even for r', if we make these changes for n and ρ.

7. DETERMINATION OF d(2, n, p)

If a p x p matrix A has the $W_p(A|n|I)$ distribution, we have proved (Lemma 9 of Chapter 3) that $E(A^{-2}) = d(2, n, p)I$. However, we deferred finding the numerical value of d(2, n, p) then. We are now in a position to evaluate it. Partition A as

$$A = \begin{bmatrix} A_{11} & A_{12} \\ A_{21} & A_{22} \end{bmatrix} \begin{matrix} p-1 \\ 1 \end{matrix} \qquad (7.1)$$

$$\qquad\quad p-1 \quad\ 1$$

Let $A^{-1} = [a^{ij}]$ and let it be partitioned as

$$A^{-1} = \left[\begin{array}{c|c} P & \underline{h} \\ \hline \underline{h}' & a^{pp} \end{array}\right]\begin{matrix} p-1 \\ 1 \end{matrix} . \qquad (7.2)$$
$$\qquad\quad p-1 \qquad 1$$

Then from (1.3.16) and observing that $A_{22\cdot 1} = A_{22} - A_{21} A_{11}^{-1} A_{12}$ is a scalar quantity, we get

$$a^{pp} = \frac{1}{A_{22\cdot 1}} . \qquad (7.3)$$

Also, from (1.3.16),

$$\underline{h} = -a^{pp} A_{11}^{-1} A_{12} . \qquad (7.4)$$

Applying theorems 1 and 2 of Section 4 to A and noting that $\Sigma = I$, for A, we get

$$E(A_{21} A_{11}^{-1} | A_{11}) = \underline{0} \qquad (7.5)$$

and

$$V(A_{21} A_{11}^{-1} | A_{11}) = I_1 \otimes A_{11}^{-1}$$
$$= A_{11}^{-1} . \qquad (7.6)$$

On account of (7.5), (7.6) can be written as

$$E(A_{11}^{-1} A_{12} A_{21} A_{11}^{-1} | A_{11}) = A_{11}^{-1} ,$$

and hence, taking expectations of both sides,

$$E(A_{11}^{-1} A_{12} A_{21} A_{11}^{-1}) = E(A_{11}^{-1}) . \qquad (7.7)$$

But by Lemma 1, Chapter 3, A_{11} has the $W_{p-1}(A_{11}|n|I_{p-1})$ distribution, and hence by Lemma 9 of Chapter 3 again

$$E(A_{11}^{-1}) = d(1, n, p-1)I_{p-1} = (n-p)^{-1} I_{p-1} \quad \text{if} \quad n-p>0.$$

Hence (7.7) becomes

$$E(A_{11}^{-1} A_{12} A_{21} A_{11}^{-1}) = \frac{1}{n-p} I_{p-1} . \tag{7.8}$$

Taking the trace of the matrices on both sides of (7.8),

$$E(\text{tr } A_{11}^{-1} A_{12} A_{21} A_{11}^{-1}) = \frac{1}{n-p} \text{tr } I_{p-1} , \quad \text{if } n-p>0$$

or

$$E(\text{tr } A_{21} A_{11}^{-1} A_{11}^{-1} A_{12}) = \frac{p-1}{n-p} , \quad \text{if } n-p>0 .$$

Hence

$$E(A_{21} A_{11}^{-1} A_{11}^{-1} A_{12}) = \frac{p-1}{n-p} , \quad \text{if } n-p>0 , \tag{7.9}$$

as $A_{21} A_{11}^{-1} A_{11}^{-1} A_{12}$ is a scalar quantity and is equal to its trace.

By Theorem 1 of Section 4, $A_{22\cdot1}$ is independently distributed of $A_{21} A_{11}^{-1}$. Therefore,

$$\begin{aligned} d(2, n, p) &= E(\text{last element of } A^{-2}) \\ &= E\{\underline{h}'\underline{h} + (a^{pp})^2\} \\ &= E\{A_{21} A_{11}^{-1} A_{11}^{-1} A_{12} \cdot (a^{pp})^2 + (a^{pp})^2\} \\ &= E(a^{pp})^2 \cdot E(A_{21} A_{11}^{-1} A_{11}^{-1} A_{12} + 1) , \end{aligned} \tag{7.10}$$

due to the independence of a^{pp} and $A_{21} A_{11}^{-1}$. But by Theorem 1 of Section 4, $A_{22\cdot1} = 1/a^{pp}$ has a χ^2 distribution with $n-(p-1)$ d.f. So

$$E(a^{pp})^2 = \frac{1}{(n-p-1)(n-p-3)} , \quad \text{if } n-p-3 > 0 . \tag{7.11}$$

Using (7.11) and (7.9) in (7.10), we finally obtain

$$\begin{aligned} d(2, n, p) &= \frac{1}{(n-p-1)(n-p-3)} \left(\frac{p-1}{n-p} + 1\right) \\ &= \frac{n-1}{(n-p)(n-p-1)(n-p-3)} , \quad \text{if } n-p-3>0. \end{aligned} \tag{7.12}$$

REFERENCES

[1] Banerjee, D. P. (1952). "On the moments of the multiple correlation coefficient in samples from normal populations", J. Ind. Soc. Agric. Stat., 4, p. 88.

[2] Bartlett, M. S. (1947). "The use of transformation", Biometrics, 3, p. 39.

[3] Daniels, H. E., and Kendall, M. G. (1958). "Short proof of Miss Harley's theorem on the correlation coefficient", Biometrika, 45, p. 471.

[4] David, F. N. (1938). Tables of the Correlation Coefficient. Cambridge University Press, London.

[5] Deemer, W. L., and Olkin, I. (1951). "The jacobians of certain matrix transformations useful in multivariate analysis", Biometrika, 38, p. 345.

[6] Fisher, R. A. (1915). "Frequency distribution of the values of the correlation coefficient in samples from an indefinitely large population", Biometrika, 10, p. 507.

[7] Fisher, R. A. (1921). "On the 'probable' error of a coefficient of correlation deduced from a small sample", Metron, 1, p. 3.

[8] Fisher, R. A. (1928). "The general sampling distribution of the multiple correlation coefficient", Proc. Roy. Soc. London, A 121, p. 654.

[9] Fisher, R. A. (1962). "The simultaneous distribution of correlation coefficients", Sankhyā A, 24, p. 1.

[10] Gayen, A. K. (1951). "The frequency distribution of the product-moment correlation coefficient in random samples of any size drawn from non-normal universes", Biometrika, 38, p. 219.

[11] Ghosh, B. K. (1966). "A symptotic expansions for the moments of the distribution of correlation coefficients", Biometrika, 53, p. 258.

[12] Gurland, J. (1968). "A relatively simple form of the distribution of the multiple correlation coefficient", J. Roy. Stat. Soc. B, 30, p. 276.

[13] Gurland, J., and Milton, R. C. (1970). "Further consideration of the distribution of the multiple correlation coefficient", J. Roy. Stat. Soc. B, p. 32.

[14] Hodgson, V. (1967). "On the sampling distribution of the multiple correlation coefficient", Presented at the IMS Meetings in Washington, D. C., December 1967.

[15] Harley, Betty I. (1954). "A note on the probability integral of the correlation coefficient", Biometrika, 41, p. 278.

[16] Harley, Betty I. (1956). "Some properties of an angular transformation for the correlation coefficient", Biometrika, 43, p. 219.

[17] Hotelling, H. (1953). "New light on the correlation coefficient and its transforms", J. Roy. Stat. Soc. B, 15, p. 193.

[18] Khatri, C. G. (1966). "A note on the large sample distribution of a transformed multiple correlation coefficient", Ann. Inst. Stat. Math., 18, p. 375.

[19] Khatri, C. G. (1968). "A note on exact moments of arc sine correlation coefficient with the help of characteristic functions", Ann. Inst. Stat. Math., 20, p. 143.

[20] Kramer, K. H. (1963). "Tables for constructing confidence limits on the multiple correlation coefficient", J. Am. Stat. Assoc., 58, p. 1082.

[21] Nabeya, S. (1951). "Note on the moments of the transformed correlation", Ann. Inst. Stat. Math., 3, p. 1.

[22] Pearson, E. S. (1971). Tables for Multivariate Analysis. Cambridge University Press, Cambridge.

[23] Ruben, H. (1966). "Some new results on the distribution of the sample correlation coefficient", J. Roy. Stat. Soc. B, 28, p. 513.

[24] Wijsman, R. A. (1959). "Applications of a certain representation of the Wishart matrix", Ann. Math. Statist., 30, p. 597.

CHAPTER 5

HOTELLING'S T^2 AND ITS APPLICATIONS

1. HOTELLING'S T^2 AND ITS DISTRIBUTION

Let a p-component random vector $\underline{u}$ have a p-variate nonsingular normal distribution, with variance-covariance matrix Σ and let a positive-definite symmetric matrix D have an independent Wishart distribution with $f(f > p)$ d.f. If the associated matrix in the distribution of D is the same as Σ, the variance-covariance matrix of $\underline{u}$, the variable $\underline{u}'(1/f\ D)^{-1}\ \underline{u}$ is known as Hotelling's T^2 based on f d.f. [10]. In other words, if $\underline{u}$ has the $N_p(\underline{u}|f|\Sigma)$ distribution, and D has an independent $W_p(D|f|\Sigma)$ distribution,

$$\frac{T^2}{f} = \underline{u}'\ D^{-1}\ \underline{u}\ . \tag{1.1}$$

Sometimes, we use T_p^2 instead of T^2 to emphasize the fact that the normal variable $\underline{u}$ has p-components. Observe that T^2 is obtained from $\underline{u}'\ \Sigma^{-1}\ \underline{u}$ by replacing Σ by its unbiased estimate $(1/f)\,D$, which is independently distributed of $\underline{u}$. In particular, when $p = 1$, T reduces to Student's t with f d.f. Hotelling's T is thus the multivariate generalization of Student's t. This will also be clear from its applications, which are the multivariate analogues of the corresponding univariate applications of Student's t. Student's t was originally intended to test the significance of an observed mean from a normal population with unknown variance but, later on, mainly due to Fisher, it was extended to numerous applications such as tests of significance of the difference between two means, tests of significance of regression and correlation coefficients (total and partial), etc. Hotelling's T^2 is useful in all similar problems in multivariate analysis and in addition in some special situations for which there is no univariate counterpart.

The distribution of T^2 when $E(\underline{u}) = \underline{\mu} = 0$ is known as the null distribution of Hotelling's T^2, while, if $\underline{\mu} \neq 0$, the distribution of T^2 is known as the non-null distribution. The merit of Student's t is that its null distribution does not involve the unknown and hence nuisance parameter σ^2, the variance of the normal distribution. Similarly, the null distribution of T^2 does not involve the unknown variance-covariance matrix Σ. This can be seen as follows: Let C be a lower triangular matrix such that $\Sigma = CC'$. Let

$$\underline{u}^* = C^{-1}\underline{u}\,; \quad D^* = C^{-1}\,DC'^{-1}\,. \tag{1.2}$$

Then it is easy to see that

$$\frac{T^2}{f} = \underline{u}'D^{-1}\underline{u} = (C\underline{u}^*)'\,(CD^*C')^{-1}\,(C\underline{u}^*) = \underline{u}^{*\prime}\,D^{*-1}\,\underline{u}^*\,, \tag{1.3}$$

i.e., T^2/f is invariant for a linear transformation of $\underline{u}$. Note that $\underline{u}^*$ has the $N_p(\underline{u}^*|\underline{\mu}^*|I_p)$ distribution, where $\underline{\mu}^* = C^{-1}\underline{\mu}$, and by Lemma 1, Section 5 of Chapter 3, D^* has the $W_p(D^*|f|I_p)$ distribution, as $C^{-1}\Sigma C'^{-1} = I$. Also D^* and u^* are independently distributed. Thus, we observe that T^2 is a function of $\underline{u}^*$ and

D^*, the distributions of which do not involve Σ, if $\underline{\mu}$ and hence $\underline{\mu}^* = \underline{0}$. Hence, the null distribution of T^2 will be independent of Σ.

Let H be an orthogonal matrix of order p, such that its last row is $\underline{u}^{*\prime}/(\underline{u}^{*\prime}\ \underline{u}^*)^{1/2}$. H is a random orthogonal matrix, as $\underline{u}^*$ are random variables. Keep $\underline{u}^*$ fixed, so that H is also fixed and then make the transformation

$$A = H\ D^*\ H' \ . \tag{1.4}$$

The Jacobian of the transformation from D^* to A is (see [4])

$$J(D^* \rightarrow A) = |H^{-1}|^{p+1} = 1 \ ,$$

as H is orthogonal. The conditional distribution of D^* when H is fixed is the same as the unconditional one, viz., $W_p(D^*|f|I)$, as D^* and $\underline{u}^*$ are independent. Making the transformation (1.4), in the distribution of D^*, the distribution of A comes out as $W_p(A|f|I)\,d\,A$, as $|D^*| = |A|$ and tr D^* = tr A, H being orthogonal. But this conditional distribution $W_p(A|f|I)\,d\,A$ of A does not involve the conditioning variates $\underline{u}^*$, and hence it is also the unconditional distribution of A, and further A is independently distributed of $\underline{u}^*$. Now

$$\begin{aligned} \frac{T^2}{f} = u^{*\prime}\ D^{*-1}\ u^* &= (H\underline{u}^*)'\ (H\,D^*\,H')^{-1}\ (H\ \underline{u}^*) \\ &= (H\underline{u}^*)'\ A^{-1}\ (H\ \underline{u}^*) \ . \end{aligned} \tag{1.5}$$

But, the last row of H is proportional to $\underline{u}^{*\prime}$, and the remaining (p-1) rows of H are orthogonal to $\underline{u}^{*\prime}$, thus

$$\underline{H}\underline{u}^* = \begin{bmatrix} 0 \\ 0 \\ \vdots \\ 0 \\ (\underline{u}^{*\prime}\ \underline{u}^*)^{1/2} \end{bmatrix} \tag{1.6}$$

Equation (1.5) therefore reduces to

$$\frac{T^2}{f} = \left[0,0,\ldots,0,\ (\underline{u}^{*\prime}\ \underline{u}^*)^{1/2}\right] A^{-1}\left[0,0,\ldots,0,\ (\underline{u}^{*\prime}\ \underline{u}^*)^{1/2}\right]'$$

$$= (\underline{u}^{*\prime}\ \underline{u}^*) \cdot a^{pp} , \tag{1.7}$$

where a^{pp} is the element in the pth row and pth column of A^{-1}. Since $\underline{u}^*$ has the $N_p(\underline{u}^*|\underline{\mu}^*|I)$ distribution, $\underline{u}^{*\prime}\ \underline{u}^*$ has the noncentral χ^2 distribution with p d.f. and noncentrality parameter (see 2.5.7) given by

$$\lambda^2 = \underline{\mu}^{*\prime}\ \underline{\mu}^* = \underline{\mu}'(CC')^{-1}\underline{\mu} = \underline{\mu}'\ \Sigma^{-1}\ \underline{\mu} . \tag{1.8}$$

By Lemma 4, Section 5 of Chapter 3, $1/a^{pp}$ is distributed as a χ^2 with f - (p-1) d.f., as $A \sim W_p(A|f|I)$. Hence from (1.7), we observe that T^2/f is distributed as the ratio V/W, where $V \sim \chi_p'^2(V|\lambda^2)$ and W has an independent χ^2 distribution with f - (p-1) d.f. Therefore, we can write the joint distribution of V and W as

$$\chi_p'^2(V|\lambda^2)\chi^2_{f-p+1}(w)\ dv\, dw ,$$

and then make a transformation,

$$T^2 = \frac{fV}{W} ,$$

from V to T^2 and integrate out W, to obtain the distribution of T^2. This non-null distribution of T^2 is given by

$$H_p'(T^2|f,\ \lambda^2)\,dT^2 = e^{-\lambda^2/2} \sum_{r=0}^{\infty} \frac{(\lambda^2/2)^r}{r!} \frac{1}{B\left(\frac{p}{2}+r,\ \frac{f-p+1}{2}\right)} \cdot \frac{(T^2/f)^{(p/2)+r-1}}{\left(1+\frac{T^2}{f}\right)^{(1/2)(f+1)+r}}\, d\left(\frac{T^2}{f}\right), \quad (1.9)$$

$$T^2 > 0 .$$

Sometines it is more convenient to use

$$x = \frac{T^2}{f + T^2} . \quad (1.10)$$

This has the same distribution as that of $V/(W+V)$ and it is given by

$$g(x|\lambda^2)dx = e^{-\lambda^2/2} \sum_{r=0}^{\infty} \frac{(\lambda^2/2)^r}{r!} B\left(x\middle|\ \frac{p}{2},\ \frac{f-p+1}{2} + r\right) dx , \quad (1.11)$$

where $B(z|\ell,\ m)$ is the p.d.f. of a beta distribution and was defined in (4.2.25). $g(x|\lambda^2)$ can alternatively be expressed as

$$g(x|\lambda^2) = e^{-\lambda^2/2}\, B\left(x\middle|\ \frac{p}{2},\ \frac{f-p+1}{2}\right)\, {}_1F_1\left(\frac{p}{2},\ \frac{f+1}{2},\ \lambda^2(1-x)\right). \quad (1.12)$$

The null distribution of T^2 can be obtained by putting $\underline{\mu} = \underline{0}$, which is the same as $\lambda^2 = 0$ in (1.9). It is

$$H_p(T^2|f)\,dT^2 = H_p'(T^2|f,\ \lambda^2 = 0)\,dT^2$$
$$= \frac{1}{B\left(\frac{p}{2},\ \frac{f-p+1}{2}\right)} \frac{(T^2/f)^{(p/2)-1}}{\left(1+\frac{T^2}{f}\right)^{(f+1)/2}}\, d\left(\frac{T^2}{f}\right) ;\ T^2 \geq 0. \quad (1.13)$$

From this, or directly from $T^2/f = V/W$, where the distribution of V reduces to that of a χ^2 distribution when $\lambda^2 = 0$, we observe that

$$\frac{f-p+1}{p} \cdot \frac{T^2}{f} = F_{p,\, f-p+1} , \tag{1.14}$$

i.e., $(f-p+1)T^2/p\, f$ has an F distribution with p and f-p+1 d.f., when $\lambda^2 = 0$. Hence we have the following theorems:

<u>Theorem 1</u>: If $\underline{u}$ has a $N_p(\underline{u}|\underline{\mu}|\Sigma)$ distribution and D has a $W_p(D|f|\Sigma)$ distribution, independent of $\underline{u}$, then $T^2 = f\, \underline{u}'\, D^{-1}\, \underline{u}$, which is known as Hotelling's T^2 based on f d.f., has the p.d.f. $H_p'(T^2|f, \lambda^2)\, dT^2$, given by (1.19), where $\lambda^2 = \underline{\mu}'\, \Sigma^{-1}\, \underline{\mu}$ is known as the noncentrality parameter of the distribution.

<u>Theorem 2</u>: If $\underline{u}$ has a $N_p(\underline{u}|\underline{0}|\Sigma)$ distribution and D has an independent $W_p(D|f|\Sigma)$ distribution, the p.d.f. of Hotelling's $T^2 = f\, \underline{u}'\, D^{-1}\, u$ is $H_p(T^2|f)$ given by (1.13) and

$$\frac{f-p+1}{p} \frac{T^2}{f}$$

has an F distribution with p and f-p+1 d.f.

Suppose $\underline{u}$ has a $N_p(\underline{u}|\underline{\mu}|k_1\Sigma)$ distribution and D has an independent $W_p(D|f|k_2^{-1}\Sigma)$distribution, and k_1, k_2 are known. In such a case, it is easy to see that Hotelling's T^2 will be

$$\frac{f}{k_1 k_2}\, \underline{u}'\, D^{-1}\, \underline{u}$$

and the noncentrality parameter will be

$$\frac{f}{k_1 k_2}\, \underline{\mu}'\, \Sigma^{-1}\, \underline{\mu} .$$

In this connection it should also be noticed that the noncentrality parameter is the same function of $E(\underline{u})$ and $V(\underline{u})$ as T^2/f is of u and D.

Another form in which T^2/f is sometimes expressed is

$$x = \frac{1}{1 + T^2/f} = \frac{|D|}{|D + \underline{u}\underline{u}'|} . \tag{1.15}$$

This follows from

$$|D + \underline{u}\underline{u}'| = |D| \, |I_p + D^{-1} \underline{u}\underline{u}'|$$

$$= |D| \, (1 + \underline{u}' D^{-1} \underline{u}) ,$$

on using (1.4.5).

We have seen above that $(f-p+1)T^2/p\,f$ is distributed as

$$\frac{V/p}{W/(f-p+1)} ,$$

i.e., as the ratio of a noncentral χ^2 variable to an independent χ^2 variable; in other words, it has a noncentral F distribution with p and f-p+1 d.f. In fixed-effects analysis of variance, the error sum of squares is always proportional to a χ^2 variable and the other sums of squares in the analysis of variance table are, in general, proportional to noncentral χ^2 variables, so that the F ratio used in an analysis of variance test has a noncentral F distribution and it is thus needed in obtaining the power function of the analysis of variance test. Tang [28] has prepared tables of this function, and these can be used for the non-null distribution of T^2. Actually Tang's tables do not give the percentage points or c.d.f. of T^2 but that of $(T^2/f)/(1+T^2/f)$, of (1.10). Moreover, he uses $\nu = \lambda/\sqrt{p+1}$ as the parameter and not λ^2. Lachenbruch [14] has extended these tables; Pearson and Hartley [20] have given charts corresponding to Tang's tables, while Bargmann and Ghosh [1] have prepared a computer program for the p.d.f. and c.d.f. of a noncentral F distribution. Other tables and charts that can be mentioned in this connection are by Lehmer [16], Ura [30], Tiku [29], Fox [6], and Patnaik [19].

In the applications of Hotelling's T^2 to multivariate problems, which we shall consider in this chapter, our object will be to find a vector $\underline{u}$ having a p-variate normal distribution and an independent Wishart matrix D based on f d.f. and an associated matrix which is the same as $V(\underline{u})$. The vector $\underline{u}$ should be such that the null hypothesis of interest in the original problem must be equivalent to the hypothesis $E(\underline{u}) = \underline{0}$. This hypothesis can be tested by the F-test of (1.14). In other words, we obtain the value of

$$T^2 = f\, \underline{u}'\, D^{-1}\, \underline{u}$$

and reject or do not reject the hypothesis under consideration, according as this T^2 is significantly large or not. This T^2 measures the departure of the null hypothesis from the true value of $E(\underline{u})$, and so we reject the null hypothesis if T^2 is too large. The yardstick for measuring the largeness of T^2 is provided by the function

$$T^2(f, p, \alpha),$$

where f is the d.f. of D, p is the number of variables in $\underline{u}$, and α is the significance level. α is to be determined from the distribution of T^2 in such a way that

$$\text{Prob(an error of the first kind)} = \text{Prob(rejecting the null hypothesis when it is true)}$$

$$= \text{Prob}(T^2 > T^2(f, p, \alpha) \mid E(\underline{u})=\underline{0}) = \alpha , \tag{1.16}$$

a preassigned quantity, which is usually very small (such as 0.05, 0.01, or 0.1) since it is the probability of making a wrong decision. We control this to the level α. However, from (1.14), when $E(\underline{u}) = \underline{0}$,

$$\text{Prob}(T^2 > T^2(f, p, \alpha) \mid E(\underline{u}) = \underline{0}) = \text{Prob}\left(F_{p,f-p+1} > F(p, f\text{-}p\text{+}1, \alpha)\right)$$
$$= \alpha \tag{1.17}$$

where

$$F(p, f\text{-}p\text{+}1, \alpha) = \frac{f-p+1}{p}\,\frac{T^2(f, p, \alpha)}{f} \tag{1.18}$$

or

$$T^2(f, p, \alpha) = \frac{p\,f\,F(p, f\text{-}p\text{+}1, \alpha)}{(f-p+1)} . \tag{1.19}$$

Since $F_{p,\, f-p+1}$ has an F distribution, the value $F(p, f\text{-}p\text{+}1, \alpha)$ can be determined from the tables of F distribution such that

$$\text{Prob}\left(F_{p,\, f-p+1} > F(p, f\text{-}p\text{+}1, \alpha)\right) = \alpha$$

and from this, using (1.19), $T^2(f, p, \alpha)$ will be determined. The null hypothesis is rejected when

$$T^2 > T^2(f, p, \alpha) ,$$

and this is known as the "critical region" of the Hotelling's

T^2 test. α, the chosen significance level, is known as the "size" of this critical region. The complementary region $T^2 \leq T^2(f, p, \alpha)$ is known as the "acceptance region", as the null hypothesis is not rejected when the observed value of T^2 is not large enough. The observations are then in agreement of the hypothesis.

When this test is employed, we run the risk of another error, viz., the error of accepting the hypothesis $E(\underline{u}) = \underline{0}$ when, in fact, it is not true. The chance of committing such an error is obviously,

$$\beta = \text{Prob}(T^2 \leq T^2(f, p, \alpha) \,|\, E(\underline{u}) \neq 0) \ . \tag{1.20}$$

The smaller this error, the better the test is and hence $1 - \beta$ or

$$\text{Prob}(T^2 > T^2(f, p, \alpha) \,|\, E(\underline{u}) \neq 0) \tag{1.21}$$

is called the "power" of Hotelling's T^2 test. As (1.9) represents the distribution of T^2 when $E(\underline{u}) \neq 0$, it is given by

$$\int_{T^2(f, p, \alpha)}^{\infty} H_p'(T^2 \,|\, f, \lambda^2)\, dT^2 \tag{1.22}$$

where

$$\lambda^2 = \{E(\underline{u})\}' \{V(\underline{u})\}^{-1} \{E(\underline{u})\} \ . \tag{1.23}$$

The power, therefore, depends on the true value of $E(\underline{u})$ only through λ^2. From (1.10) and (1.11), the power function is

$$1 - \beta = 1 - \int_0^{x(f, p, \alpha)} g(x \,|\, \lambda^2) dx \ , \tag{1.24}$$

where

$$x(f, p, \alpha) = \frac{T^2(f, p, \alpha)}{f + T^2(f, p, \alpha)} = \frac{pF/(f-p+1)}{1 + p\,F(p,\ f-p+1,\ \alpha)/(f-p+1)} . \tag{1.25}$$

The value of (1.24) can be obtained, for different values of λ^2, from Tang's tables.

Before we proceed to consider the applications of Hotelling's T^2, we shall, in the next section, show how this test can also be obtained from the "union-intersection" principle of test construction - a heuristic principle for obtaining multivariate tests due to S. N. Roy [23, 24]. This process helps in obtaining simultaneous confidence bounds on certain parametric functions associated with this distribution of $\underline{u}$.

2. APPLICATION OF THE UNION-INTERSECTION PRINCIPLE TO OBTAIN HOTELLING'S T^2 TEST

As before, we set up the null hypothesis

$$H_0: \quad \underline{\mu} = \underline{0} . \tag{2.1}$$

where $\underline{\mu} = E(\underline{u})$, against the alternative that $\underline{\mu} \neq \underline{0}$. We are assuming that $\underline{u}$ has the $N_p(\underline{u}|\underline{\mu}|\Sigma)$ distribution. Σ is unknown but an independent matrix D having the $W_p(D|f|\Sigma)$ distribution is available. Let $\underline{a}$ be any arbitrary non-null vector of p constant elements. It is obvious that, when H_0 is true, $\underline{a}'\underline{\mu} = 0$ for every $\underline{a}$ and, conversely, if $\underline{a}'\underline{\mu} = 0$ for every $\underline{a}$, H_0 will be true. Thus H_0 will be rejected if at least one of the hypotheses

$$H_a: \ \underline{a}'\underline{\mu} = 0 , \quad \underline{a} \in A . \tag{2.2}$$

(A is the set of all p-component non-null vectors of constants) is rejected. Hence

$$H_o = \bigcap_a H_a \tag{2.3}$$

where $\bigcap$ denotes "intersection". Let ω_a denote the "critical region" for the hypothesis H_a. If our sample observations are such that the "sample point" falls in the critical region for at least one of the hypotheses $H_a (\underline{a} \in A)$, H_a and hence H_o will be rejected. It is therefore evident that the critical region for H_o will be formed by the union of these critical regions for the separate H_a. Thus ω, the critical region for H_o, is

$$\omega = \bigcup_a \omega_a . \tag{2.4}$$

This is the union-intersection principle of S. N. Roy. The sizes of the ω_a should be such that the 'final' size of ω turns out to be α, a preassigned quantity. The merit of this procedure is that H_a is about a "scalar" quantity $\underline{a}'\underline{\mu}$ which is the mean of a univariate random variable $\underline{a}'\underline{u}$, and if we have an optimum test for this univariate hypothesis, we may be able to obtain a good test (in some sense) for the multivariate hypothesis H_o. Nandi [18] has investigated the properties of Roy's union-intersection test in general and showed that the test is consistent if the component tests (univariate) are so, unbiased under certain conditions, and admissible if, again, the component tests are admissible. Details of these properties are beyond the scope of this book, and the reader may refer to papers by Roy [23, 24] and Nandi [18].

We now illustrate how this union-intersection principle again leads to Hotelling's T^2 test. Since $\underline{a}'\underline{u}$ has a normal distribution with mean $\underline{a}'\underline{\mu}$ and variance $\underline{a}'\Sigma\underline{a}$ and by Lemma 2 of Chapter 3, $\underline{a}'D\underline{a}/\underline{a}'\Sigma\underline{a}$ has a χ^2 distribution with f d.f., independent of $\underline{a}'\underline{u}$, the hypothesis H_a can be tested by the Student's t-test

$$t_a = \frac{\underline{a}'\underline{u}}{(\underline{a}'D\underline{a}/f)^{1/2}} \tag{2.5}$$

based on f d.f. The t-test has the usual optimum properties, which are well known from univariate statistical theory. The critical region ω_a for H_a is thus

$$|t_a| > \text{a suitable constant, say } t', \tag{2.6}$$

depending on f and the size of ω_a. Alternatively, it is also equivalent to

$$t_a^2 > t'^2 . \tag{2.7}$$

Now, if each of the t_a's is less than t'^2, none of the hypothesis H_a will be rejected and so H_o will not be rejected. If any of the t_a^2's is however greater than or equal to t'^2, the corresponding H_a and hence H_o will be rejected. Thus H_o will not be rejected if each $t_a^2 \leq t'^2$. But this will be so if the maximum value of t_a^2, with respect to $\underline{a}$, is less than t'^2. Hence, our procedure will be simpler if we find this maximum value instead of undertaking the impossible task of verifying whether t_a^2 is less than t'^2 or not for each and every $\underline{a}$ in A.

Now

$$t_a^2 = \frac{f(\underline{a}'\underline{u})^2}{\underline{a}'D\underline{a}} \tag{2.8}$$

Taking partial derivatives of t_a^2 with respect to the elements of $\underline{a}$ and equating them to zero, we find that t_a^2 is maximum when $\underline{a}$ is proportional to $D^{-1}\underline{u}$. Substituting this value of $\underline{a}$ in t_a^2, we obtain as

$$\operatorname*{Max}_a \frac{f(\underline{a}'\underline{u})^2}{\underline{a}'D\underline{a}} = f\ \underline{u}'D^{-1}\underline{u}\ , \tag{2.9}$$

which is nothing but Hotelling's T^2, based on f d.f. The critical region ω for H_o is thus $T^2 > t'^2$, where t' is to be chosen such that the size of ω is a predetermined constant α. But this needs the distribution of T^2 when H_o is true and we have, in the previous section, already seen how to obtain this t'^2, which we called $T^2(f, p, \alpha)$.

Let us now see how the T^2 test can be used for obtaining some simultaneous confidence bounds. Since $E(\underline{u} - \underline{\mu}) = \underline{0}$, it follows from Theorem 1 that

$$T_o^2 = f(\underline{u} - \underline{\mu})'\ D^{-1}\ (\underline{u} - \underline{\mu}) \tag{2.10}$$

has the null distribution of Hotelling's T^2, given by (1.13), and so, from (1.17),

$$\text{Prob}(T_o^2 \leq T^2(f, p, \alpha)) = 1 - \alpha\ . \tag{2.11}$$

But, from (2.9) we have

$$f(\underline{u} - \underline{\mu})'\ D^{-1}\ (\underline{u} - \underline{\mu}) = \operatorname*{Max}_a \frac{f\{\underline{a}(\underline{u} - \underline{\mu})\}^2}{\underline{a}'D\underline{a}}\ . \tag{2.12}$$

Substituting this in (2.11), we obtain

$$\text{Prob}\left\{\underset{a}{\text{Max}} \left| \frac{\underline{a}'(\underline{u} - \underline{\mu})}{(\underline{a}'D\underline{a})^{1/2}} \right| \leq \frac{1}{\sqrt{f}} T(f, p, \alpha)\right\} = 1 - \alpha ,$$

which is the same as

$$\text{Prob}\left\{\underline{a}'\underline{u} - \left(\frac{\underline{a}'D\underline{a}}{f}\right)^{1/2} T(f, p, \alpha) \leq \underline{a}'\underline{\mu} \leq \underline{a}'\underline{u} + \left(\frac{\underline{a}'D\underline{a}}{f}\right)^{1/2} T(f, p, \alpha)\right\}$$

$$\text{for every } \underline{a} \text{ in } A$$

$$= 1 - \alpha . \tag{2.13}$$

Better still, we use (1.19) and obtain

$$\underline{a}'\underline{u} \pm (\underline{a}'D\underline{a})^{1/2} \left\{\frac{p\ F(p,\ f-p+1,\ \alpha)}{f-p+1}\right\}^{1/2} \tag{2.14}$$

as the "simultaneous confidence bounds" for all linear parametric functions $\underline{a}'\underline{\mu}$. The confidence coefficient associated with all such intervals, generated for different $\underline{a}$, is $1 - \alpha$. If one is interested in some particular linear functions of $\underline{\mu}$, one can obtain the corresponding confidence intervals from (2.14), using the appropriate $\underline{a}$'s. Then the assurance is $1 - \alpha$ that these intervals, and all other intervals corresponding to other $\underline{a}$'s in A, contain the corresponding $\underline{a}'\underline{\mu}$'s.

3. DISTRIBUTION OF RAO'S U STATISTIC

Let, as before, $\underline{u}$ have a $N_p(\underline{u}|\underline{\mu}|\Sigma)$ distribution and D have an independent $W_p(D|f|\Sigma)$ distribution. Let $\underline{u}$, $\underline{\mu}$, Σ, D be partitioned as

$$\underline{u} = \begin{bmatrix} \underline{u}(1) \\ \hline \underline{u}(2) \end{bmatrix} \begin{matrix} k \\ p-k \end{matrix}, \quad \underline{\mu} = \begin{bmatrix} \underline{\mu}(1) \\ \hline \underline{\mu}(2) \end{bmatrix} \tag{3.1}$$

$$\Sigma = \underset{\quad k \qquad p-k}{\left[\begin{array}{c|c} \Sigma_{11} & \Sigma_{12} \\ \hline \Sigma_{21} & \Sigma_{22} \end{array}\right]} \begin{matrix} k \\ p-k \end{matrix}, \quad D = \underset{\quad k \qquad p-k}{\left[\begin{array}{c|c} D_{11} & D_{12} \\ \hline D_{21} & D_{22} \end{array}\right]} \begin{matrix} k \\ p-k \end{matrix} \tag{3.2}$$

We also set

$$\beta = \Sigma_{21} \Sigma_{11}^{-1}, \quad B = D_{21} D_{11}^{-1} \tag{3.3}$$

$$\Sigma_{22\cdot 1} = \Sigma_{22} - \Sigma_{21} \Sigma_{11}^{-1} \Sigma_{12}, \quad D_{22\cdot 1} = D_{22} - D_{21} D_{11}^{-1} D_{12}, \tag{3.4}$$

and

$$\lambda_p^2 = \underline{\mu}' \Sigma^{-1} \underline{\mu}, \quad \lambda_k^2 = \underline{\mu}'(1) \Sigma_{11}^{-1} \underline{\mu}(1) . \tag{3.5}$$

Hotelling's T^2 based on all the variables in $\underline{u}$ and the one based on only $\underline{u}(1)$ are, respectively,

$$T_p^2 = f \underline{u}' D^{-1} u, \quad T_k^2 = f \underline{u}'(1) D_{11}^{-1} \underline{u}(1) . \tag{3.6}$$

From (2.4.5),

$$\Sigma^{-1} = \left[\begin{array}{c|c} \Sigma_{11}^{-1} + \beta' \Sigma_{22\cdot 1}^{-1} \beta & -\beta' \Sigma_{22\cdot 1}^{-1} \\ \hline - \Sigma_{22\cdot 1}^{-1} \beta & \Sigma_{22\cdot 1}^{-1} \end{array}\right] \tag{3.7}$$

and

$$D^{-1} = \left[\begin{array}{c|c} D_{11}^{-1} + B' D_{22\cdot 1}^{-1} B & - B D_{22\cdot 1}^{-1} \\ \hline - D_{22\cdot 1}^{-1} B & D_{22\cdot 1}^{-1} \end{array}\right] \tag{3.8}$$

Using (3.1) and (3.7) in λ_p^2, given by (3.5), we find

$$\lambda_p^2 = \lambda_k^2 + (\underline{\mu}(2) - \beta \underline{\mu}(1))' \Sigma_{22\cdot 1}^{-1} (\underline{\mu}(2) - \beta \underline{\mu}(1)) . \tag{3.9}$$

Similarly, using (3.1) and (3.8) in T_p^2, given by (3.6), we get

$$\begin{aligned} T_p^2 &= T_k^2 + f(\underline{u}(2) - B \underline{u}(1))' D_{22\cdot 1}^{-1} (\underline{u}(2) - B \underline{u}(1)) \\ &= T_k^2 + f z' D_{22\cdot 1}^{-1} \underline{z} , \end{aligned} \tag{3.10}$$

where

$$\underline{z} = \underline{u}(2) - B \underline{u}(1) . \tag{3.11}$$

Rao's [22] U statistic is defined as

$$U = \frac{1 + T_k^2/f}{1 + T_p^2/f} . \tag{3.12}$$

Let us consider its distribution. We recall from Theorems 1 and 2 of Chapter 4, that

$$D_{22\cdot 1} \text{ has the } W_{p-k}(D_{22\cdot 1}|f-k|\Sigma_{22\cdot 1}) \text{ distribution} \tag{3.13}$$

and is independent of B. The elements of B have a multivariate normal distribution when D_{11} is fixed. The mean and variance-covariance matrix of this conditional distribution of B are

$$E(B|D_{11}) = B, \quad V(B|D_{11}) = \Sigma_{22\cdot 1} \otimes D_{11}^{-1} . \tag{3.14}$$

We use these results to consider the conditional distribution of the (p-k)-component vector $\underline{z}$, when D_{11} and $\underline{u}(1)$ are fixed. We first observe that

$$\begin{aligned} E(\underline{z}|D_{11}, \underline{u}(1)) &= E(\underline{u}(2) - B\,\underline{u}(1)|D_{11}, \underline{u}(1)) \\ &= E(\underline{u}(2)|\underline{u}(1)) - E(B|D_{11})\underline{u}(1) \\ &= \underline{\mu}(2) + \beta(\underline{u}(1) - \underline{\mu}(1)) - \beta\,\underline{u}(1) \\ &= \underline{\mu}(2) - \beta\,\underline{\mu}(1) , \end{aligned} \tag{3.15}$$

where we have used (2.4.11) for obtaining $E(\underline{u}(2)|\underline{u}(1))$. Also, as D and $\underline{u}$ are independently distributed,

$$\begin{aligned} V(\underline{z}|D_{11}, \underline{u}(1)) &= V(\underline{u}(2) - B\,\underline{u}(1)|D_{11}, \underline{u}(1)) \\ &= V(\underline{u}(2)|\underline{u}(1)) + V(B\,\underline{u}(1)|D_{11}, \underline{u}(1)) \\ &= \Sigma_{22\cdot 1} + \Sigma_{22\cdot 1}\,\underline{u}'(1)\,D_{11}^{-1}\,\underline{u}(1) . \end{aligned} \tag{3.16}$$

The second term in the right-hand side of the last step follows from applying (1.2.7) to (3.14). Using (3.6), we finally obtain

$$V(\underline{z} | D_{11}, \underline{u}(1)) = \Sigma_{22\cdot 1}(1 + T_k^2/f) , \tag{3.17}$$

which is a nonsingular matrix. Now, when $\underline{u}(1)$ and D_{11} are fixed, $\underline{z}$ is a linear function of the normal variables $\underline{u}(2)$ and B, and so by Theorem 2 of Chapter 2, the conditional distribution of $\underline{z}$, when $\underline{u}(1)$ and D_{11} are fixed, is (p-k) variate normal with mean and variance-covariance matrix given by (3.15) and (3.17). But this conditional distribution of $\underline{z}$ involves the conditioning variates $\underline{u}(1)$ and D_{11} only through $(1 + T_k^2/f)$, occurring in (3.17). Hence, it is the conditional distribution of $\underline{z}$ when T_k^2 is fixed. We have already seen that $D_{22\cdot 1}$ is independently distributed of B and so it is independently distributed of $\underline{z}$, as $\underline{z}$ is a function of B and $\underline{u}$ only. We have thus the following two independent distributions:

(a) when T_k^2 is fixed,

$$\underline{z} \sim N_{p-k}(\underline{z} | \underline{\mu}(2) - \beta\, \underline{\mu}(1) | (1 + T_k^2/f)\, \Sigma_{22\cdot 1})$$

and

(b) $D_{22\cdot 1} \sim W_p(D_{22\cdot 1} | f-k | \Sigma_{22\cdot 1})$.

Hence using Theorem 1, conditional on T_k^2 being fixed,

$$\frac{(f - k)}{1 + T_k^2/f}\, \underline{z}'\, D_{22\cdot 1}^{-1}\, \underline{z} \tag{3.18}$$

is a Hotelling's T^2 based on f-k d.f., and noncentrality parameter

$$\frac{1}{1 + T_k^2/f}\, (\underline{\mu}(2) - \beta\, \underline{\mu}(1))'\, \Sigma_{22\cdot 1}^{-1}\, (\underline{\mu}(2) - \beta\, \underline{\mu}(1)) . \tag{3.19}$$

But from (3.10) and (3.9), (3.18) and (3.19) reduce to

$$T^2_{2\cdot 1} = \frac{(f-k)(T^2_p - T^2_k)}{f + T^2_k} \tag{3.20}$$

and

$$\frac{(\lambda^2_p - \lambda^2_k)}{1 + T^2_k/f}, \tag{3.21}$$

respectively. We thus have the following theorem:

Theorem 3: If $\underline{u}$ has a $N_p(\underline{u}|\underline{\mu}|\Sigma)$ distribution and D has an independent $W_p(D|f|\Sigma)$ distribution, and if $\underline{u}$, $\underline{\mu}$, Σ, D are partitioned as in (3.1) and (3.2), the conditional distribution of $T^2_{2\cdot 1} = (f-k)(T^2_p - T^2_k)/(f + T^2_k)$, when T^2_k is fixed, is $H'_{p-k}(T^2_{2\cdot 1}|f-k, (\lambda^2_p - \lambda^2_k)f/(f + T^2_k))\, dT^2_{2\cdot 1}$, where T^2_p, T^2_k λ^2_p, and λ^2_k are defined by (3.5) and (3.6).

In particular, when $\lambda^2_p = \lambda^2_k$, we observe by use of Theorem 2, that the conditional distribution of

$$\frac{(f-k) - (p-k) + 1}{(p-k)} \; \frac{T^2_{2\cdot 1}}{f-k}, \tag{3.22}$$

when T^2_k is fixed, is an F distribution with p-k and (f-k) - (p-k) + 1 d.f. However this conditional distribution does not involve T^2_k, and so it is the unconditional distribution also and $T^2_{2\cdot 1}$ is independently distributed of T^2_k. Further (3.22) reduces to

$$\frac{f-p+1}{p-k}\left(\frac{1}{U} - 1\right) \tag{3.23}$$

where U, a statistic due to C. Radhakrishna Rao [22], is already defined in (3.12). We have, therefore,

Theorem 4: With $\underline{u}$, $\underline{\mu}$, Σ, D, U as defined earlier,

$$\frac{f-p+1}{p-k}\left(\frac{1}{U} - 1\right) \sim F_{p-k,\, f-p+1}\,,$$

$$\underline{\text{if } \lambda_p^2 = \lambda_k^2}\,,$$

and U is independently distributed of T_k^2.

It should be noted that λ_p^2 will be equal to λ_k^2 in any of the following cases:

(a) $\underline{\mu}(2) = \beta\, \underline{\mu}(1)$ (this follows from 3.9),

(b) $\underline{\mu}(2) = \underline{0}$, if it is already known that $\underline{u}(1)$ and $\underline{u}(2)$ are independent (because in that case $\beta = 0$ and this condition reduces to (a)),

(c) $\underline{\mu}(2) = \underline{0}$, if it is already known that $\underline{\mu}(1) = \underline{0}$,

(d) any linear function of $\underline{u}$ uncorrelated with $\underline{u}(1)$ has zero mean value. (3.24)

Condition (d) can be seen as follows: Let $\underline{a}'\underline{u}$ be any linear function of $\underline{u}$. It can be written as

$$\underline{a}'\underline{u} = \underline{a}'(1)\, \underline{u}(1) + \underline{a}'(2)\, \underline{u}(2)\,,$$

if $\underline{a}'$ is partitioned as

$$[\underline{a}'(1) | \underline{a}'(2)]\,.$$

Since $\underline{a}'\underline{u}$ is to be uncorrelated with $\underline{u}(1)$,

$$\underline{0} = \mathrm{Cov}(\underline{a}'\underline{u}, \underline{u}(1)) = \mathrm{Cov}(\underline{a}'(1)\, \underline{u}(1) + \underline{a}'(2)\, \underline{u}(2), \underline{u}(1))$$
$$= \underline{a}'(1)\, \Sigma_{11} + \underline{a}'(2)\, \Sigma_{21}\,.$$

Hence,

$$a'(1) = - a'(2)\, \Sigma_{21}\, \Sigma_{11}^{-1}$$

$$= - a'(2)\, \beta$$

and, therefore,

$$E(\underline{a}'\underline{u}) = \underline{a}'\underline{\mu} = \underline{a}'(1)\, \underline{\mu}(1) + \underline{a}'(2)\, \underline{\mu}(2)$$

$$= - \underline{a}'(2)\, \underline{\beta}\, \underline{\mu}(1) + \underline{a}'(2)\, \underline{\mu}(2)$$

$$= \underline{a}'(2)\, (\underline{\mu}(2) - \underline{\beta}\, \underline{\mu}(1)) .$$

If this mean value is zero for every $\underline{a}$, it is obvious that $\underline{\mu}(2) - \beta\, \underline{\mu}(1)$ is null and in that case $\lambda_p^2 = \lambda_k^2$.

The F-test provided by Theorem 4 can thus be used to test any of the four hypotheses (a), (b), (c), or (d).

4. APPLICATIONS OF HOTELLING'S T^2

(A_1) Test of Significance of A Hypothetical Mean Vector

Given a random sample of size n from a $N_p(\underline{x}|\underline{\mu}|\Sigma)$ population, we may wish to test the hypothesis

$$H_o:\quad \underline{\mu} = \underline{\mu}_o \quad \text{(specified)} . \tag{4.1}$$

When Σ is known, we have already seen in (3.4.13) that the hypothesis can be tested by the statistic

$$n(\bar{\underline{x}} - \underline{\mu}_o)'\, \Sigma^{-1}(\bar{\underline{x}} - \underline{\mu}_o) , \tag{4.2}$$

which has a χ^2 distribution with p.d.f. if H_o is true. But when Σ is unknown, we use the following two distributions:

(i) $\sqrt{n}\,\bar{\underline{x}}$ has the $N_p(\sqrt{n}\,\bar{\underline{x}}|\sqrt{n}\,\underline{\mu}|\Sigma)$ distribution,

(ii) S has an independent $W_p(S|n-1|\Sigma)$ distribution, where $\bar{\underline{x}}$ and S are, respectively, the vector of the sample means and the matrix of the corrected s.s. and s.p. of the sample observations. These distributions were derived in Section 4 of Chapter 3. Let, therefore,

$$\underline{u} = \sqrt{n}(\bar{\underline{x}} - \underline{\mu}_o) . \tag{4.3}$$

Its distribution is

$$N_p(\underline{u}|\sqrt{n}(\underline{\mu} - \underline{\mu}_o)|\Sigma) , \tag{4.4}$$

and it is independent of S. Hence, by Theorem 1 of Section 1,

$$\begin{aligned} T^2 &= f\,\underline{u}'\,S^{-1}\,\underline{u} \\ &= (n-1)n\,(\bar{\underline{x}} - \underline{\mu}_o)'\,S^{-1}(\bar{\underline{x}} - \underline{\mu}_o) \end{aligned} \tag{4.5}$$

is a Hotelling's T^2 based on f = n-1 d.f. and noncentrality parameter

$$\lambda^2 = n(\underline{\mu} - \underline{\mu}_o)'\,\Sigma^{-1}(\underline{\mu} - \underline{\mu}_o) . \tag{4.6}$$

Observe that T^2 as given by (4.5) is nothing but (4.2) with Σ replaced by its unbiased estimate $[1/(n-1)]\Sigma$, i.e., T^2 is obtained by "Studentizing" (4.2). When the null hypothesis H_o is true, $\lambda^2 = 0$ or $E(\underline{u}) = \underline{0}$, and by Theorem 2 of Section 1,

$$\frac{f-p+1}{p}\,\frac{T^2}{f} = \frac{n-p}{p}\,n(\bar{\underline{x}} - \underline{\mu}_o)'\,S^{-1}(\bar{\underline{x}} - \underline{\mu}_o) \tag{4.7}$$

has an F distribution with p and n-p d.f. This F test can, therefore, be used to test the null hypothesis H_o. The critical region will be $T^2 > T^2(f, p, \alpha)$, as discussed in Section 1, with T^2 given by (4.5) and $f = n-1$, α being the level of significance. The power of the test will be (1.24), with λ^2 given by (4.6). Confidence bounds (simultaneous) for linear parametric functions of the type $\underline{a}'\underline{\mu}$ will be

$$\underline{a}'\underline{\bar{x}} \pm (\underline{a}'S\underline{a})^{1/2} \left\{\frac{p\, F(p, n-p, \alpha)}{n(n-p)}\right\}^{1/2} \tag{4.8}$$

and can be obtained in the same manner as (2.14), from $\underline{u} = \sqrt{n}(\underline{\bar{x}} - \underline{\mu})$.

(A_2) Test of Equality of the Mean Vectors of Two Normal Populations, with the Same but Unknown Variance-Covariance Matrix and other Associated Tests

Let X and Y be two $p \times n_1$ and $p \times n_2$ matrices of sample observations, in samples of sizes n_1 and n_2 from two independent populations $N_p(\underline{x}|\underline{\mu}|\Sigma)$ and $N_p(\underline{x}|\underline{\nu}|\Sigma)$, respectively. Consider the hypothesis

$$H_0: \underline{\mu} - \underline{\nu} = \underline{\delta}_0, \text{ a given vector.} \tag{4.9}$$

For testing this hypothesis, we consider

$\underline{\bar{x}} = X\, E_{n_1,1}/n_1$, the vector of sample means for the first sample,

$\underline{\bar{y}} = Y\, E_{n_2,1}/n_2$, the vector of sample means for the second sample,

$S_x = X(I - (1/n_1)E_{n_1,n_1})X'$, the matrix of corrected s.s. and s.p. of the observations from the first sample,

$S_y = Y(I - (1/n_2)E_{n_2,n_2})Y'$, the matrix of the corrected s.s. and s.p. of the observations from the second sample. (4.10)

We also let

$$\underset{-}{\delta} = \underset{-}{\mu} - \underset{-}{\nu}, \quad \underset{-}{d} = \underset{-}{\bar{x}} - \underset{-}{\bar{y}} \tag{4.11}$$

$$S = S_x + S_y , \tag{4.12}$$

which is called the "pooled" matrix of corrected s.s. and s.p. from both the samples

$$f = n_1 + n_2 - 2 , \quad C^2 = \frac{n_1 n_2}{n_1 + n_2} . \tag{4.13}$$

From Section 4 of Chapter 3, we have the following four independent distributions:

$$\text{(i)} \quad \underset{-}{\bar{x}} \sim N_p(\underset{-}{\bar{x}}|\underset{-}{\mu}|\ n_1^{-1}\ \Sigma),$$

$$\text{(ii)} \quad \underset{-}{\bar{y}} \sim N_p(\underset{-}{\bar{y}}|\nu|\ n_2^{-1}\ \Sigma),$$

$$\text{(iii)} \quad S_x \sim W_p(S_x|n_1 - 1|\Sigma),$$

$$\text{(iv)} \quad S_y \sim W_p(S_y|n_2 - 1|\Sigma).$$

By Theorem 2 of Chapter 2,

$$\underset{-}{u} = C(\underset{-}{d} - \underset{-}{\delta}_o) \quad \text{has the} \quad N_p(\underset{-}{u}|C(\underset{-}{\delta} - \underset{-}{\delta}_o)|\Sigma) \text{ distribution.} \tag{4.14}$$

Also, by the additive property of Wishart distributions proved in Section 6 of Chapter 3,

$$S \quad \text{has the } W_p(S|f|\Sigma) \quad \text{distribution,} \tag{4.15}$$

independent of $\underset{-}{u}$. Hence, by Theorem 1 of Section 1,

$$\begin{aligned} T^2 &= f\ \underset{-}{u}'\ S^{-1}\ \underset{-}{u} \\ &= C^2 f\ (\underset{-}{d} - \underset{-}{\delta}_o)'\ S^{-1}(\underset{-}{d} - \underset{-}{\delta}_o) \\ &= \left(\frac{n_1 n_2}{n_1 + n_2}\right) (n_1 + n_2 - 2)\ (\underset{-}{\bar{x}} - \underset{-}{\bar{y}} - \underset{-}{\delta}_o)'\ S^{-1}(\underset{-}{\bar{x}} - \underset{-}{\bar{y}} - \underset{-}{\delta}_o) \end{aligned} \tag{4.16}$$

is a Hotelling's T^2 based on $f = n_1 + n_2 - 2$ d.f. and noncentrality parameter

$$\lambda^2 = C^2(\underline{\delta} - \underline{\delta}_o)' \Sigma^{-1}(\underline{\delta} - \underline{\delta}_o)$$
$$= \frac{n_1 n_2}{n_1+n_2} (\underline{\mu} - \underline{\nu} - \underline{\delta}_o)' \Sigma^{-1}(\underline{\mu} - \underline{\nu} - \underline{\delta}_o). \qquad (4.17)$$

when the null hypothesis is true, $\underline{\delta} = \underline{\delta}_o$ and $E(\underline{u}) = \underline{0}$, and so by Theorem 2 of Section 1,

$$\frac{f-p+1}{p} \frac{T^2}{f} = \frac{n_1+n_2-p-1}{p} \cdot \frac{T^2}{n_1+n_2-2} \qquad (4.18)$$

has an F distribution with p and n_1+n_2-p-1 d.f. This F test is therefore used to test H_o. The critical region will be $T^2 > T^2(f, p, \alpha)$ with T^2 and f as given by (4.16) and (4.13) and α is the significance level. The power of the test will be given by (1.24), with λ^2 defined by (4.17). Simultaneous confidence bounds for linear parametric functions $\underline{a}'\underline{\delta}$ will be provided by

$$\underline{a}'\underline{d} \pm (1/C)(\underline{a}' S \underline{a})^{1/2} \left\{\frac{p\, F(p,\ f-p+1,\ \alpha)}{f-p+1}\right\}^{1/2}, \qquad (4.19)$$

in accordance with (2.14).

A particular case that is of importance in discriminant analysis (which will be considered in a subsequent chapter) is $\underline{\delta}_o = \underline{0}$, i.e., the test of equality of the mean vectors of the two normal populations. The appropriate T^2 for this test will be obtained by putting $\underline{\delta}_o = \underline{0}$ in (4.16) and is, therefore,

$$T^2 = C^2 f\, \underline{d}'\, S^{-1}\, \underline{d}$$
$$= \left(\frac{n_1 n_2}{n_1+n_2}\right) (n_1+n_2-2)\ (\bar{\underline{x}} - \bar{\underline{y}})'\, S^{-1}(\bar{\underline{x}} - \bar{\underline{y}}). \qquad (4.20)$$

The noncentrality parameter in the distribution of this Hotelling's

T^2 is, on putting $\underline{\delta}_o = \underline{0}$ in (4.17),

$$\lambda^2 = C^2 \, \underline{\delta}' \, \Sigma^{-1} \, \underline{\delta} = \frac{n_1 n_2}{n_1+n_2} \, (\underline{\mu} - \underline{\nu})' \, \Sigma^{-1}(\underline{\mu} - \underline{\nu}) \, . \quad (4.21)$$

If the null hypothesis $\underline{\mu} = \underline{\nu}$ is true, (4.18) will provide an F test as before, with T^2 given by (4.20) and $f = n_1+n_2-2$. The quantity $\underline{\delta}' \, \Sigma^{-1} \, \underline{\delta}$ in (4.21) is denoted by Δ^2 and was proposed by Mahalanobis [17] as a measure of the divergence or distance between the two normal populations, $N_p(\underline{x}|\underline{\mu}|\Sigma)$ and $N_p(\underline{y}|\underline{\nu}|\Sigma)$. The corresponding sample quantity, obtained by replacing $\underline{\delta}$ and Σ in Δ^2 by their estimates, is denoted by D^2, which is given by

$$D^2 = \underline{d}'(\tfrac{1}{f} \, S)^{-1} \, \underline{d} \quad \text{or} \quad f \, \underline{d}' \, S^{-1} \, \underline{d} \qquad (4.22)$$

and is known as Mahalanobis's Studentized D^2. From (4.20), it is obvious that

$$T^2 = C^2 \, D^2 \, , \qquad (4.23)$$

or that T^2 and D^2 are almost the same, except for the constant C^2. The distribution of D^2, when $\underline{\mu} \neq \underline{\nu}$, is therefore, by Theorem 1 of Section 1,

$$H_p'(C^2 \, D^2|f, \, C^2 \, \Delta^2) \, d(C^2 \, D^2) \, , \qquad (4.24)$$

and is obtained from the distribution of T^2 by using (4.23) and noting that $\lambda^2 = C^2 \, \Delta^2$; f is, of course, n_1+n_2-2. Expression (4.24) is known as the non-null distribution of D^2 and was first obtained by Bose and Roy [3]. The null distribution of D^2 is obtained by putting $\Delta^2 = 0$ in (4.24) and reduces to

$$H_p(C^2 \, D^2|f) \, d(C^2 \, D^2). \qquad (4.25)$$

When $\Delta^2 = 0$, i.e., $\underline{\mu} = \underline{\nu}$, the F test of (4.18) can be expressed in terms of D^2, by using $T^2 = c^2 D^2$, as

$$\frac{n_1+n_2-p-1}{p} \frac{n_1 n_2}{n_1+n_2} \frac{D^2}{n_1+n_2-2} = F_{p,n_1+n_2-p-1} . \qquad (4.26)$$

Let us now partition $\underline{x}$, $\underline{d}$, $\underline{\delta}$, Σ and S as

$$\underline{x} = \left[\begin{array}{c}\underline{x}(1)\\ \hline \underline{x}(2)\end{array}\right], \quad \underline{d} = \left[\begin{array}{c}\underline{d}(1)\\ \hline \underline{d}(2)\end{array}\right]\begin{array}{l}k\\ p\text{-}k\end{array}, \quad \underline{\delta} = \left[\begin{array}{c}\underline{\delta}(1)\\ \hline \underline{\delta}(2)\end{array}\right]\begin{array}{l}k\\ p\text{-}k\end{array}$$

$$\Sigma = \underset{\quad k \qquad p\text{-}k}{\left[\begin{array}{c|c}\Sigma_{11} & \Sigma_{12}\\ \hline \Sigma_{21} & \Sigma_{22}\end{array}\right]}\begin{array}{l}k\\ p\text{-}k\end{array}, \quad S = \underset{\quad k \qquad p\text{-}k}{\left[\begin{array}{c|c}S_{11} & S_{12}\\ \hline S_{21} & S_{22}\end{array}\right]}\begin{array}{l}k\\ p\text{-}k\end{array}. \qquad (4.27)$$

Also let

$$T_k^2 = c^2 f \, \underline{d}'(1) \, S_{11}^{-1} \, \underline{d}(1) = c^2 D_k^2 , \qquad (4.28)$$

$$\lambda_k^2 = c^2 \, \underline{\delta}'(1) \, \Sigma_{11}^{-1} \, \underline{\delta}(1) = c^2 \Delta_k^2 . \qquad (4.29)$$

In other words, Δ_k^2 is the (distance)^2 between the two normal populations when only the first k characters $\underline{x}(1)$ are used. D_k^2 is the corresponding Studentized (distance)^2. To emphasize the fact that we use all the components of $\underline{x}$ in D^2 or T^2, we shall now rewrite D^2 and T^2 as D_p^2 and T_p^2, respectively. We now apply Theorem 3 of Section 3 to C $\underline{d}$ which has the $N_p(C\,\underline{d}|C\,\delta|\Sigma)$ distribution and S which has the $W_p(S|f|\Sigma)$ distribution, independent of $\underline{d}$. We obtain, by using this theorem, the following result: The conditional distribution of $c^2 D_{2\cdot1}^2 = (f-k)\,c^2(D_p^2 - D_k^2)/(f + D_k^2)$, when D_k^2 is fixed is

$$H'_{p-k}\left(c^2 D_{2\cdot1}^2 \,\middle|\, f-k, \frac{c^2 f(\Delta_p^2 - \Delta_k^2)}{f + c^2 D_k^2}\right) d(c^2 D_{2\cdot1}^2) , \qquad (4.30)$$

where $f = n_1+n_2-2$ and the function H' is defined by (1.9).

Hence, or by using (3.23) and (3.12), we can test the hypothesis $\Delta_p^2 = \Delta_k^2$ by the statistic

$$U = \frac{1 + C^2 D_k^2/f}{1 + C^2 D_p^2/f}, \tag{4.31}$$

the actual test being

$$\frac{f-p+1}{p-k}\left(\frac{1}{U} - 1\right) = F_{p-k,\, f-p+1}. \tag{4.32}$$

The relevance and usefulness of this test in the context of discriminant analysis will be discussed later.

(A_3) Fisher-Behrens Problem in the Multivariate Case

Let us now consider the problem of testing the hypothesis

$$H: \underline{\mu} = \underline{\nu}, \tag{4.33}$$

where $\underline{\mu}$ is the mean vector of a p-variate normal population with variance-covariance matrix Σ_1 and $\underline{\nu}$ is the mean vector of an independent normal population with variance-covariance matrix Σ_2. Σ_1 and Σ_2 are unknown and unequal. The corresponding problem in the univariate case is known as the Fisher-Behrens problem. The solution of Fisher and Behrens, however, is controversial, and so we consider here only the multivariate extension (due to Bennet [2]) of a noncontroversial solution due to Scheffe' [25] in the univariate case. Let us suppose that random samples of sizes n_1 and n_2 are available from the two normal populations. Let X and Y be, respectively, the $p \times n_1$ and $p \times n_2$ matrices of the sample observations. We assume $n_1 < n_2$ and partition Y as

$$Y = \underset{\quad n_1 \quad m}{[Y_1 | Y_2]p}, \tag{4.34}$$

where $m = n_2 - n_1$. Each column of X has a p-variate normal distribution with mean $\underline{\mu}$ and variance-covariance matrix Σ_1 and the columns of X are independently distributed. Hence [remembering (2.2.2)],

$$E(X) = \underline{\mu}\, E_{1n_1}, \quad V(X) = \Sigma_1 \otimes I_{n_1} . \tag{4.35}$$

Similarly,

$$E(Y_1) = \underline{\nu}\, E_{1n_1}, \quad E(Y_2) = \underline{\nu}\, E_{1m},$$

$$V(Y_1) = \Sigma_2 \otimes I_{n_1}, \quad V(Y_2) = \Sigma_2 \otimes I_m, \tag{4.36}$$

and, further,

$$\mathrm{Cov}(Y_1, Y_2) = 0, \tag{4.37}$$

as all the columns of Y are independent. Define a $p \times n_1$ matrix Z as

$$Z = X - \left(\frac{n_1}{n_2}\right)^{1/2} Y_1\left(I - \frac{1}{n_1} E_{n_1 n_1}\right) - \frac{1}{n_2} Y E_{n_2 n_1} . \tag{4.38}$$

From (4.35) and (4.36), it is easy to see that

$$E(Z) = (\underline{\mu} - \underline{\nu})E_{1n_1} . \tag{4.39}$$

Using (4.34), Z can be rewritten as

$$Z = X - \left(\frac{n_1}{n_2}\right)^{1/2} Y_1\left(I - \frac{1}{n_1} E_{n_1 n_1}\right) - \frac{1}{n_2}\left(Y_1 E_{n_1 n_1} + Y_2 E_{mn_1}\right)$$

$$= X - Y_1 P - \frac{1}{n_2} Y_2\, E_{mn_1}, \tag{4.40}$$

where

$$P = \left(\frac{n_1}{n_2}\right)^{1/2}\left(I - \frac{1}{n_1} E_{n_1 n_1}\right) + \frac{1}{n_2} E_{n_1 n_1} . \tag{4.41}$$

Then, from (4.35), (4.36), (4.37), and (1.2.7), we obtain

$$V(Z) = \Sigma_1 \otimes I_{n_1} + \Sigma_2 \otimes (P'I_{n_1}P) + \frac{1}{n_2^2}\Sigma_2 \otimes (E_{n_1 m} I E_{m n_1})$$
$$= \Sigma_1 \otimes I_{n_1} + \Sigma_2 \otimes (P'P + \frac{m}{n_2^2} E_{n_1 n_1})$$
$$= (\Sigma_1 + \frac{n_1}{n_2}\Sigma_2) \otimes I_{n_1}$$
$$= \Sigma \otimes I_{n_1}, \qquad (4.42)$$

where

$$\Sigma = \Sigma_1 + (n_1/n_2)\Sigma_2 \qquad (4.43)$$

is non-singular. Since Z is a linear combination of the normal variables X and Y, from (4.39), (4.42), and Theorem 2 of Chapter 2, we see that each column of Z has a p-variate normal distribution with mean $\underline{\mu} - \underline{\nu}$ and variance-covariance matrix Σ given by (4.43). Further, from (4.42), all the columns of Z are independently distributed. In other words, Z can be regarded as a random sample of size n_1 from a p-variate normal population with mean $\underline{\mu} - \underline{\nu}$ and variance-covariance matrix Σ, which is unknown. The hypothesis H_o is then the hypothesis that the mean of Z is null, and the problem therefore reduces to the one we already considered in (A_1). The appropriate statistic is the Hotelling's T^2, based on n_1-1 d.f. and given by

$$\frac{T^2}{n_1 - 1} = n_1 \bar{\underline{z}}' S_z^{-1} \bar{\underline{z}}, \qquad (4.44)$$

where

$$\bar{\underline{z}} = \frac{1}{n_1} Z E_{n_1 1}, \text{ the mean vector of the observations in Z} \qquad (4.45)$$

and

$$S_z = Z(I - \frac{1}{n_1} E_{n_1 n_1})Z' , \quad (4.45b)$$

the matrix of the corrected s.s. and s.p. of the observations Z. The test is (from 4.7)

$$\frac{n_1 - p}{p} \frac{T^2}{n_1 - 1} = F_{p,\, n_1 - p} . \quad (4.46)$$

In actual application, however, the task of calculating (4.44) is made much simpler due to the following results:

$$\begin{aligned} \underline{\bar{z}} &= \frac{1}{n_1} Z\, E_{n_1 1} \\ &= \frac{1}{n_1} X\, E_{n_1 1} - \frac{1}{n_2} Y\, E_{n_2 1} \\ &= \underline{\bar{x}} - \underline{\bar{y}} , \end{aligned} \quad (4.47)$$

where $\underline{\bar{x}}$ is the vector of means obtained from the first sample X and $\underline{\bar{y}}$ is the vector of the means from the second sample. Equation (4.47) can be proved by substituting (4.38) for Z. Similarly,

$$S_z = U(I - \frac{1}{n_1} E_{n_1 n_1})U' = S_u \quad (4.48)$$

where

$$U = X - \left(\frac{n_1}{n_2}\right)^{1/2} Y_1 . \quad (4.49)$$

It is therefore easier to form the matrix U by subtracting from every column of X, $(n_1/n_2)^{1/2}$ times the corresponding column of Y and then, if one forms the matrix S_u of the corrected s.s. and s.p. of these observations U, one gets S_z from (4.48). Thus, for the purpose of calculations, (4.44) can be written as

$$\frac{T^2}{n_1 - 1} = n_1(\underline{\bar{x}} - \underline{\bar{y}})' S_u^{-1}(\underline{\bar{x}} - \underline{\bar{y}}) . \quad (4.50)$$

It is worth noting that all the n_2 observations are utilized in finding $\bar{\underline{y}}$, but only n_1 of them are used in S_u. Which $n_2 - n_1$ columns of Y are ignored is arbitrary in this test and is a weakness of the test.

If H_o is not true, the noncentrality parameter in the distribution of T^2 will be (see 4.6)

$$\lambda^2 = n_1(\underline{\mu} - \underline{\nu})' \Sigma^{-1}(\underline{\mu} - \underline{\nu}) . \tag{4.51}$$

When n_1 and n_2 are equal, say equal to n, the test becomes much simpler. We then have

$$Z = X - Y , \tag{4.52}$$

and the Hotelling's T^2 is given by

$$\frac{T^2}{n-1} = n(\bar{\underline{x}} - \bar{\underline{y}})' S_z^{-1}(\bar{\underline{x}} - \bar{\underline{y}}) . \tag{4.53}$$

where

$$S_z = (X - Y)(I - \frac{1}{n} E_{nn})(X - Y)' . \tag{4.54}$$

(A_4) Test of Significance of Contrasts

Suppose, we wish to test the hypothesis

$$H_o: \mu_1 = \mu_2 = \ldots = \mu_p \text{ (= } \gamma \text{ unspecified)}, \tag{4.55}$$

on the basis of a random sample of size n from the p-variate normal population $N_p(\underline{x}|\underline{\mu}|\Sigma)$, where $\mu_1, \ldots, \mu_p$ are the elements of $\underline{\mu}$. Let, as before, X denote the $p \times n$ matrix of the sample observations, $\bar{\underline{x}}$ the vector of the sample means, and S the matrix of the corrected s.s. and s.p. of the sample observations.

We call a $(p-1) \times p$ matrix C a contrast matrix if it is of rank p-1 and is such that

$$C\, E_{p1} = \underline{0}. \tag{4.56}$$

It is obvious that the sum of the elements in any row of C is zero. Following are some typical contrast matrices:

$$C_1 = \begin{bmatrix} 1 & -1 & 0 & \cdots & 0 & 0 \\ 0 & 1 & -1 & \cdots & 0 & 0 \\ \vdots & & & & & \\ 0 & 0 & 0 & \cdots & 1 & -1 \end{bmatrix} \tag{4.57}$$

$$C_2 = \begin{bmatrix} 1 & 0 & \cdots & 0 & -1 \\ 0 & 1 & \cdots & 0 & -1 \\ \vdots & & & & \\ 0 & 0 & \cdots & 1 & -1 \end{bmatrix} \tag{4.58}$$

$$\begin{bmatrix} \frac{1}{\sqrt{(1)(2)}} & \frac{-1}{\sqrt{(1)(2)}} & 0 & \cdots & 0 & 0 \\ \frac{1}{\sqrt{(2)(3)}} & \frac{1}{\sqrt{(2)(3)}} & \frac{-2}{\sqrt{(2)(3)}} & \cdots & 0 & 0 \\ \vdots & & & & & \\ \frac{1}{\sqrt{(p-1)p}} & \frac{1}{\sqrt{(p-1)p}} & \frac{1}{\sqrt{(p-1)p}} & \cdots & \frac{1}{\sqrt{(p-1)(p)}} & \frac{-(p-1)}{\sqrt{(p-1)p}} \end{bmatrix}. \tag{4.59}$$

Observe that in C_3, the row vectors are mutually orthogonal and are of unit length, so that if one more row $(1/\sqrt{p})E_{1p}$ is added to C_3, it will be an orthogonal matrix of order p. This orthogonal matrix is called Helmert's orthogonal matrix. Since there can be at most (p-1) row vectors which are linearly independent and orthogonal to E_{p1}, it is obvious that two contrast matrices,

C_α and C_β, must be related by

$$C_\alpha = D\, C_\beta \ , \tag{4.60}$$

where D is of order (p-1) x (p-1) and rank p-1. In other words, the contrasts $C_\alpha \underline{x}$ among $x_1, \ldots, x_p$ must be linearly independent linear functions of any other set of contrasts $C_\beta \underline{x}$.

Now the hypothesis H_o can be written as

$$H_o: \quad \underline{\mu} = \gamma\, E_{p1} \ , \tag{4.61}$$

where γ is unknown, and hence, if H_o is true,

$$\underline{\nu} = E(C\underline{x}) = \underline{0} \ , \tag{4.62}$$

where C is any contrast matrix. Conversely, if $E(C\underline{x}) = \underline{0}$ for some contrast matrix C, we have $C\underline{\mu} = 0$ but, on account of (4.60), $C = D\, C_2$, with C_2 given by (4.58), and hence $\underline{0} = C\,\underline{\mu} = D\, C_2\, \underline{\mu}$ yields $C_2\, \underline{\mu} = 0$, as D is nonsingular and hence $\mu_1 = \ldots = \mu_p$, i.e., H_o is true.

But $C\underline{x}$ has the $N_{p-1}(C\underline{x}|C\underline{\mu}|C\,\Sigma\, C')$ distribution by Theorem 2 of Chapter 2 and CX is a random sample of size n from this (p-1) variate normal population. The problem of testing H_o, given by (4.55), thus reduces to testing the significance of the mean of $N_{p-1}(C\underline{x}|C\underline{\mu}|C\,\Sigma\, C')$ on the basis of the sample CX, as H_o is equivalent to $C\underline{\mu} = \underline{0}$. This is the same problem as (A_1), and the appropriate Hotelling's T^2 is thus given by (d.f. n-1)

$$\frac{T^2}{n-1} = n(C\bar{\underline{x}})'\ (CSC')^{-1}\ (C\bar{\underline{x}}) \tag{4.63}$$

and the test for H_o is provided by

$$\frac{(n-1) - (p-1) + 1}{p-1} \frac{T^2}{n-1} = F_{p-1,\ n-p+1}, \qquad (4.64)$$

as CSC' has the $W_{p-1}(CSC'|n-1|C\Sigma C')$ distribution independent of $C\underline{\bar{x}}$. It should be observed that (4.63) does not depend on the choice of the contrast matrix C. In other words, if C_α and C_β are two contrast matrices,

$$(C_\alpha\underline{\bar{x}})'(C_\alpha SC'_\alpha)^{-1}(C_\alpha\underline{\bar{x}}) = (C_\beta\underline{\bar{x}})'(C_\beta SC'_\beta)^{-1}(C_\beta\underline{\bar{x}}), \qquad (4.65)$$

because of (4.60) and the nonsingularity of D. It should also be remembered that $(CSC')^{-1}$ in (4.63) cannot be further simplified as C is not a square matrix, it is of order $(p-1) \times p$ and, as such, has no inverse. In practice, it is easier to choose C_1 or C_2 as the contrast matrix for calculating T^2. If H_o is not true, the distribution of T^2 will have

$$\lambda^2 = n(C\underline{\mu})'(C\Sigma C')^{-1}(C\underline{\mu}) \qquad (4.66)$$

as the noncentrality parameter. Simultaneous confidence intervals for all contrasts among the means $\mu_1, \mu_2, \ldots, \mu_p$, i.e., linear functions of the type $\underline{a}'\underline{\mu}$. Satisfying $\underline{a}'E_{p1} = 0$, can be obtained by the method outlined in Section 2 and will be [from (2.14) or (4.8)]

$$\underline{a}'\underline{\bar{x}} \pm (\underline{a}'s\underline{a})^{1/2}\left\{\frac{(p-1)\ F(p-1,\ n-p+1,\ \alpha)}{n(n-p+1)}\right\}^{1/2}. \qquad (4.67)$$

In certain cases, it is known, a priori, that some k contrasts are significant, i.e., the expected values of some k contrasts among $x_1, \ldots, x_p$ are not null. In these cases one is interested in knowing whether every other contrast uncorrelated with these k contrasts has zero mean. In other words, one wishes to know whether correlation with these k contrasts is the only reason for the remaining p-1-k contrasts to have significant mean values. Rao's U test can be used to test this. Let $B_1\underline{x}$, where $B_1 E_{p1} = 0$ and B_1 is a $k \times p$ matrix of rank k, denote the k contrasts assigned by prior information to be significant. Then we can take our contrast matrix C of (4.62) to be one in which the first k rows are the same as those of B_1, so that

$$C\,\bar{\underline{x}} = \begin{bmatrix} B_1\bar{\underline{x}} \\ \hline B_2\bar{\underline{x}} \end{bmatrix}, \quad \text{say,} \tag{4.68}$$

and $B_2\bar{\underline{x}}$ represent the remaining p-1-k contrasts. Then applying Theorem 3 to the $N_{p-1}(\sqrt{n}\,C\,\bar{\underline{x}}|\sqrt{n}\,C\,\underline{\mu}|C\Sigma C')$ and $W_{p-1}(CSC'|n-1|C\Sigma C')$ distributions of $C\,\bar{\underline{x}}$ and CSC', and noting condition (d) of (3.24), we find that the hypothesis under consideration, of significance of contrasts uncorrelated with $B_1\bar{\underline{x}}$, can be tested by the F test:

$$\frac{n-p+1}{p-k-1}\left(\frac{1}{U} - 1\right) = F_{p-k-1,\ n-p+1}\ , \tag{4.69}$$

where

$$U = \frac{1 + T_k^2/(n-1)}{1 + T^2/(n-1)}\ ,$$

$$T_k^2 = n(n-1)(B_1\bar{\underline{x}})'(B_1SB_1')^{-1}(B_1\bar{\underline{x}})\ . \tag{4.70}$$

is the Hotelling T^2 based on $B_1\bar{\underline{x}}$, the assigned contrasts, and T^2 given by (4.63) is the Hotelling's T^2 based on all the contrasts $C\,\bar{\underline{x}}$.

C. Radhakrishna Rao [22] has considered a practical application of these tests. In his illustration, $p = 4$ and x_1, x_2, x_3, x_4 represent the thickness of cork borings on trees in the North, South, East and West directions. The hypothesis $E(x_1) = E(x_2) = E(x_3) = E(x_4)$ is thus the hypothesis of equal bark deposits in every direction. He takes the contrast matrix

$$C = \begin{bmatrix} 1 & 1 & -1 & -1 \\ 1 & -1 & 0 & 0 \\ 0 & 0 & 1 & -1 \end{bmatrix} \tag{4.71}$$

so that the contrasts $C\,\bar{\underline{x}}$ measure, respectively, the difference in bark deposits of (i) North, South versus East, West; (ii) North versus South, and (iii) East versus West directions. There was reason to believe that the first contrast will be significant and hence the question whether differences in the North and South or East and West directions were solely due to their correlation with the first contrast can be resolved by using U test of (4.69), with $k = 1$ and B_1 consisting of the single row vector [1, 1, -1, -1], the first row of the matrix C.

Another application of the test (4.64) is given by Graybill [9] who considers a randomized block experiment in which some of the usual assumptions are not valid.

If we consider a randomized block experiment with b blocks and v treatments, the model is

$$E(y_{ir}) = \mu + t_i + b_r\,, \quad (i = 1, 2, \ldots, v;\ r = 1, 2, \ldots, b) \tag{4.72}$$

where y_{ir} is the yield of the ith treatment in the rth block, μ is a general constant, t_i is the effect of the ith treatment, and b_r is the effect of the rth block; μ, t_i, b_r are all fixed effects. The yields are usually assumed to be independent normal variables with a common variance. In some situations, it so happens that yields in the same block only are correlated (this may happen if the b_r are not fixed but random, or it is a split-plot experiment), and the variance of a yield is not constant but depends on the treatment applied. In other words, the yields y_{ir} have the following variances and covariances:

$$\operatorname{Cov}(y_{ir}, y_{i'r'}) = \begin{cases} 0, & r \neq r' \\ \sigma_{ii'}, & r = r' \end{cases}$$

$$V(y_{ir}) = \sigma_{ii} \ . \qquad (4.73)$$

Alternatively, if Y denotes the $v \times b$ matrix of the yields y_{ir}, we can write (4.73) as (see 2.2.2)

$$V(Y) = \Sigma \otimes I_b \ , \qquad (4.74)$$

where Σ is the $v \times v$ matrix of the elements $\sigma_{ii'}(i, i' = 1, 2, \ldots, v)$. The hypothesis under consideration is the usual hypothesis

$$H_o: t_1 = t_2 = \ldots = t_v \qquad (4.75)$$

of the homogeneity of treatments, consider any $(v-1) \times v$ contrast matrix C. To be definite, we shall choose C_2 of (4.58) with p replaced by v. Then, from (4.72)

$$E(C_2Y) = \begin{bmatrix} t_1 - t_v \\ t_2 - t_v \\ t_{v-1} - t_v \end{bmatrix} E_{1b} \tag{4.76}$$

and reduces to $\underline{0}$ if H_o is true. Further,

$$V(C_2Y) = (C_2 \Sigma C_2') \otimes I_p , \tag{4.77}$$

since C_2Y are linear functions of normal variables Y, and as $V(C_2Y)$ is nonsingular, we see from (4.76), (4.77), and Theorem 2 of Chapter 2 that every column of C_2Y has a (v-1)-variate normal distribution with mean vector $\underline{0}$ (if H_o is true) and variance-covariance matrix $C_2 \Sigma C_2'$; further the columns are independently distributed. In other words, CY is a random sample of size b from (v-1)-variate normal population, and the test of H_o is the same as the test of significance of the mean of this normal population, which we already have considered in (A_1). The test is thus

$$\frac{(b-1) - (v-1) + 1}{v-1} \cdot \frac{T^2}{b-1} = F_{v-1,\ b-v+1} \tag{4.78}$$

where

$$T^2 = b(b-1)\ (C_2\bar{\underline{y}})'\ (C_2SC_2')^{-1}\ (C_2\bar{\underline{y}}) ,$$

$$\bar{\underline{y}} = \frac{1}{b} Y E_{b1} , \tag{4.79}$$

$$S = Y(I - \frac{1}{b} E_{bb})Y' . \tag{4.80}$$

From the d.f. of F in (4.78) it is obvious that this test can be used only if $b > v$. If H_o is not true, the noncentrality parameter in the distribution of T^2 will be

$$\lambda^2 = b[t_1-t_v,\ldots,t_{v-1}-t_v](C_2\Sigma C_2')^{-1}[t_1-t_v,\ldots,t_{v-1}-t_v]' \,. \tag{4.81}$$

Simultaneous confidence intervals for contrasts among $t_1, \ldots, t_v$, of the type $\underline{h}'\underline{t} = h_1t_1 + \ldots + h_vt_v$ with $h_1 + \ldots + h_v = 0$, are, from (4.67),

$$\underline{h}'\bar{\underline{y}} \pm (\underline{h}'S\underline{h})^{1/2} \left\{\frac{(v-1)\ F(v-1,\ b-v+1,\ \alpha)}{b(b-v+1)}\right\}^{1/2}. \tag{4.82}$$

(A_5) Profile Analysis

If a battery of p psychological tests is administered to a group and $\mu_1, \mu_2, \ldots, \mu_p$ are the true expected scores in the p tests, the profile of the group is defined as the graph obtained by joining the points $(1, \mu_1), (2, \mu_2), \ldots, (p, \mu_p)$ successively. In practive, the μ's are estimated from the average scores in a sample.

If we have two possibly different groups, we will have two profiles, one determined by the points (i, μ_i) $(i = 1,2,\ldots,p)$ and the other by (i, ν_i), where $\nu_1, \ldots, \nu_p$ are the true mean scores of the other group. The profiles are said to be similar if the segment of line joining (i, μ_i) to $(i+1, \mu_{i+1})$ is parallel to the corresponding segment of line joining (i, ν_i) to $(i+1, \nu_i+1)$ of the other profile, for each i. Obviously, this will be so if the hypothesis

$$H_o\colon\ \mu_i - \mu_{i-1} = \nu_i - \nu_{i-1},\quad i = 2, 3, \ldots, p \tag{4.83}$$

is true. Let us assume that the test scores $\underline{x}$ are normally distributed and that their variance-covariance matrix is the same for both the groups. Let X be a $p \times n_1$ matrix of scores of n_1 individuals from the first group, and similarly let Y be a $p \times n_2$ matrix of the scores of n_2 individuals from the second group. Then X and Y represent samples of sizes n_1 and n_2 from the two independent normal populations $N_p(\underline{x}|\underline{\mu}|\Sigma)$ and $N_p(\underline{x}|\underline{\nu}|\Sigma)$, where $\underline{\mu}$, $\underline{\nu}$ are vectors of μ_i, ν_i, respectively. The hypothesis H_o of (4.83) is then the same as

$$H_o: E(C_1\bar{\underline{x}}) \text{ is the same in both the groups,} \qquad (4.84)$$

where C_1 is the contrast matrix defined in (4.57). However, $C_1\bar{\underline{x}}$ has the $N_{p-1}(C_1\underline{x}|C_1\underline{\mu}|C_1\Sigma C_1')$ distribution in the first group and $N_{p-1}(C_1\underline{x}|C_1\underline{\nu}|C_1\Sigma C_1')$ distribution in the second group, and C_1X, C_1Y are samples from these. The problem of testing H_o of (4.84) is then the same as (A_2), which we have already considered, and so a test will be provided by (see 4.26)

$$\frac{n_1+n_2-(p-1)-1}{(p-1)} \frac{n_1n_2}{n_1+n_2} \frac{D^2}{n_1+n_2-2} = F_{p-1,\, n_1+n_2-(p-1)-1}, \qquad (4.85)$$

where

$$D^2 = (n_1+n_2-2)(C_1\underline{d})'(C_1SC_1')^{-1}(C_1\underline{d}), \qquad (4.86)$$

$$\underline{d} = \bar{\underline{x}} - \bar{\underline{y}}, \quad \bar{\underline{x}} = \frac{1}{n_1} X E_{n_1 1}, \quad \bar{\underline{y}} = \frac{1}{n_2} Y E_{n_2 1}, \qquad (4.87)$$

$$S = X(I - \frac{1}{n_1} E_{n_1 n_1})X' + Y(I - \frac{1}{n_2} E_{n_2 n_2})Y' .$$

If we decide to accept the hypothesis of similarity of the two profiles on the basis of the F test in (4.85), the next logical question will be whether the average true score of the two groups is the same, i.e., whether

$$H_1: \frac{1}{p}(\mu_1 + \dots + \mu_p) = \frac{1}{p}(\nu_1 + \dots + \nu_p)$$

$$\text{or } E_{1p}\underline{\mu} = E_{1p}\underline{\nu} . \tag{4.88}$$

Observe that $E_{1p}\,\underline{x}$ has a univariate normal distribution with variance $E_{1p}\,\Sigma\,E_{p1}$ and means $E_{1p}\,\underline{\mu}$ and $E_{1p}\,\underline{\nu}$ in the two groups, and hence the hypothesis H_1 can be tested by the usual t test for testing the equality of the means of two normal populations with the same variance. If we square this t, we shall get the F test:

$$(n_1 + n_2 - 2)\,\frac{n_1 n_2}{n_1+n_2}\,(E_{1p}\underline{d})^2/(E_{1p}\,S\,E_{p1})$$

$$= F_{1,\, n_1+n_2-2} . \tag{4.89}$$

If H_o is true,

$$\mu_1 - \nu_1 = \mu_2 - \nu_2 = \dots = \mu_p - \nu_p ,$$

and if in addition H_1 is also true,

$$\mu_1 + \dots + \mu_p = \nu_1 + \dots + \nu_p .$$

From these two equations it is easy to see that, when H_o and H_1 hold,

$$\mu_1 = \nu_1,\ \ \mu_2 = \nu_2,\ \dots,\ \mu_p = \nu_p ,$$

i.e., both the profiles are exactly the same. If so, we now ask the question: Are the common values of μ_i, ν_i the same? That is, does

$$H_2:\ \ \mu_1 = \mu_2 = \dots = \mu_p \text{ hold?} \tag{4.90}$$

If H_o, H_1, H_2 all hold, the distribution of $\underline{x}$ is the same in both the groups and X and Y are therefore samples from the same population. They can therefore be pooled together yielding the grand mean vector

$$\underline{g} = \frac{n_1\bar{\underline{x}} + n_2\bar{\underline{y}}}{n_1 + n_2} \tag{4.91}$$

as the estimate of the common value of $\underline{\mu}$, $\underline{\nu}$ and its distribution is $N_p(\underline{g}|\underline{\mu}|\frac{1}{N}\Sigma)$, where $N = n_1 + n_2$. Obviously $\underline{g}$ is independently distributed of the $W_p(S|n_1 + n_2 - 2|\Sigma)$ distribution of S. The hypothesis H_2 is equivalent to the hypothesis

$$E(C\underline{g}) = 0 ,$$

where C is any contrast matrix, and so the problem reduces again to the one we considered in (A_4). The test will be

$$\frac{(n_1+n_2-2) - (p-1) + 1}{p-1} \frac{T^2}{n_1+n_2-2} = F_{p-1,\ n_1+n_2-p} , \tag{4.92}$$

where

$$\frac{T^2}{n_1+n_2-2} = N(C\underline{g})' (CSC')^{-1} (C\underline{g}) . \tag{4.93}$$

It should be noticed that, if the hypothesis H_2 is accepted, the common profile of the two groups consists of a line parallel to the x-axis, at a distance equal to the common value of $\mu_1, \ldots, \mu_p; \nu_1, \ldots, \nu_p$, estimated by $\underline{g}'E_{p1}/p$.

(A_6) Test of Symmetry of Organs

In many biological and anthropological problems, $x_1, x_2, \ldots, x_p$ represent measurements or characteristics on the left side and $x_{p+1}, x_{p+2}, \ldots, x_{2p}$ represent the same measurements on the right side or x_i $(i = 1, \ldots, p)$ represents characteristics of one of a twin pair and x_{p+i} $(i = 1, 2, \ldots, p)$ are those for the other member and we are interested in testing the hypothesis of symmetry of the left and right sides or equality of the measurements of the twins. The hypothesis can be expressed mathematically as

$$H_o: \ \mu_i = \mu_{p+i} \ (i = 1, 2, \ldots, p) \ . \tag{4.94}$$

where $\mu_1, \ldots, \mu_{2p}$ are the true means of $x_1, \ldots, x_{2p}$. Let $\bar{\underline{x}}$, S represent the mean vector and the matrix of corrected s.s. and s.p. of n sample observations in a sample of size n from this 2p-variate normal population of $x_1, \ldots, x_{2p}$. Partition $\bar{\underline{x}}$ and S as

$$\bar{\underline{x}} = \left[\begin{array}{c} \bar{\underline{x}}(1) \\ \hline \bar{\underline{x}}(2) \end{array}\right] \begin{array}{c} p \\ p \end{array} , \quad S = \left[\begin{array}{c|c} S_1 & S_2 \\ \hline S_2' & S_3 \end{array}\right] \begin{array}{c} p \\ p \end{array} \ . \tag{4.95}$$
$$\qquad\qquad\qquad\qquad\qquad p \quad p$$

Then, $\sqrt{n}\ \bar{\underline{x}}$ has $N_{2p}(\sqrt{n}\ \bar{\underline{x}} | \sqrt{n}\ \underline{\mu} | \Sigma)$ distribution, and S has an independent $W_{2p}(S | n-1 | \Sigma)$ distribution, where $\underline{\mu}$ is the vector of $\mu_1, \ldots, \mu_{2p}$ and Σ is the $2p \times 2p$ variance-covariance matrix of the 2p variables $\underline{x}$. Hence

$$\underline{y} = \sqrt{n}(\bar{\underline{x}}(1) - \bar{\underline{x}}(2)) = \sqrt{n}\ A\bar{\underline{x}} \tag{4.96}$$

has the $N_p(\underline{y} | \sqrt{n}\ (\underline{\mu}(1) - \underline{\mu}(2)) | A \Sigma A')$ distribution, where

$$\underline{\mu} = \left[\begin{array}{c} \underline{\mu}(1) \\ \hline \underline{\mu}(2) \end{array}\right] \begin{array}{c} p \\ p \end{array} \tag{4.97}$$

and

$$\underset{p \times 2p}{A} = [I_p | - I_p] \; . \tag{4.98}$$

The distribution of $ASA' = S_1 - S_2' - S_2 + S_3$ is, by Lemma 1 of Chapter 3,

$$W_p(ASA' | n-1 | A \Sigma A') \; , \tag{4.99}$$

independent of $\underline{y}$, and hence, by Theorem 2, if H_o is true, i.e., $\underline{\mu}(1) = \underline{\mu}(2)$,

$$\frac{(n-1) - p+1}{p} \cdot \frac{T^2}{n-1} \text{ has the } F_{p,\ n-p} \text{ distribution}, \tag{4.100}$$

where

$$\frac{T^2}{n-1} = \underline{y}'(ASA')^{-1}\,\underline{y}$$

$$= n(\bar{\underline{x}}(1) - \bar{\underline{x}}(2))'\,(S_1 - S_2' - S_2 + S_3)^{-1}\,(\bar{\underline{x}}(1) - \bar{\underline{x}}(2)). \tag{4.101}$$

(A_6) Application of Hotelling's T^2 to Test the Significance of a Vector of Regression Coefficients

We shall use the same notation and symbols as in Section 4 of Chapter 4, where we considered the distribution of the matrix of regression coefficients B in the regression of $\underline{x}(2)$ on $\underline{x}(1)$. By Theorem 2 of that chapter, we know that the conditional distribution of B, when the observations on $\underline{x}(1)$ are fixed, is $N_{(p-k)k}(B|\boldsymbol{\beta}|\Sigma_{22.1} \otimes S_{11}^{-1})$. Let $\underline{b}_r$ and $\underline{\boldsymbol{\beta}}_r$ denote, respectively, the rth columns of B and $\boldsymbol{\beta}$. Then it is obvious that the distribution of $\underline{b}_r$ is

$$N_{p-k}(\underline{b}_r | \underline{\boldsymbol{\beta}}_r | s^{rr}\,\Sigma_{22.1})\, d\,\underline{b}_r \quad (r = 1,\ 2,\ \ldots,\ k), \tag{4.102}$$

where s^{ij} are the elements of S_{11}^{-1}. By Theorem 1, $\underline{b}_r$ is independently distributed of $S_{22\cdot1}$, the matrix of residual s.s. and s.p., and has the distribution

$$W_{p-k}(S_{22\cdot1}|n-1-k|\Sigma_{22\cdot1})\, d\,S_{22\cdot1} \;. \tag{4.103}$$

If we wish to test the hypothesis

$$H_o:\ \underline{\beta}_r = \text{a specified vector } \underline{\gamma}\,, \tag{4.104}$$

we observe that, under the null hypothesis,

$$\underline{u} = \frac{1}{\sqrt{s^{rr}}}(\underline{b}_r - \underline{\gamma}) \text{ has the } N_{p-k}(\underline{u}|\underline{0}|\Sigma_{22\cdot1}) \text{ distribution} \tag{4.105}$$

independent of $S_{22\cdot1}$, and hence by Theorem 2 of this chapter,

$$\frac{(n-1-k) - (p-k) + 1}{p-k}\,\frac{T^2}{n-1-k} \text{ has the } F_{p-k,\ n-p} \text{ distribution,} \tag{4.106}$$

where

$$\frac{T^2}{n-1-k} = \underline{u}'\, S_{22\cdot1}^{-1}\, \underline{u}$$

$$= \frac{1}{s^{rr}}(\underline{b}_r - \underline{\gamma})'\, S_{22\cdot1}^{-1}(\underline{b}_r - \underline{\gamma})\;. \tag{4.107}$$

The F test (4.106) can thus be used to test H_o. If H_o is not true, by Theorem 1 of this chapter, we know that the noncentrality parameter in the distribution of T^2 will be

$$\lambda^2 = \frac{1}{s^{rr}}(\underline{\beta}_r - \underline{\gamma})'\, S_{22\cdot1}^{-1}(\underline{\beta}_r - \underline{\gamma})\;, \tag{4.108}$$

and the power of our test will be a function of this λ^2. The simultaneous confidence intervals for linear functions $\underline{a}'\underline{\beta}_r$ will be, by (2.14),

$$\underline{a}'\underline{b}_r \pm (s^{rr}\, \underline{a}'\, S_{22\cdot1}\, \underline{a})^{1/2} \left\{\frac{(p-k)\ F(p-k,\ n-p,\ \alpha)}{n-p}\right\}^{1/2}. \tag{4.109}$$

(A_7) Application of Hotelling's T^2 to Test the Independence of x_p and $x_1, x_2, \ldots, x_{p-1}$

Let S be the matrix of the corrected s.s. and s.p. of sample observations in a sample of size n from the p-variate normal population $N_p(\underline{x}|\underline{\mu}|\Sigma)$, where $x_i (i = 1, 2, \ldots, p)$ are the elements of $\underline{x}$. Partition Σ and S as

$$\Sigma = \left[\begin{array}{c|c} \Sigma_{p-1} & \underline{\sigma} \\ \hline \boldsymbol{\sigma}' & \boldsymbol{\sigma}_{pp} \end{array}\right]\begin{array}{l} p-1 \\ 1 \end{array}, \quad S = \left[\begin{array}{c|c} S_{p-1} & \underline{s} \\ \hline \underline{s}' & s_{pp} \end{array}\right], \qquad (4.110)$$

$$\qquad\quad p-1 \qquad 1$$

Let

$$S^* = S_{p-1} - \underline{s}(s_{pp})^{-1}\underline{s}' = S_{p-1} - \frac{\underline{s}\underline{s}'}{s_{pp}} \qquad (4.111)$$

and

$$\Sigma^* = \Sigma_{p-1} - \frac{\underline{\sigma}\underline{\sigma}'}{\boldsymbol{\sigma}_{pp}} . \qquad (4.112)$$

By application of Theorems 1 and 2 of Chapter 4, to the $W_p(S|n-1|\Sigma)$ distribution of S, we have

$$S^* \text{ has the } W_{p-1}(S^*|(n-1) - 1|\Sigma^*) \text{ distribution} \qquad (4.113)$$

and is independent of $\underline{b} = \underline{s}(s_{pp})^{-1}$, i.e., $\frac{1}{s_{pp}}\,\underline{s}$, the conditional distribution of which, when s_{pp} is fixed, is

$$N_{p-1}(\underline{b}|\underline{\boldsymbol{\beta}}|\frac{1}{s_{pp}}\Sigma^*), \qquad (4.114)$$

where $\boldsymbol{\beta} = \underline{\sigma}(\sigma_{pp})^{-1}$.

Applying Theorem 1 of this chapter to these distributions of $\underline{b}$ and S^*, we find that the conditional distribution of

$$T^2 = s_{pp}\, \underline{b}'\, S^{*-1}\, \underline{b} \cdot (n-2) , \qquad (4.115)$$

when s_{pp} is fixed, is that of a Hotelling's T^2 based on (n-2) d.f. and noncentrality parameter

$$\lambda^2 = s_{pp}\, \underline{\beta}'\, \Sigma^{*-1}\, \underline{\beta}\ . \tag{4.116}$$

We can use this Hotelling's T^2 to test the independence of x_p and $(x_1, x_2, \ldots, x_{p-1})$. As $\underline{x}$ has a p-variate normal distribution, x_p and $(x_1, \ldots, x_{p-1})$ will be independent if and only if $\underline{\sigma}$, the covariance vector of x_p with $(x_1, \ldots, x_{p-1})$, is null. When $\underline{\sigma} = \underline{0}$, $\underline{\beta}$ is also null and then $\lambda^2 = 0$ and so, by Theorem 2 of this chapter,

$$\frac{(n-2) - (p-1) + 1}{p-1}\, \frac{T^2}{n-2} = F_{p-1,\ n-p}\ , \tag{4.117}$$

and this F test can be used to test the hypothesis under consideration.

We shall now examine this T^2 more closely. By (1.4.16),

$$\begin{aligned} S^{*-1} &= (I - \frac{1}{s_{pp}} S_{p-1}^{-1}\, \underline{s}\underline{s}')^{-1}\, S_{p-1}^{-1} \\ &= S_{p-1}^{-1} + \frac{S_{p-1}^{-1}\, \underline{s}\underline{s}'\, S_{p-1}^{-1}}{s_{pp} - \underline{s}'\, S_{p-1}^{-1}\, \underline{s}}\ , \end{aligned} \tag{4.118}$$

and so, from (4.115), we obtain

$$\begin{aligned} \frac{T^2}{n-2} &= s_{pp}\, \underline{b}'\, S^{*-1}\, \underline{b} \\ &= \underline{s}'\, S^{*-1}\, \underline{s}/s_{pp} \\ &= \underline{s}'\left\{S_{p-1}^{-1} + \frac{S_{p-1}^{-1}\, \underline{s}\underline{s}'\, S_{p-1}^{-1}}{s_{pp} - \underline{s}'\, S_{p-1}^{-1}\, \underline{s}}\right\} \underline{s} \cdot (s_{pp})^{-1} \\ &= \underline{s}'\, S_{p-1}^{-1}\, \underline{s}/(s_{pp} - \underline{s}'\, S_{p-1}^{-1}\, \underline{s}) \\ &= R^2/(1 - R^2)\ , \end{aligned} \tag{4.119}$$

where R is the multiple correlation coefficient between x_p and $x_1, \ldots, x_p$ in a sample of size n. This follows from (4.2.1). Similarly,

$$\lambda^2 = s_{pp} \cdot \gamma^2/(1 - \gamma^2) , \tag{4.120}$$

where γ is the population multiple correlation coefficient between x_p and $x_1, \ldots, x_{p-1}$. Thus Hotelling's T^2 is related to R^2. The distribution of T^2 when s_{pp} is fixed is thus the distribution of $R^2/(1 - R^2)$ when s_{pp} is fixed and is, by Theorem 1,

$$H'_{p-1}\left(\frac{(n-2)R^2}{1 - R^2} \,\middle|\, (n-2), \lambda^2\right) \; d\left(\frac{(n-2)R^2}{1 - R^2}\right) . \tag{4.121}$$

Remember that in Chapter 4, Section 2, we derived the distribution of R^2 when b_{11}^2, a quantity defined in that section, is fixed. The distribution is $\phi(R^2|b_{11}^2)dR^2$ and is given by (4.2.21). Note that s_{pp} is fixed when x_p is fixed and b_{11}^2 is fixed when x_1, x_2, $\ldots$, x_{p-1} are fixed. Thus (4.121) gives the conditional distribution of R^2 when x_p is fixed and $x_1, \ldots, x_{p-1}$ are random, and (4.2.21) gives the conditional distribution of R^2 when $x_1, \ldots,$ x_{p-1} are fixed and x_p is random. It is a remarkable feature <u>that both these distributions are exactly the same</u>, the only difference being that the noncentrality parameter in one case is $\lambda^2 = s_{pp}\, \gamma^2/(1 - \gamma^2)$ and in the other case (see 4.2.18) is $\lambda^2 = b_{11}^2\, \gamma^2/(1 - \gamma^2)$, i.e., in one case it is a function of the observations on x_p (the fixed variable) and in the other case it is a function of the observations on $x_1, x_2, \ldots, x_{p-1}$, the fixed variables. In particular, when $\gamma = 0$, i.e., there is no true relation between x_p and $x_1, \ldots, x_{p-1}$, both the distributions are identical and, as observed in (4.117) or (4.2.30), $(n-p)R^2/(p-1)(1-R^2)$ has an $F_{p-1,\, n-p}$ d.f. This is known as the "duality" in the relationship between x_p and $x_1, \ldots, x_{p-1}$. In the absence of any true correlations between x_p and $(x_1, \ldots, x_{p-1})$ the conditional distribution of R^2 is the same, whether x_p is fixed and $x_1, \ldots, x_{p-1}$ are random or $x_1, \ldots, x_{p-1}$ are fixed and x_p is random.

5. APPLICATIONS OF HOTELLING'S T^2 IN PROBLEMS INVOLVING LINEAR HYPOTHESES

Consider a p-variate normal distribution

$$N_p(\underline{y} \mid A\underline{\theta} \mid \Sigma) \, , \tag{5.1}$$

where the matrix A is of order $p \times k$ and rank r, and $\underline{\theta}$ is a $k \times 1$ vector of unknown parameters. It is assumed that A is known and that a matrix S, having the distribution

$$W_p(S \mid f \mid \Sigma) \tag{5.2}$$

independent of $\underline{y}$, is also available. In such a case, the first problem is to test whether the model

$$H_o\colon \; E(\underline{y}) = A\underline{\theta} \tag{5.3}$$

is adequate, and, if so, the next problem is to test a linear hypothesis of the type

$$H_1\colon \; B\underline{\theta} = \underline{0} \, , \tag{5.4}$$

where B is some given matrix of order $m \times k$ and rank $m \leq r$.

When Σ is known, this problem is simple and has been solved (see C. R. Rao [22]) by transforming from $\underline{y}$ to $\underline{z}$ by $z = C\underline{y}$, where C is such that $C\Sigma C' = I$. Then the z_i's in $\underline{z}$ are all independent normal and both H_o and H_1 can be tested by the usual least squares theory (see Chapter 2, Section 5). The tests come out as follows: If H_o is true,

$$t_1 = \underset{\theta}{\text{Min}}(\underline{y} - A\,\underline{\theta})'\,\Sigma^{-1}(\underline{y} - A\,\underline{\theta}) \sim \chi^2_{p-r} \tag{5.5}$$

and provides a test of the adequacy of the model, while for testing H_1, we employ

$$t_2 = \underset{\theta}{\text{Min}}(\underline{y} - A\,\underline{\theta})'\,\Sigma^{-1}(\underline{y} - A\,\underline{\theta}), \text{ subject to } B\,\underline{\theta} = \underline{0} \tag{5.6}$$

and test H_1 by the χ^2 test

$$t_2 - t_1 \sim \chi^2_m \,. \tag{5.7}$$

This test is valid only if B satisfies the condition that there exists an $m \times p$ matrix L such that

$$B = L\,A\,, \tag{5.8}$$

and this is known as the condition of estimability of $B\,\underline{\theta}$. However, we are not assuming Σ to be known here, and we wish to test H_o and H_1. We would naturally expect our statistics to be obtained from t_1, t_2 by "Studentizing", i.e., by replacing Σ by its independent estimate $(1/f)S$. The tests, therefore, will not be χ^2 tests but will be based on Hotelling's T^2 [see Rao [21]].

Since the rank of A is r, we can find a matrix D of order $(p-r) \times p$ such that the rows of D are orthogonal to those of A' or such that

$$D\,A = 0\ . \tag{5.9}$$

Then, from (5.1), (5.2), Theorem 2 of Chapter 2, and Lemma 1 of Chapter 3,

$$\underline{z}(1) = D\,\underline{y} \text{ has the } N_{p-r}(\underline{z}(1)\,|\,DA\underline{\theta}\,|\,D\Sigma D') \text{ distribution} \tag{5.10}$$

and is independent of the

$$W_{p-r}(DSD'\,|\,f\,|\,D\Sigma D') \text{ distribution of } DSD'\ . \tag{5.11}$$

Hence,

$$T^2_{p-r} = f\ \underline{z}'(1)\ (DSD')^{-1}\ \underline{z}(1) \tag{5.12}$$

is a Hotelling's T^2 based on f d.f., and if our model H_o is true, $E(\underline{z}(1)) = DA\underline{\theta} = 0$ because of (5.9). Therefore,

$$\frac{f-(p-r)+1}{p-r}\,\frac{T^2_{p-r}}{f} = F_{p-r,\ f-(p-r)+1}\ . \tag{5.13}$$

The adequacy of the model (5.3) can thus be tested by this F test. In practice, however, instead of obtaining the matrix D such that $DA = 0$, it is sometimes easier to use

$$\frac{T^2_{p-r}}{f} = \underset{\theta}{\text{Min}}\ (\underline{y} - A\underline{\theta})'\ S^{-1}(\underline{y} - A\underline{\theta})\ . \tag{5.14}$$

This relation can be proved by considering the $p \times p$ nonsingular matrix P, partitioned as

$$P = \underset{p}{\begin{bmatrix} D \\ \hline E \end{bmatrix}}\begin{matrix} p-r \\ r \end{matrix}\ , \tag{5.15}$$

where E is so chosen that $DSE' = 0$. Then, because of (5.9) and (5.15),

$$
\begin{aligned}
(\underline{y} - A\underline{\theta})' S^{-1}(\underline{y} - A\underline{\theta}) &= (P\underline{y} - PA\underline{\theta})' (PSP')^{-1} (P\underline{y} - PA\underline{\theta}) \\
&= (D\underline{y}) (DSD')^{-1} (D\underline{y}) \\
&+ (E\underline{y} - EA\underline{\theta})'(ESE')^{-1}(E\underline{y} - EA\underline{\theta}),
\end{aligned} \tag{5.16}
$$

as $DSE' = 0$. The minimum value of (5.16) obviously occurs when $E\underline{y} = EA\underline{\theta}$, as the first term on the right side of the last equation in (5.16) does not involve θ. The required minimum value is thus

$$(D\underline{y})(DSD')^{-1}(D\underline{y}) = \underline{z}'(1) (DSD')^{-1} \underline{z}(1) ,$$

as $\underline{z}(1) = D\underline{y}$, and this proves the result (5.14).

We now assume (5.3) to be adequate and test H_1, assuming (5.8), the condition of estimability, to hold. If so, let

$$\underline{z} = \left[\frac{\underline{z}(1)}{\underline{z}(2)}\right] = \left[\frac{D}{L}\right] \underline{y} = G\underline{y} \text{ say .} \tag{5.17}$$

Then

$$\underline{z} \text{ has the } N_{p-r+m}(\underline{z} | GA\theta | G\Sigma G') \text{ distribution} \tag{5.18}$$

and is independent of the

$$W_{p-r+m}(GSG' | f | G\Sigma G') \text{ distribution.} \tag{5.19}$$

But $E(\underline{z}(1)) = DA\underline{\theta} = 0$ and

$$
\begin{aligned}
E(\underline{z}(2)) &= LA\underline{\theta} \\
&= B\underline{\theta}
\end{aligned} \tag{5.20}
$$

is equal to zero only if H_1 is true. Thus we wish to test $E(\underline{z}(2)) = 0$, given $E(\underline{z}(1)) = 0$, with $\underline{z}$ having the distribution (5.18) and GSG' having the distribution (5.19).

We now apply Theorem 4 of this chapter to these distributions and note that condition (c) of (3.24) is valid. The required test is, therefore, provided by Rao's U statistic

$$U = \frac{1 + T^2_{p-r}/f}{1 + T^2_{p-r+m}/f}, \tag{5.21}$$

where

$$\frac{T^2_{p-r+m}}{f} = \underline{z}'(GSG')^{-1}\underline{z} = (G\underline{y})'\,(GSG')^{-1}\,(G\underline{y}) . \tag{5.22}$$

T^2_{p-r} is defined by (5.12). The test for H_1 is

$$\frac{f - (p-r+m) + 1}{m}\left(\frac{1}{U} - 1\right) = F_{m,\, f - (p-r+m) + 1} . \tag{5.23}$$

In practice, however, instead of finding G to calculate (5.22), it is easier to use the relation,

$$\frac{T^2_{p-r+m}}{f} = \min_{\theta}\,(\underline{y} - A\underline{\theta})'\,S^{-1}(\underline{y} - A\underline{\theta}),\ \text{subject to the condition } B\underline{\theta} = \underline{0} . \tag{5.24}$$

This relation can be proved in exactly the same way as (5.14). Sometimes it is easier to substitute the condition $B\underline{\theta} = \underline{0}$ in the model (5.3) itself and have a revised model $E(\underline{y}) = A^*\underline{\theta}$. Then $T^2_{p-r+m/f}$ is $\min_{\theta}\,(y - A^*\underline{\theta})'\,S^{-1}(y - A^*\underline{\theta})$. In such cases, it is also useful to note that $m = \text{rank } A - \text{rank } A^*$.

If we differentiate

$$\phi = (\underline{y} - A\underline{\theta})'\,S^{-1}(y - A\underline{\theta}) , \tag{5.25}$$

with respect to $\underline{\theta}$ and, equate the result to $\underline{0}$, for minimization, we get the normal equations (replacing $\underline{\theta}$ by $\hat{\underline{\theta}}$)

$$A'S^{-1}\underline{y} = (A'S^{-1}A)\hat{\underline{\theta}} . \tag{5.26}$$

Let $\hat{\underline{\theta}}$ be any solution of these equations, then it is easy to see that [from (5.14)]

$$T^2_{p-r/f} = \text{Min } \phi = \underline{y}'S^{-1}\underline{y} - \hat{\underline{\theta}}' A' S^{-1}\underline{y} . \tag{5.27}$$

Similarly,

$$T^2_{p-r+m/f} = \underline{y}'S^{-1}\underline{y} - \hat{\underline{\theta}}^{*\prime} A^{*\prime} S^{-1}\underline{y} ,$$

where $\hat{\underline{\theta}}^*$ is any solution of $A^{*\prime}S^{-1}\underline{y} = (A^{*\prime}S^{-1}A^*)\hat{\underline{\theta}}^*$. It can also be shown, by a little algebra, that

$$\frac{T^2_{p-r+m}}{f} = (B\hat{\theta})'(HSH')^{-1}(B\hat{\theta}) + \frac{T^2_{p-r}}{f} , \tag{5.28}$$

where the matrix H must be obtained by expressing $B\hat{\underline{\theta}}$ in terms of $\underline{y}$ as

$$B\hat{\underline{\theta}} = H\underline{y} , \tag{5.29}$$

then, $V(B\hat{\theta}) = H\Sigma H'$, and hence HSH', occurring in (5.28), is f times the estimated variance of $B\hat{\underline{\theta}}$.

It should be noted that it is not essential that the rank of the matrix B and its number of rows be the same as we assumed. If they are not, m in (5.21), (5.22), (5.23), or (5.28) must be replaced by the rank of B. However, it is essential that the condition (5.8) of estimability be satisfied. As an illustration, consider the hypothesis

$$A\underline{\theta} = \underline{0} . \tag{5.30}$$

Here, A has p rows but its rank is r. $A\theta$ is estimable, as $E(\underline{y}) = A\theta$. Hence, for this hypothesis, the test will be (asm = r)

$$\frac{f-p+1}{r}\left(\frac{1}{U} - 1\right) = F_{r,\ f-p+1} , \tag{5.31}$$

where

$$U = \frac{1 + T^2_{p-r}/f}{1 + T^2_p/f}, \tag{5.32}$$

$$\frac{T^2_p}{f} = \underset{\theta}{\text{Min}}\ (\underline{y} - A\underline{\theta})'\ S^{-1}(\underline{y} - A\underline{\theta}) \text{ subject to } A\underline{\theta} = 0$$

$$= \underline{y}'S^{-1}\underline{y}\ , \tag{5.33}$$

and T^2_{p-r} is the same as before.

If the hypothesis $B\underline{\theta} = \underline{0}$ of (5.4) is modified as

$$H:\quad B\underline{\theta} = \underline{\xi}\ , \tag{5.34}$$

where $\underline{\xi}$ is a specified vector, it is easy to see that (5.23) will still be able to test this hypothesis if we modify T^2_{p-r+m}/f as the conditional minimum of ϕ when $B\underline{\theta} = \underline{\xi}$. Equation (5.28) will then be changed to

$$\frac{T^2_{p-r+m}}{f} = (B\hat{\underline{\theta}} - \underline{\xi})'\ (HSH')^{-1}\ (B\hat{\underline{\theta}} - \underline{\xi}) + \frac{T^2_{p-r}}{f}, \tag{5.35}$$

and hence

$$\frac{T^2_{p-r+m}}{f} - \frac{T^2_{p-r}}{f} = (B\hat{\underline{\theta}} - \underline{\xi})'\ (HSH')^{-1}\ (B\hat{\underline{\theta}} - \underline{\xi})$$

$$= \underset{\underline{g}}{\text{Max}}\ \frac{(\underline{g}'B\hat{\underline{\theta}} - \underline{g}'\underline{\xi})^2}{\underline{g}'(HSH')\underline{g}}\ . \tag{5.36}$$

The last step can be verified by actually maximizing with respect to the arbitrary vector $\underline{g}$. Hence, from (5.23),

$$\text{Prob}\left\{\frac{f-(p-r+m)+1}{m}\ \frac{1}{1+T^2_{p-r}/f}\ \underset{g}{\text{Max}}\ \frac{(\underline{g}'B\hat{\underline{\theta}} - \underline{g}'\underline{\xi})^2}{\underline{g}'\ (HSH')\ \underline{g}}\right.$$

$$\left. \leq F(m,\ f-p+r-m+1,\ \alpha)\right\} = 1 - \alpha\ . \tag{5.37}$$

($F(a, b, \alpha)$ is defined in Section 1). After a little manipulation of (5.37), following the same line of argument as we employed in deriving (2.14), we obtain

$$\underline{g}'B\hat{\underline{\theta}} \pm \left\{\left(1 + \frac{T^2_{p-r}}{f}\right) \frac{vm}{f - (p-r+m) + 1} F(m,\ f-p+r-m+1,\ \alpha)\right\}^{1/2} \tag{5.38}$$

as the simultaneous confidence bounds on all parametric functions of the type $\underline{g}'B\hat{\underline{\theta}}$. Here v is given by,

$$\frac{1}{f} v = \frac{1}{f} \underline{g}'(HSH')\underline{g}$$

$$= \text{estimate of } \underline{g}'\ H\Sigma H'\ \underline{g}$$

$$= \text{estimate of } V(\underline{g}'B\hat{\underline{\theta}}). \tag{5.39}$$

In particular, if one needs confidence bounds for linear parametric functions of the type $\underline{g}'A\underline{\theta}$, we use (5.30), (5.31), and (5.32) and obtain the confidence bounds as (putting $m = r$)

$$\underline{g}'A\hat{\underline{\theta}} \pm \left\{\left(1 + \frac{T^2_{p-r}}{f}\right) \frac{vr}{f-p+1} F(r,\ f-p+1,\ \alpha)\right\}^{1/2}. \tag{5.40}$$

In particular, when B consists of a single row vector, say $\underline{b}'$, we can obtain a confidence interval for the parametric function $\underline{b}'\underline{\theta}$, by putting $m = 1$ in (5.38), as

$$\underline{b}'\hat{\underline{\theta}} \pm \left\{\left(1 + \frac{T^2_{p-r}}{f}\right) \frac{v}{f - (p-r)} F(1,\ f-p+r,\ \alpha)\right\}^{1/2}. \tag{5.41}$$

Here again,

$$\frac{1}{f} v = \text{the estimate of the variance of } \underline{b}'\hat{\underline{\theta}}, \text{ obtained by replacing } \Sigma \text{ by } \frac{1}{f} S. \tag{5.42}$$

It is essential that $\underline{b}'\underline{\theta}$ be estimable.

We shall now illustrate the tests derived in this section by two problems that arise in practice.

(I_1) Growth Curves

If y_t denotes the observation on a character at time t for an individual, $y_{t_1}, y_{t_2}, \ldots, y_{t_p}$ will be correlated and we assume a p-variate normal distribution, with Σ as the variance-covariance matrix for them. If we wish to test whether a model of the type

$$E(y_t) = \alpha + \beta_1 t + \beta_2 t^2 + \ldots + \beta_{r-1} t^{r-1} \qquad (5.43)$$

will be adequate to represent the behavior of E(y) over time, we can use the tests developed in this section. Equation (5.43) is known as a growth curve, and such a problem arises in experiments on animals or plants, while investigating the effect of diets or fertilizers. Let us assume that we have these observations y_t for n individuals, at times $t_1, t_2, \ldots, t_p$. Usually, the observations are taken at regular intervals of times and, in that case, by a change of scale and original, we can take $t_1, \ldots, t_p$ to be 0, 1, ..., p-1; but this is not essential. Let y_{tr} be the observation on the rth individual at time t ($t = t_1, \ldots, t_p$) and let Y denote the p x n matrix of these observations. Let

$$\underline{\bar{y}} = \frac{1}{n} Y E_{n1} \; ; \quad S = Y(I - \frac{1}{n} E_{nn})Y' . \qquad (5.44)$$

Then, on account of (5.43),

$$\sqrt{n}\, \underline{\bar{y}} \text{ has the } N_p(\sqrt{n}\, \underline{\bar{y}} | \sqrt{n}\, A\, \underline{\theta} | \Sigma) \text{ distribution,} \qquad (5.45)$$

where

$$A = \begin{bmatrix} 1 & t_1 & t_1^2 \cdots t_1^{r-1} \\ 1 & t_2 & t_2^2 \cdots t_2^{r-1} \\ \vdots & & \\ 1 & t_p & t_p^2 \cdots t_p^{r-1} \end{bmatrix} \qquad (5.46)$$

and

$$\underline{\theta} = \begin{bmatrix} \alpha \\ \beta_1 \\ \beta_2 \\ \vdots \\ \beta_{r-1} \end{bmatrix} \qquad (5.47)$$

Also,

$$S \text{ has the } W_p(S|n-1|\Sigma) \text{ distribution, independent of } \bar{\underline{y}} . \qquad (5.48)$$

We can, therefore, use the test (5.13) to test the adequacy of the specification (5.43), with $f = n - 1$, $p = p$, $r = \text{rank } A = r$, and

$$\frac{T^2_{p-r}}{n-1} = n \underset{\theta}{\text{Min}} \; (\bar{\underline{y}} - A\underline{\theta})' \; S^{-1}(\bar{\underline{y}} - A\underline{\theta}) \; . \qquad (5.49)$$

Let $\hat{\alpha}, \hat{\beta}_1, \ldots, \hat{\beta}_{r-1}$ denote the values of $\alpha, \beta_1, \ldots, \beta_{r-1}$ that minimize the expression in (5.49). These will be the point estimates of $\alpha, \beta_1, \ldots, \beta_{r-1}$ (subject to estimability). Confidence intervals for any of them can be obtained from (5.41). However, if one needs confidence bounds to the polynomial

$$\psi(t) = \alpha + \beta_1 t + \ldots + \beta_{r-1} \, t^{r-1}$$

for all values of t, one has to use (5.40). Although vectors of the form $[1, t, t^2, \ldots, t^{r-1}]$ do not cover all the points of an r-dimensional space, a probability statement similar to (5.37) applicable to all combinations of α, β_1, ..., β_{r-1} is still true as an inequality, and so the confidence bounds for $\psi(t)$ will be given by

$$\hat{\psi}(t) \pm \left\{\left(1 + \frac{T^2_{p-r}}{n-1}\right) \frac{vr}{n-p} F(r, n-p, \alpha)\right\}^{1/2} , \qquad (5.50)$$

where

$$\hat{\psi}(t) = \hat{\alpha} + \hat{\beta}_1 t + \dots + \hat{\beta}_{r-1} t^{r-1} \qquad (5.51)$$

and

$$\frac{1}{f} v = \text{estimate of the variance of } \hat{\psi}(t) , \qquad (5.52)$$

which can be obtained by expressing $\hat{\psi}(t)$ as a linear function of $\bar{\underline{y}}$ first and then using the fact that $V(\bar{\underline{y}}) = (1/n)\Sigma$ and that $[1/(n-1)]S$ estimates Σ.

One can thus obtain confidence bounds for the $\psi(t_i)$ $(i = 1, 2, \dots, p)$, and from these, the limits to other values of t can be obtained by graduation, if necessary. The confidence bounds are valid only when the model (5.43) is supported by the data, this model, therefore, must first be tested by (5.13), using (5.49).

(I_2) Coplanarity of the Mean Vectors of Several Normal Populations

Let $\underline{\mu}(1)$, $\underline{\mu}(2)$, ..., $\underline{\mu}(k)$ be the mean vectors of k independent p-variate normal populations, with the same variance-covariance matrix Σ. We assume the $\underline{\mu}$'s to be known but Σ to be unknown. But an estimate of Σ based on a Wishart matrix S of f d.f. is assumed to be available. Let $\underline{x}$ be the vector of observations from a new p-variate normal population with the same variance-covariance matrix. The mean vectors $\underline{\mu}(1), \ldots, \underline{\mu}(k)$ determine a hyperplane in p-dimensional geometry, and we wish to test whether E(x), the new mean vector, also lies on the same hyperplane. In other words, we wish to know whether there exist arbitrary constants $\theta_1, \theta_2, \ldots, \theta_k$ such that

$$E(\underline{x}) = \theta_1 \underline{\mu}(1) + \theta_2 \underline{\mu}(2) + \ldots + \theta_k \underline{\mu}(k) , \qquad (5.53)$$

where $\theta_1 + \theta_2 + \ldots + \theta_k = 1$. If (5.53) is valid, we can rewrite it as (using $\theta_k = 1 - \theta_1 - \ldots - \theta_{k-1}$)

$$\begin{aligned} E(\underline{x} - \underline{\mu}(k)) &= \theta_1(\underline{\mu}(1) - \underline{\mu}(k)) + \ldots + \theta_{k-1} (\underline{\mu}(k-1) - \underline{\mu}(k)) \\ &= [\underline{\mu}(1) - \underline{\mu}(k), \ldots, \underline{\mu}(k-1) - \underline{\mu}(k)] \begin{bmatrix} \theta_1 \\ \vdots \\ \theta_{k-1} \end{bmatrix} \\ &= A\underline{\theta}, \text{ say.} \end{aligned} \qquad (5.54)$$

We therefore determine

$$T^2_{p-r}/f = \min_{\theta} (\underline{x} - \underline{\mu}(k) - A\underline{\theta})' S^{-1}(\underline{x} - \underline{\mu}(k) - A\underline{\theta}) , \qquad (5.55)$$

as required to apply (5.13). In this case,

$$r = \text{rank of } [\underline{\mu}(1) - \underline{\mu}(k), \ldots, \underline{\mu}(k-1) - \underline{\mu}(k)].$$

If the F test of (5.13) turns out to be insignificant, the specification (5.53) and hence the hypothesis of coplanarity will be tenable.

Having ascertained the validity of (5.53), we now wish to test whether the new observations $\underline{x}$ comes not from any new population but from one of the k given populations, say the kth population only. In other words, we wish to test

$$H_1: \theta_1 = 0, \theta_2 = 0, \ldots, \theta_{k-1} = 0 \tag{5.56}$$

(obviously $\theta_k = 1$, as $\Sigma\theta_i = 1$). If we substitute these values of H_1 in (5.54), the revised model is

$$E(\underline{x} - \underline{\mu}(k)) = \underline{0}$$
$$= A^*\underline{\theta}, \text{ say},$$

so that rank $A^* = 0$, and hence m = d.f. associated with H_1 are rank A - rank $A^* = r - 0 = r$, and obviously

$$\frac{T^2_{p-r+m}}{f} = \frac{T^2_p}{f} = \text{Min } (\underline{x} - A\underline{\theta})' \, S^{-1}(\underline{x} - A\underline{\theta}) \,, \quad \text{subject to } H_1$$
$$= (\underline{x} - \underline{\mu}(k))' \, S^{-1}(\underline{x} - \underline{\mu}(k)) \,. \tag{5.57}$$

The test for H_1 is thus

$$\frac{f-p+1}{r} \frac{T^2_p - T^2_{p-r}}{f + T^2_{p-r}} = F_{r,\, f-p+1} \,. \tag{5.58}$$

6. OPTIMUM PROPERTIES OF HOTELLING'S T^2 TESTS

In Section 4 (A_1), we considered application of Hotelling's T^2 to test the null hypothesis $H_o: \underline{\mu} = \underline{0}$, against $H_1: \underline{\mu}'\Sigma^{-1}\underline{\mu} =$ say δ, at a level of significance α. $\underline{\mu}$ and Σ were the mean and variance-covariance matrix of a p-variate normal population and $\bar{\underline{x}}$, S are, respectively, the mean and corrected s.s. and s.p. matrix from a sample of size n from the population. The test statistic is $T^2 = n(n-1)\, \bar{\underline{x}}' \, S^{-1} \, \bar{\underline{x}}$.

James and Stein [12] have shown that, among procedures based on the sufficient statistics ($\bar{\underline{x}}$, S), the T^2 test is the best invariant procedure under the real linear group. (The reader who is unfamiliar with this and other nomenclature in this section should refer to Lehman [15] or Ferguson [5]. Simaika [26] proved the T^2 test to be uniformly most powerful of level α, among tests whose power function depends only on $\underline{\mu}'\Sigma^{-1}\underline{\mu}$, for testing H_o against H_1': $\underline{\mu}'\Sigma^{-1}\underline{\mu} > 0$. Stein [27] showed that the T^2 test is admissible for testing H_o against H_1'. Hsu [11] showed that the T^2 test maximizes a certain integral over H_1 of the power. Giri, et.al. [7] have proved that the T^2 test maximizes, among all α-level tests, the minimum power under H_1, when $p = 2$, $n = 3$ and for each possible choice of δ and α. Kiefer and Schwartz [13] have considered the admissible Bayes character of T^2 and other fully invariant tests for classical multivariate normal problems. They also investigated the behavior of T^2 closer to H_o, unlike Stein [27] whose method insures that no other test of the same size is superior to T^2, "far" from H_o. Giri and Kiefer [8] have proved certain local and asymptotic minimax properties of the T^2 test. A simple proof of the monotonocity of the probability ratio for T^2 is available in Wijsman [31], if one notes the relation between T^2 and R^2, established in (A_7) of Section 4.

REFERENCES

[1] Bargmann, R. E., and Ghosh, S. P. (1964). "Noncentral statistical distribution programs for a computer language", Report No. RC-1231, IBM Watson Research Center, Yorktown Heights, New York.

[2] Bennett, B. M. (1951). "Note on a solution of the generalized Behrens-Fisher problem", Ann. Inst. Stat. Math., 2, p. 87.

[3] Bose, R. C., and Roy, S. N. (1938). "The distribution of studentized D^2-statistic", Sankhyā, 4, p. 19.

[4] Deemer, Walter L., and Olkin, I. (1951). "The jacobians of certain matrix transformations useful in multivariate analysis", Biometrika, 38, p. 345.

[5] Ferguson, Thomas S. (1967). Mathematical Statistics. Academic Press, New York.

[6] Fox, M. (1956). "Charts of the power of the F-test", Ann. Math. Statist., 27, p. 484.

[7] Giri, N., Kiefer, J., and Stein, C. (1963). "Minimax character of Hotelling's T^2 test in the simplest case", Ann. Math. Statist., 34, p. 1524.

[8] Giri, N., and Kiefer, J. (1964). "Local and asymptotic minimax properties of multivariate tests", Ann. Math. Statist., 35, p. 21.

[9] Graybill, F. (1954). "Variance heterogeneity in a randomized block design", Biometrics, 10, p. 516.

[10] Hotelling, H. (1931). "The generalization of Student's ratio", Ann. Math. Statist., 2, p. 360.

[11] Hsu, P. L. (1945). "On the power function of the E^2-test and the T^2-test", Ann. Math. Statist., 16, p. 278.

[12] James, W., and Stein, C. (1960). "Estimation with quadratic loss", Proc. Fourth Berkeley Symp. Math. Stat. Prob., 1, p. 361.

[13] Kiefer, J., and Schwartz, R. (1965). "Admissible Bayes character of T^2, $-R^2$, and other fully invariant tests for classical multivariate normal problems", Ann. Math. Statist., 36, p. 747.

[14] Lachenbruch, P. A. (1967). "The noncentral F distribution-extension of Tang's tables", University of North Carolina Mimeo Series, No. 531.

[15] Lehman, E. L. (1959). Testing Statistical Hypotheses. John Wiley and Sons, New York.

[16] Lehmer, E. (1944). "Inverse tables of probabilities of errors of the second kind", Ann. Math. Statist., 15, p. 388.

[17] Mahalanobis, P. C. (1936). "On the generalized distance in statistics", Proc. Nat. Inst. Soc. India, 12, p. 49.

[18] Nandi, H. K. (1965). "On some properties of Roy's union-intersection tests", Cal. Stat. Ass. Bull., 14, p. 9.

[19] Patnaik, P. B. (1949). "The noncentral χ^2 and F-distributions and their applications", Biometrika, 36, p. 202.

[20] Pearson, E. S., and Hartley, H. O. (1951). "Charts of the power function for analysis of variance tests, derived from the noncentral F-distribution", Biometrika, 38, p. 112.

[21] Rao, C. Radhakrishna (1959). "Some problems involving linear hypotheses in multivariate analysis", Biometrika, 46, p. 49.

[22] Rao, C. Radhakrishna (1965). Linear Statistical Inference and its Applications. John Wiley and Sons, New York.

[23] Roy, S. N. (1953). "On a heuristic method of test construction and its use in multivariate analysis", Ann. Math. Statist., 24, p. 220.

[24] Roy, S. N. (1957). Some Aspects of Multivariate Analysis. John Wiley and Sons, New York.

[25] Scheffe', H. (1943). "On solutions of the Behrens-Fisher problem based on the t-distribution", Ann. Math. Statist., 14, p. 35.

[26] Simaika, J. B. (1941). "An optimum property of two statistical tests", Biometrika, 32, p. 70.

[27] Stein, C. (1956). "The admissibility of Hotelling's T^2-test", Ann. Math. Statist., 27, p. 616.

[28] Tang, P. C. (1938). "The power function of the analysis of variance tests with tables and illustrations of their use", Stat. Research Memoirs, 2, p. 126.

[29] Tiku, M. L. (1967). "Tables of the power of the F-test", J. Am. Stat. Assoc., 62, p. 525.

[30] Ura, S. (1954). "A table of the power function of the analysis of variance test", Reports of Statistical Application Research, JUSE 3, p. 23.

[31] Wijsman, R. (1959). "Applications of a certain representation of the Wishart matrix", Ann. Math. Statist., 30, p. 597.

CHAPTER 6

DISCRIMINANT ANALYSIS

1. PROCEDURE FOR DISCRIMINATION

Given two populations π_1 and π_2, how do we decide whether an individual belongs to π_1 or π_2 on the basis of measurements on a p-component vector of variables $\underline{x}$? Table 1 gives a list of several examples where such a problem arises in practice. The table is prepared from an exhaustive list of case studies of discriminant analysis in the literature compiled by Hodges [14].

TABLE 1

Variables measured $\underline{x}$	Populations π_1 and π_2
(1) Scores in different psychological tests	Successful and unsuccessful salesmen
(2) Price of car, down payment, income, length of loan	Good and bad auto loan risks
(3) Measures of anxiety, dependence, guilt, perfectionism	Nonulcer dyspeptics and controls
(4) Anthropological measurements on skulls	Male and female
(5) Stature, sitting height, head length, breadth, nasal length	Two ethnic groups, e.g., Aryans and Dravidians

Cont.

TABLE 1 (continued)

	Variables measured $\underset{-}{x}$	Populations π_1 and π_2
(6)	Length, amplitude, rate of change in price cycle	Consumer's goods and producer's goods
(7)	B.T.U., volatile matter, fixed carbon, percentage of ash of coal	Coal from two mines
(8)	Ph, N, P contents of soil	Soils with and without Azotobacter
(9)	Sepal and petal length and width	Two species of flowers, e.g., Iris Setosa and Iris Versicolor
(10)	Scores in tests of intelligence and coordination	Engineers and air pilots
(11)	Frequencies of different words	Authorship of a disputed article, e.g., whether written by Shakespere or not
(12)	Different medical tests, such as blood, urine, blood pressure	Persons with and without a particular disease, e.g., diabetic and nondiabetic
(13)	Width of shoulder, thickness of flank, width of flank, length of leg	Two different grades of lamb carcasses, e.g., Royal and Tallarook

Let R denote the entire p-dimensional space in which the point of observation $\underset{-}{x}$ falls. We then have to divide this region R into two, say R_1 and R_2, by some suitable optimum method and lay down a rule of procedure, such as

$$\text{If } \underset{-}{x} \text{ falls in } R_1, \text{ assign the individual to } \pi_1;$$
$$\text{If } \underset{-}{x} \text{ falls in } R_2, \text{ assign to } \pi_2. \tag{1.1}$$

Obviously, with any such procedure, an error of misclassification is inevitable, i.e., the rule may assign an individual to π_2, when he really belongs to π_1, and vice versa. A good rule should control this error of discrimination. Let us assume that $f_1(\underline{x})$ and $f_2(\underline{x})$ are the p.d.f.'s of $\underline{x}$ in the two populations π_1 and π_2. Then the total chance of misclassification, say α, consists of two components,

$$\alpha_1 = \text{Prob (an individual belonging to } \pi_1 \text{ is misclassified as belonging to } \pi_2) \quad (1.2)$$

and

$$\alpha_2 = \text{Prob (an individual belonging to } \pi_2 \text{ is misclassified).} \quad (1.3)$$

Obviously,

$$\alpha_1 = \int_{R_2} f_1(\underline{x})d\underline{x} \ , \quad \alpha_2 = \int_{R_1} f_2(\underline{x})d\underline{x} \ , \quad (1.4)$$

as the individual with measurements $\underline{x}$ is assigned to π_2 when $\underline{x}$ falls in R_2 and to π_1 when $\underline{x}$ falls in R_1, and the p.d.f. of $\underline{x}$ is f_1 in π_1 and f_2 in π_2. The total chance of misclassification is $\alpha = \alpha_1 + \alpha_2$. If there is no reason why we should prefer α_1 to be smaller or bigger than α_2, we shall naturally give equal importance to avoid the two errors, viz., misclassification of an individual in π_1 and in π_2, and shall therefore set $\alpha_1 = \alpha_2$. But there may be situations in which the two are not equally important, as, for example, it is more important to avoid hanging an innocent person as compared to letting a murderer free. In such situations, it is the responsibility of the user of discriminant analysis to assess precisely the relative importance of α_1 and α_2. It will naturally depend on the penalties of misclassifications of the two kinds. For example, if a potentially good candidate for admission to a medical school is rejected, the nation will suffer a shortage in medical persons, but, on the contrary, if a bad candidate is admitted, he may not be able to complete the course successfully and money, resources, equipment used by him will be a waste. When the penalties associated with α_1, α_2 are unequal, the relation $\alpha_1 = \alpha_2$ will have to be replaced by some other relation, such as $\alpha_1 = k\,\alpha_2$, where k is to be assessed by the user and supplied to the statistician. For the sake of simplicity, we assume $k = 1$ and the change in what follows, when $k \neq 1$, is obvious and the rule can be suitably modified.

The problem, therefore, reduces to dividing R into R_1 and R_2 such that

$$\alpha = \int_{R_2} f_1(\underline{x})d\underline{x} + \int_{R_1} f_2(\underline{x})d\underline{x}$$

$$= \int_{R_2} (f_1(\underline{x}) - f_2(\underline{x}))d\underline{x} + 1 \tag{1.5}$$

[$1 = \int_R f_2(x)dx = \int_{R_1} f_2(\underline{x})dx + \int_{R_2} f_2(\underline{x})d\underline{x}$ is utilized] is minimized, subject to the condition,

$$\alpha_1 = \alpha_2 . \tag{1.6}$$

A solution of this "minimization" problem is provided by the famous Neyman-Pearson lemma, a proof which is easily available in Rao [21]. The solution is as follows:

On the boundary separating R_1 from R_2,

$$\frac{f_1(\underline{x})}{f_2(\underline{x})} = \text{a constant ,} \tag{1.7}$$

the constant being so chosen that the condition (1.6) is satisfied. We assumed in this derivation that $\underline{x}$ has a continuous distribution and admits p.d.f.'s f_1 and f_2 in π_1 and π_2, but this is not essential and the final solution (1.7) is the same even for discrete distributions. $f_1(\underline{x})/f_2(\underline{x})$ is the likelihood ratio for the observation $\underline{x}$, and the solution (1.7) is, therefore, the likelihood ratio solution.

If we assume that π_1 and π_2 are, respectively, the $N_p(\underline{x}|\underline{\mu}(1)|\Sigma)$ and $N_p(\underline{x}|\underline{\mu}(2)|\Sigma)$ populations, we obtain

$$2\log_e \frac{f_1(\underline{x})}{f_2(\underline{x})} = (\underline{x}-\underline{\mu}(2))'\Sigma^{-1}(\underline{x}-\underline{\mu}(2))-(\underline{x}-\underline{\mu}(1))'\Sigma^{-1}(\underline{x}-\underline{\mu}(1))$$

$$= \underline{\delta}'\,\Sigma^{-1}\underline{x} - (1/2)(\underline{\mu}(1)+\underline{\mu}(2))'\Sigma^{-1}\,\underline{\delta}$$

$$= \underline{a}'\underline{x} - (1/2)(\underline{\mu}(1)+\underline{\mu}(2))'\underline{a}\ , \tag{1.8}$$

where

$$\underline{\delta} = \underline{\mu}(1) - \underline{\mu}(2) = E(\underline{x}|\pi_1) - E(\underline{x}|\pi_2) \tag{1.9}$$

and

$$\underline{a} = \Sigma^{-1}\,\underline{\delta}\ . \tag{1.10}$$

From (1.7), therefore, it follows that, on the boundary separating R_1 and R_2,

$$\underline{a}'\underline{x} = \text{a suitable constant, say } h. \tag{1.11}$$

Let us, therefore, define

$$R_1 \text{ as the region, where } \underline{a}'\underline{x} \geq h,$$

$$R_2 \text{ as the region, where } a'x < h. \tag{1.12}$$

We have arbitrarily decided to include the boundary $\underline{a}'\underline{x} = h$ in R_1, but we can make a "randomized" decision instead, by tossing a coin when $\underline{x}$ falls on the boundary $\underline{a}'\underline{x} = h$ and deciding whether the individual should be assigned to π_1 or π_2 according as the coin shows head or tail. From (1.4), we have

$$\alpha_1 = \int_{\underline{a}'\underline{x}<h} N_p(\underline{x}|\underline{\mu}(1)|\Sigma)d\underline{x}, \quad \alpha_2 = \int_{a'x\geq h} N_p(\underline{x}|\underline{\mu}(2)|\Sigma)d\underline{x}. \tag{1.13}$$

Make a transformation from $\underline{x}$ to $u = \underline{a}'\underline{x}$ and p-1 other suitable variables in (1.13). Note that the range of integration depends only on u, so that these other variables are integrated out, reducing α_1 and α_2 to

$$\alpha_1 = \int_{-\infty}^{h} \phi_1(u)du, \quad \alpha_2 = \int_{h}^{\infty} \phi_2(u)du, \tag{1.14}$$

where ϕ_1, ϕ_2 are the p.d.f.'s of u in π_1 and π_2 respectively. Since u is a linear function of the normal variables $\underline{x}$, it has a normal distribution, with

$$\begin{aligned} E(u) &= \underline{a}'\underline{\mu}(1) \text{ in } \pi_1, \\ &= \underline{a}'\underline{\mu}(2) \text{ in } \pi_2, \end{aligned} \tag{1.15}$$

$$\begin{aligned} V(u) &= \underline{a}'\Sigma\underline{a} \\ &= \underline{\delta}'\Sigma^{-1}\underline{\delta} \\ &= \Delta_p^2, \text{ in both } \pi_1 \text{ and } \pi_2. \end{aligned} \tag{1.16}$$

Hence

$$\phi_i(u) = \frac{1}{\Delta_p\sqrt{2\pi}} e^{-\frac{1}{2\Delta_p^2}(u - \underline{a}'\underline{\mu}(i))^2}, \quad (i = 1, 2)$$

and so, by making a transformation from u to $\{u - E(u)\}/\Delta_p$, we obtain

$$\alpha_1 = \Phi\left(\frac{h - \underline{a}'\underline{\mu}(1)}{\Delta_p}\right), \quad \alpha_2 = \Phi\left(\frac{\underline{a}'\underline{\mu}(2) - h}{\Delta_p}\right), \tag{1.17}$$

where Φ is the c.d.f. of a standard normal variable, i.e.,

$$\Phi(z_o) = \int_{-\infty}^{z_o} \frac{1}{\sqrt{2\pi}} e^{-z^2/2} dz . \tag{1.18}$$

But $\alpha_1 = \alpha_2$, and so, from (1.17) and the symmetry of a normal curve,

$$\frac{\underline{a}'\underline{\mu}(2) - h}{\Delta_p} = \frac{h - \underline{a}'\underline{\mu}(1)}{\Delta_p} \tag{1.19}$$

or

$$\underline{h} = \frac{1}{2} a'\{\underline{\mu}(1) + \underline{\mu}(2)\} . \tag{1.20}$$

Substituting this back in (1.17), we get (on using $a = \Sigma^{-1} \underline{\delta}$)

$$\alpha_1 = \alpha_2 = \Phi(-\frac{1}{2} \Delta_p) . \tag{1.21}$$

The discriminating procedure in the case of two p-variate normal populations with different means $\underline{\mu}(1)$, $\underline{\mu}(2)$ but the same variance-covariance matrix Σ is as follows:

(i) Compute $u = \underline{a}'\underline{x} = \underline{\delta}' \Sigma^{-1} \underline{x}$.

(ii) Compute $h = \frac{1}{2} \underline{a}'\{\underline{\mu}(1) + \underline{\mu}(2)\}$.

(iii) Assign the individual with measurements $\underline{x}$ to π_1 if $u \geq h$ and to π_2 if $u < h$.

(iv) The chance of misclassification is $\alpha = \alpha_1 + \alpha_2$, where $\alpha_1 = \alpha_2 = \Phi(-\frac{1}{2} \Delta_p)$, $\Delta_p^2 = \underline{\delta}' \Sigma^{-1} \underline{\delta} = \underline{a}'\underline{\delta}$. (1.22)

The chance of misclassification is a monotonic decreasing function of Δ_p. In Chapter 5, Section 4, we defined this Δ_p^2 as the square of the "distance" between the two populations π_1 and π_2, based on the p characters $\underline{x}$. The function $\underline{a}'\underline{x} = u$ is called Fisher's "discriminant function". Fisher [13] however obtained this discriminant function from the following considerations.

When we have two univariate populations (normal) with means μ_1, μ_2 and a common variance σ^2, the distance between the means, in units of the standard deviation σ, is $(\mu_1 - \mu_2)/\sigma$, if $\mu_1 > \mu_2$. It is intuitively clear that an observation x will be classified to belong to π_1 if it is nearer to μ_1 and to π_2 if it is nearer to μ_2. In this, there is a risk of misclassification and that risk is smaller if $(\mu_1 - \mu_2)/\sigma$ is larger, because, in that case the two normal curves are farther apart, the overlap is smaller. On the contrary, if $(\mu_1 - \mu_2)/\sigma$ is smaller, there is considerable overlapping of the areas under the two normal curves, and the chance of misclassification will be increased. So, when dealing with p-variate normal populations, Fisher suggested that the p characters be combined linearly, in such a way that the distance $(\mu_1 - \mu_2)/\sigma$ for this linear combination is maximum and the classification rule be based on this "optimum" linear combination. In other words, the p-dimensional classification procedure is reduced to a one-dimensional one, by choosing $\underline{a}'\underline{x}$ such that

$$\frac{|\text{Mean of } \underline{a}'\underline{x} \text{ in } \pi_1 - \text{Mean of } \underline{a}'\underline{x} \text{ in } \pi_2|}{\text{Standard deviation of } \underline{a}'\underline{x}} = \frac{|\underline{a}'\underline{\delta}|}{(\underline{a}'\Sigma\underline{a})^{1/2}} \tag{1.23}$$

is maximized with respect to $\underline{a}$. Differentiating (1.23) with respect to $\underline{a}$ and equating the result to zero, we obtain

$$\underline{\delta} = \left(\frac{\underline{a}'\underline{\delta}}{\underline{a}'\Sigma\underline{a}}\right)\Sigma\underline{a}. \tag{1.24}$$

This equation in $\underline{a}$ has no unique solution. A solution is

$$\underline{a} = \Sigma^{-1} \underline{\delta} , \tag{1.25}$$

but any vector proportional to $\underline{a}$ will also satisfy (1.24). So the optimum linear combination is $\underline{a}'\underline{x}$ or any linear function of $\underline{x}$, proportional to $\underline{a}'\underline{x}$. Since classification of an individual with p measurements $\underline{x}$ is thus based on only one single linear function $\underline{a}'\underline{x}$, it is called a discriminant function. As we have already seen, the coefficients of a discriminant function, viz., $a_1, \ldots, a_p$, are not unique but their ratios are, as not only $\underline{a}'\underline{x}$ but any function of the form $c\,\underline{a}'\underline{x}$, where c is a constant, can also be a discriminant function. The classification procedure based on $\underline{a}'\underline{x} = \underline{\delta}'\,\Sigma^{-1}\,\underline{x}$ is (1.22) as before. The value of (1.23), when $\underline{a} = \Sigma^{-1}\underline{\delta}$ is substituted, i.e., the "distance" between π_1 and π_2 based on $\underline{a}'\underline{x}$, reduces to Δ_p, the Mahalanobis distance defined earlier.

2. THE SAMPLE DISCRIMINANT FUNCTION

In practice, we seldom know $\underline{\mu}(1)$, $\underline{\mu}(2)$, and Σ. What are usually available are two samples of sizes n_1 and n_2 from π_1 and π_2, respectively. In this case, an obvious method will be to replace $\underline{\delta}$ and Σ by their estimates in the classification procedure (1.22). Let X and Y be the $p \times n_1$ and $p \times n_2$ matrices of the sample observations from π_1, which is a $N_p(\underline{x}|\underline{\mu}(1)|\Sigma)$ population, and π_2, which is a $N_p(\underline{x}|\underline{\mu}(2)|\Sigma)$ population. As in (A_2) of Section 4, Chapter 5, we define

$$\underline{\bar{x}} = \frac{1}{n_1} X E_{n_1 1}, \quad \underline{\bar{y}} = \frac{1}{n_2} Y E_{n_2 1},$$

$$S_x = X(I - \frac{1}{n_1} E_{n_1 n_1})X', \quad S_y = Y(I - \frac{1}{n_2} E_{n_2 n_2})Y',$$

$$S = S_x + S_y, \quad f = n_1 + n_2 - 2,$$

$$\underline{d} = \underline{\bar{x}} - \underline{\bar{y}}, \quad c^2 = \frac{n_1 n_2}{n_1 + n_2}. \tag{2.1}$$

Then

$$c\, \underline{d} \text{ has the } N_p(c\, \underline{d} | c\, \underline{\delta} | \Sigma) \text{ distribution} \tag{2.2}$$

and is independent of S, which has the $W_p(S|f|\Sigma)$ distribution. Therefore, $\underline{d}$ is an unbiased estimate of $\underline{\delta}$ and $(1/f)S$ is an unbiased estimate of Σ. Using these, instead of $\underline{\delta}$ and Σ, Fisher's discriminant function becomes

$$\begin{aligned} \underline{\ell}'\underline{x} &= \underline{d}'(\frac{1}{f} S)^{-1} \underline{x} \\ &= f\, \underline{d}' S^{-1} \underline{x}. \end{aligned} \tag{2.3}$$

This is known as the sample discriminant function. The classification procedure (1.22) now becomes:

(i) Compute $\underline{\ell}'\underline{x} = f\, \underline{d}' S^{-1} \underline{x}$, $\underline{\ell} = f S^{-1} \underline{d}$.

(ii) Compute $\frac{1}{2} \underline{\ell}'(\underline{\bar{x}} + \underline{\bar{y}})$.

(iii) Assign the individual with measurements $\underline{x}$ to π_1 or π_2, according as (2.4)

$$\underline{\ell}'\underline{x} - \frac{1}{2} \underline{\ell}'(\underline{\bar{x}} + \underline{\bar{y}}) \tag{2.5}$$

is ≥ 0, or < 0. The statistic (2.5) is called Anderson's [1] classification statistic. The chance of misclassification, when this procedure is employed, is, of course, not $\Phi(-\frac{1}{2}\Delta_p)$, nor can it be obtained in a similar manner, because $\underline{\ell}'\underline{x}$ does not have a normal distribution, as $\underline{a}'\underline{x}$. $\underline{\ell}$ is not a constant vector but is a random variable, being a function of $\underline{d}$ and S. $\underline{\ell}'\underline{x}$ is not a linear function of normal variables. We shall consider this point later in the chapter.

We could have obtained this sample discriminant function $\underline{\ell}'\underline{x}$ by maximizing

$$\frac{|\underline{\ell}'\underline{d}|}{(f^{-1}\ \underline{\ell}'\ S\underline{\ell})^{1/2}}, \tag{2.6}$$

an expression analogous to (1.23), with $\underline{\delta}$, Σ replaced by their estimates. Observe that the square of this "distance", based on sample quantities, is proportional to

$$\frac{\frac{n_1 n_2}{n_1+n_2}(\bar{z}_1 - \bar{z}_2)^2}{\frac{1}{n_1+n_2-2}(s_1^2 + s_2^2)}, \tag{2.7}$$

where $z = \underline{\ell}'\underline{x}$, $\bar{z}_1 = \underline{\ell}'\bar{\underline{x}}$, $\bar{z}_2 = \underline{\ell}'\bar{\underline{y}}$, $s_1^2 = \underline{\ell}'S_x\underline{\ell}$, and $s_2^2 = \underline{\ell}'S_y\underline{\ell}$. But

$$\frac{n_1 n_2}{n_1+n_2}(\bar{z}_1 - \bar{z}_2)^2 = c^2\ (\underline{\ell}'\underline{d})^2 \tag{2.8}$$

is the "between populations" s.s. and

$$s_1^2 + s_2^2 = \underline{\ell}'\ S\ \underline{\ell} \tag{2.9}$$

is the "within populations" s.s. in an analysis of variance carried on a variable z, the observations on which are $\underline{\ell}'X$ in π_1 and $\underline{\ell}'Y$ in π_2. The usual analysis of variance table for z is

TABLE 2

Analysis of Variance for $z = \underline{\ell}'\underline{x}$

Source	d.f.	Sum of squares
Between populations	1	$c^2(\underline{\ell}'\underline{d})^2 = c^2\underline{\ell}'\,\underline{d}\underline{d}'\,\underline{\ell}$
Within populations	$n_1 + n_2 - 2$	$\underline{\ell}'\,S\,\underline{\ell}$
Total	$n_1 + n_2 - 1$	$\underline{\ell}'(S + c^2\,\underline{d}\underline{d}')\underline{\ell}$

Maximizing (2.6) is thus the same as maximizing the F ratio

$$\frac{\text{"between populations" s.s./d.f. 1}}{\text{"within populations" s.s./d.f. f}}$$

in the analysis of variance of the variable $\underline{\ell}'\underline{x}$. From (1.25), it is easy to see that the F ratio is maximized when $\underline{\ell}$ is proportional to $S^{-1}\underline{d}$. Substituting this value of $\underline{\ell}$ in (2.6), we find the "sample" distance to be

$$D_p = (f\ \underline{d}'\ S^{-1}\ \underline{d})^{1/2}\ . \tag{2.10}$$

$D_p^2 = f\ \underline{d}'\ S^{-1}\ \underline{d}$ is the "Studentized" D^2 of Mahalanobis, based on the p characters $\underline{x}$, mentioned in Chapter 5. Its distribution was seen to be

$$H_p'(c^2\ D_p^2 | f,\ c^2\ \Delta_p^2)\ d(c^2\ D_p^2)\ , \tag{2.11}$$

from which it can be deduced that

$$E(D_p^2) = \frac{f}{f-p-1}\ (\Delta_p^2 + \frac{p}{c^2})\ . \tag{2.12}$$

D_p^2 is thus not unbiased for Δ_p^2, the square of the distance between π_1 and π_2. An unbiased estimate of Δ_p^2 is

$$\frac{f-p-1}{f}\ D_p^2 - \frac{p}{c^2}\ . \tag{2.13}$$

In order to find out the contribution of $x_{k+1}, x_{k+2}, \ldots, x_p$ to the distance between π_1 and π_2 in addition to the one based on $x_1, \ldots, x_k$ we shall partition $\underline{x}$, $\underline{\delta}$, Σ, $\underline{d}$, S, $\underline{a}$ as

$$\underline{x} = \left[\begin{array}{c} \underline{x}(1) \\ \hline \underline{x}(2) \end{array}\right]\begin{array}{l} k \\ p-k \end{array}, \qquad \underline{\delta} = \left[\begin{array}{c} \underline{\delta}(1) \\ \hline \underline{\delta}(2) \end{array}\right]\begin{array}{l} k \\ p-k \end{array}$$

$$\underline{a} = \Sigma^{-1}\underline{\delta} = \left[\begin{array}{c} \underline{a}(1) \\ \hline \underline{a}(2) \end{array}\right]\begin{array}{l} k \\ p-k \end{array}, \qquad \underline{d} = \left[\begin{array}{c} \underline{d}(1) \\ \hline \underline{d}(2) \end{array}\right]$$

$$\Sigma = \left[\begin{array}{c|c} \Sigma_{11} & \Sigma_{12} \\ \hline \Sigma_{21} & \Sigma_{22} \end{array}\right]\begin{array}{l} k \\ p-k \end{array}, \qquad S = \left[\begin{array}{c|c} S_{11} & S_{12} \\ \hline S_{21} & S_{22} \end{array}\right]\begin{array}{l} k \\ p-k \end{array}. \tag{2.14}$$
$$\begin{array}{cc} k & p-k \end{array} \qquad\qquad \begin{array}{cc} k & p-k \end{array}$$

We also let

$$\boldsymbol{\beta} = \Sigma_{21}\Sigma_{11}^{-1}, \qquad B = S_{21}S_{11}^{-1}$$

$$\Sigma_{22\cdot1} = \Sigma_{22} - \Sigma_{21}\Sigma_{11}^{-1}\Sigma_{12}, \quad S_{22\cdot1} = S_{22} - S_{21}S_{11}^{-1}S_{12}. \tag{2.15}$$

Using (5.3.7) for Σ^{-1} and a similar partitioning for S^{-1}, we can see that (see 5.3.9 also)

$$\Delta_p^2 = \Delta_k^2 + (\underline{\delta}(2) - \boldsymbol{\beta}\underline{\delta}(1))'\,\Sigma_{22\cdot1}^{-1}(\underline{\delta}(2) - \boldsymbol{\beta}\underline{\delta}(1)), \tag{2.16}$$

$$D_p^2 = D_k^2 + f(\underline{d}(2) - B\,\underline{d}(1))'\,S_{22\cdot1}^{-1}(\underline{d}(2)-B\underline{d}(1)), \tag{2.17}$$

where

$$\Delta_k^2 = \underline{\delta}'(1)\,\Sigma_{11}^{-1}\,\underline{\delta}(1), \quad D_k^2 = f\,\underline{d}'(1)\,S_{11}^{-1}\,\underline{d}(1), \tag{2.18}$$

are, respectively, the true squared distance and "Studentized" squared distance between π_1 and π_2, based on the first k variables $\underline{x}(1)$ only. The increase in the distance between π_1 and π_2, due to the characters $x_{k+1}, \ldots, x_p$, over the distance based on $x_1, \ldots, x_k$ is given by

$$\Delta_p^2 - \Delta_k^2 = (\underline{\delta}(2) - \beta\underline{\delta}(1))' \Sigma_{22\cdot 1}^{-1}(\underline{\delta}(2) - \beta\underline{\delta}(1)). \quad (2.19)$$

Its unbiased estimate, because of (2.13), is

$$\frac{1}{f}\left\{(f-p-1)D_p^2 - (f-k-1)D_k^2\right\} - \frac{1}{c^2}(p-k) . \quad (2.20)$$

3. TESTS ASSOCIATED WITH DISCRIMINANT FUNCTIONS

(B_1) Consider the hypothesis

$$H_1: \ a_{k+1} = \cdots = a_p = 0, \text{ i.e., } \ \underline{a}(2) = \underline{0} ,$$

where $\underline{a} = \Sigma^{-1}\underline{\delta}$ is the coefficient vector of the discriminant function $\underline{a}'\underline{x}$, and $\underline{a}$ is partitioned as in (2.14). From (5.3.7), we find

$$\underline{a}(2) = \Sigma_{22\cdot 1}^{-1}(\underline{\delta}(2) - \beta\underline{\delta}(1)), \quad (3.1)$$

where $\Sigma_{22\cdot 1}$, $\delta(2)$, $\underline{\delta}(1)$, and β are as defined in Section 2. H_1 is thus equivalent to the hypothesis

$$H_2: \ \underline{\delta}(2) = \beta\underline{\delta}(1) , \quad (3.2)$$

which can be written alternatively as

$$H_2: \ E(\underline{x}(2)\,|\,\underline{x}(1)) \text{ is the same in both } \pi_1 \text{ and } \pi_2. \quad (3.3)$$

Also, from (2.16), we observe that both H_1 and H_2 are equivalent to

$$H_3: \ \Delta_p^2 = \Delta_k^2 .$$

We have already derived a test for this hypothesis in Chapter 5, Section 4 (A_2). The test is [see (5.4.32)]

$$\frac{f-p+1}{p-k} \frac{c^2(D_p^2 - D_k^2)}{f + c^2 D_k^2} = F_{p-k,\, f-p+1} . \quad (3.4)$$

If H_1 (or H_2 or H_3) is accepted, it means that the variables $x_{k+1}, \ldots, x_p$ do not have any additional discriminating ability, once $x_1, x_2, \ldots, x_k$ have already been considered.

(B_2) A hypothesis of considerable interest in discriminant analysis is

$$H_4: \; E(\underline{x}(2)) \text{ is the same in } \pi_1 \text{ and } \pi_2, \text{ given that } E(\underline{x}(1)) \text{ is the same in } \pi_1 \text{ and } \pi_2$$

that is

$$H_4: \; \underline{\delta}(2) = 0, \text{ given } \underline{\delta}(1) = 0.$$

This alternatively can be expressed as

$$H_5: \; \Delta_p^2 = 0, \text{ given } \Delta_k^2 = 0$$

or as

$$H_6: \; \Delta_{p-k}^2 = 0, \text{ given } \Delta_k^2 = 0,$$

where Δ_{p-k}^2 is the (distance)^2 based on $\underline{x}(2)$, viz., $\underline{\delta}'(2)\, \Sigma_{22}^{-1}\, \underline{\delta}(2)$.

From (2.16), it is obvious that, whenever H_4 (or H_5, H_6) is true, $\Delta_p^2 = \Delta_k^2$, and so the F test given in (3.4) can be used to test H_4. Thus the same F test can be used to test any of H_1, H_2, H_3. H_4, H_5, H_6.

When it is known that $\underline{x}(1)$ has the same mean in both the populations, $\underline{x}(1)$ is known as the vector of concomitant variables or ancillary variables. They have no discriminating ability by themselves but, in the presence of other variables having discriminating ability, these ancillary variables, on account of their correlations with these variables, may provide additional discrimination. A good discussion of such concomitant variables in discriminant analysis will be found in Rao [22], Cochran and Bliss [8], and Cochran [10].

Rao [22] has suggested a different test for H_4. It is based on the statistic $c^2(D_p^2 - D_k^2)/f$. In Section 4 of Chapter 5, we obtained the conditional distribution of $c^2(D_p^2 - D_k^2)/(f + c^2 D_k^2)$, when D_k^2 is fixed. From this and the distribution of D_k^2, one can obtain the unconditional distribution of $c^2(D_p^2 - D_k^2)/f$. This distribution can then be used to test H_4. According to Rao, $c^2(D_p^2 - D_k^2)/f$ is a better statistic than $c^2(D_p^2 - D_k^2)/(f + c^2 D_k^2)$ for testing H_4, because the variance of the estimate of Δ_p^2 obtained from $c^2(D_p^2 - D_k^2)/f$ is less than the variance of the estimate of Δ_p^2, based on $c^2(D_p^2 - D_k^2)/(f + C^2 D_k^2)$. This point is investigated in considerable detail by Kathleen Subrahmaniam and Kocherlakota Subrahmaniam [26], who also have come to the same conclusion that the test based on $c^2(D_p^2 - D_k^2)/f$ is far superior. They have provided tables of percentage points of the distribution of $w = c^2(D_p^2 - D_k^2)/f$. When these tables were not available, Rao suggested the use of an approximate F test, viz.,

$$\frac{f-p+1}{p-k}\left[\frac{f-k+1}{f+1}\,\frac{c^2(D_p^2 - D_k^2)}{f}\right] = F_{p-k,\ f-p+1}\,. \qquad (3.5)$$

This approximate test is derived from (3.4) by replacing $1/(f + c^2 D_k^2)$ in (3.4) by its expected value. From the distribution of D_k^2 [given by (2.11), with p changed to k], it can be readily shown that

$$E(f + c^2 D_k^2)^{-1} = \frac{(f-k+1)}{f(f+1)} .$$

Using this for $1/(f + c^2 D_k^2)$ in (3.4), we obtain (3.5) as an approximate test.

Another approximation suggested by Rao is that

$$\frac{f-k+1}{f+1} \frac{(f-p+1)}{1} c^2(D_p^2 - D_k^2) \text{ is a } \chi^2 \text{ with p-k d.f. when } H_4 \text{ is true.}$$

The Subrahmaniams have investigated these two approximations also. According to them, this latter one is a better approximation. Both the approximations are conservative in that the actual level of significance is less than the prescribed level of significance.

(B_3) Goodness of fit of a hypothetical discriminant function:

H_o: A given function $\underline{h}'\underline{x}$ is good enough for discriminating between π_1 and π_2.

We shall now develop a test for this hypothesis. From $\underline{x}$, transform to $z_1 = \underline{h}'\underline{x}$ and $z_2, \ldots, z_p$, which are p-1 other suitable linear functions of $\underline{x}$. First we prove that a discriminant function is invariant for a linear transformation. Let the matrix of transformation from $\underline{x}$ to $\underline{z} = [z_1, \ldots, z_p]'$ be H, i.e., $\underline{z} = H\underline{x}$. Then $V(\underline{z}) = H\Sigma H'$ and difference in the means of $\underline{z}$ in π_1 and π_2 is $H\underline{\delta}$. The discriminant function based on $\underline{z}$ is therefore

$$(H\underline{\delta})'\ (H\Sigma H')^{-1}\ \underline{z} = \underline{\delta}'\ \Sigma^{-1}\ \underline{x}\ , \tag{3.6}$$

as H is nonsingular. But if the hypothesis H_o is true, $\underline{\delta}'\Sigma^{-1}\underline{x}$ must be proportional to $\underline{h}'\underline{x}$, i.e., z_1. So,

$$(H\delta)'\ (H\Sigma H')^{-1}\ \underline{z} = \gamma\ z_1\ ,$$

where γ is some constant. The coefficients of $z_2, z_3, \ldots, z_p$ in $(H\delta)'\ (H\Sigma H')^{-1}\underline{z}$ are thus all zero. The hypothesis H_o, therefore, is equivalent to the hypothesis that the variables z_2, z_3, ..., z_p do not occur in the discriminant function based on $\underline{z}$. We already know from (B_1) of this section that this hypothesis can be tested by [see (3.4)]

$$\frac{f-p+1}{p-1}\ \frac{c^2(D_p^2 - D_1^2)}{f + c^2\ D_1^2} = F_{p-1,\ f-p+1}\ , \tag{3.7}$$

where D_p^2 is D^2 based on $z_1, \ldots, z_p$ and D_1^2 is D^2 based on z_1 alone. But D^2 is invariant for a nonsingular linear transformation like $\underline{z} = H\underline{x}$, as

$$(H\underline{d})'\ (HSH')^{-1}\ (H\underline{d}) = \underline{d}'S^{-1}\underline{d}\ , \tag{3.8}$$

and so D_p^2 in (3.7) is the D^2 based on $\underline{x}$, viz., $f\ \underline{d}'\ S^{-1}\ \underline{d}$. However, as $z_1 = \underline{h}'\underline{x}$,

$$\frac{D_1^2}{f} = (\underline{h}'\underline{d})'\ (\underline{h}'S\underline{h})^{-1}\ (\underline{h}'\underline{d}) = \frac{(\underline{h}'\underline{d})^2}{\underline{h}'S\underline{h}}\ . \tag{3.9}$$

(B_4) Test for an assigned ratio of two coefficients in the discriminant function: Since only ratios of discriminant function coefficients are unique, a hypothesis of the type $a_i = \rho$ is meaningless, but a hypothesis

$$\text{H:} \quad \frac{a_i}{a_j} = \rho, \tag{3.10}$$

which specifies the ratio of two coefficients in the discriminant function $\underline{a}'\underline{x}$, is of interest. To derive a test for this hypothesis, we transform from $\underline{x}$ to new variables $\underline{z} = [z_1, \ldots, z_p]'$, by the nonsingular transformation

$$\begin{aligned} z_r &= x_r, \quad (r = 1, 2, \ldots, i\text{-}1, i+1, \ldots, p), \\ z_i &= \rho x_i + x_j . \end{aligned} \tag{3.11}$$

Since a discriminant function is invariant for a nonsingular linear transformation, the discriminant function based on $\underline{z}$ is the same as that based on $\underline{x}$, viz., $\underline{a}'\underline{x}$, which when H is true, can be written as

$$\begin{aligned} & a_1x_1 + \ldots + a_{i-1}x_{i-1} + \rho a_jx_i + a_{i+1}x_{i+1} + \ldots + a_px_p \\ & \quad = a_1z_1 + \ldots + a_{i-1}z_{i-1} + a_jz_i + a_{i+1}z_{i+1} + \ldots \\ & \qquad + a_{j-1}z_{j-1} + a_{j+1}z_{j+1} + \ldots + a_pz_p , \end{aligned} \tag{3.12}$$

showing that the coefficient of z_j is zero. H is thus equivalent to the hypothesis that the coefficient of z_j in the discriminant function based on $\underline{z}$ is null. A test for this, from (B_1) of this section, is

$$\frac{f-p+1}{1} \frac{c^2(D_p^2 - D_{p-1}^2)}{f + c^2 D_{p-1}^2} = F_{1,\, f-p+1} , \tag{3.13}$$

where $D_p^2 = D^2$ based on $\underline{z} = D^2$ based on $\underline{x}$ due to invariance, and $D_{p-1}^2 = D^2$ based on the p-1 variables $z_1, \ldots, z_{j-1}, z_{j+1}, \ldots, z_p$. Now,

$$\begin{bmatrix} z_1 \\ \vdots \\ z_{j-1} \\ z_{j+1} \\ \vdots \\ z_p \end{bmatrix} = L\underline{x} ,$$

where L is a $(p-1) \times p$ matrix, obtained from I_p by deleting the jth row and replacing the ith row by ρ times the ith row plus the jth row. Hence,

$$D^2_{p-1}, \text{ i.e., } D^2 \text{ based on } z_1, \ldots, z_{j-1}, z_{j+1}, \ldots, z_p$$
$$= f(L\underline{d})' (LSL')^{-1} (L\underline{d}) . \qquad (3.14)$$

If the problem is not to test the hypothesis $a_i/a_j = \rho$ but to obtain a confidence interval for a_i/a_j, we have to equate the left-hand side of (3.13), which will now be a function of an unknown $\rho(= a_i/a_j)$ to $F(1, f-p+1, \alpha)$ and solve the resulting quadratic equation in ρ. The roots of this quadratic equation, in general, will yield a confidence interval for ρ, with confidence coefficient $1 - \alpha$.

4. ANALOGY OF DISCRIMINANT ANALYSIS WITH REGRESSION ANALYSIS

We shall use the same notation as in Section 2. If we define a dummy variable ξ as

$$\xi = \lambda_1, \text{ if an individual belongs to population } \pi_1$$
$$= \lambda_2, \text{ if he belongs to } \pi_2 , \qquad (4.1)$$

we can write

$$E(\underline{x}) = \underline{\alpha} + \underline{\beta}\, \xi , \qquad (4.2)$$

where $\underline{x}$ is the vector of the measurements on the variables $x_1, \ldots, x_p$ for the individual, and

$$\underline{\alpha} = \frac{1}{\lambda_1 - \lambda_2} (\lambda_1 \underline{\mu}(2) - \lambda_2 \underline{\mu}(1)), \quad \underline{\beta} = \frac{1}{\lambda_1 - \lambda_2} (\underline{\mu}(1) - \underline{\mu}(2)). \tag{4.3}$$

because (4.2) reduces to $\underline{\mu}(1)$ when $\xi = \lambda_1$ and to $\underline{\mu}(2)$ when $\xi = \lambda_2$. Equation (4.2) is thus formally the regression of $\underline{x}$ on ξ. In regression analysis, the problem is to predict the value of the dependent variable from the values of the independent variables. However, in this regression (4.2), the problem is just the opposite. The dependent variate $\underline{x}$ is known and, from this, we have to decide whether to assign the individual to π_1 or π_2, i.e., we wish to predict ξ, the independent variable in (4.2). It is, therefore, natural to consider the regression of ξ on $\underline{x}$ and not of $\underline{x}$ on ξ as in (4.2). Let us find it. The observations on $\underline{x}$, from the $N = n_1 + n_2$ observations in the two samples from π_1 and π_2, and the observations on ξ, as defined in (4.1), are given in Table 3.

TABLE 3

Observations on $\underline{x}$ and ξ

Variable	Observations on the first n_1 individuals coming from π_1	Observations on the next n_2 individuals from π_2
$\underline{x}$	X	Y
ξ	λ_1 (n_1 times)	λ_2 (n_2 times)

The matrix of corrected s.s. and s.p. of all the N observations on $\underline{x}$ is

$$[X|Y]\,(I - \frac{1}{N}E_{NN})\,[X|Y]'$$
$$= X(I - \frac{1}{n_1}E_{n_1 n_1})X' + Y(I - \frac{1}{n_2}E_{n_2 n_2})Y' + \frac{n_1 n_2}{n_1+n_2}\underline{d}\underline{d}'$$
$$= S + c^2\,\underline{d}\underline{d}', \tag{4.4}$$

where $\underline{d} = (1/n_1)\,X\,E_{n_1 1} - (1/n_2)\,Y\,E_{n_2 1}$, as defined earlier. The matrix of the corrected s.p. of $\underline{x}$ with ξ is

$$[X|Y]\,(I - \frac{1}{N}E_{NN})\,[\lambda_1 E_{1n_1}|\lambda_2 E_{1n_2}]'$$
$$= c^2(\lambda_1 - \lambda_2)\,\underline{d}\,. \tag{4.5}$$

Finally, the corrected s.s. of observations on ξ is

$$\left[\frac{\lambda_1 E_{n_1 1}}{\lambda_2 E_{n_2 1}}\right]'(I - \frac{1}{N}E_{NN})\,[\lambda_1 E_{1n_1}|\lambda_2 E_{1n_2}]'$$
$$= c^2(\lambda_1 - \lambda_2)^2\,. \tag{4.6}$$

If the regression of ξ on $\underline{x}$ is

$$\text{const.} + b_1x_1 + b_2x_2 + \dots + b_px_p\,, \tag{4.7}$$

from the usual least squares theory of minimizing the sum of squares of deviations of ξ from its regression, we find that $\underline{b}$, the vector of the regression coefficients $b_1, \dots, b_p$, satisfies the normal equations

$$c^2(\lambda_1 - \lambda_2)\underline{d} = (S + c^2\,\underline{d}\underline{d}')\underline{b}\,, \tag{4.8}$$

where the left-hand side of (4.8) is obtained from (4.5) and the first matrix on the right-hand side of (4.8) is taken from (4.4). The solution of (4.8) is

$$\underline{b} = c^2(\lambda_1 - \lambda_2)\,(S + c^2\,\underline{d}\underline{d}')^{-1}\,\underline{d}\,. \tag{4.9}$$

But, from (1.4.16),

$$(S + c^2 \underline{d}\underline{d}')^{-1} = (I + c^2 S^{-1} \underline{d}\underline{d}')^{-1} S^{-1}$$
$$= \left(I - \frac{c^2 S^{-1} \underline{d}\underline{d}'}{1 + c^2 \underline{d}' S^{-1} \underline{d}}\right) S^{-1} . \qquad (4.10)$$

Hence, (4.9) reduces to

$$\underline{b} = \frac{c^2(\lambda_1 - \lambda_2)}{1 + c^2 D_p^2/f} S^{-1}\underline{d} , \qquad (4.11)$$

where $D_p^2/f = \underline{d}'S^{-1}\underline{d}$. This shows that $\underline{b}$ is proportional to $S^{-1}\underline{d}$ and $\underline{b}'\underline{x}$ is proportional to $\underline{\ell}'\underline{x}$, the sample discriminant function of (2.3). The discriminant function and the regression function are thus the same, apart from a constant of proportionality, and this regression approach will also lead to the same classification procedure (2.4) as before.

The regression s.s., in the regression of ξ on $\underline{x}$, is

$$\text{S.S.R.}(\underline{x}) = c^2(\lambda_1 - \lambda_2)\underline{b}'\underline{d} . \qquad (4.12)$$

[From the least squares theory, it is the sum of the products of the left-hand side quantities in (4.8) and the regression coefficients $\underline{b}$ of (4.9)]. Using (4.11), it reduces to

$$\text{S.S.R.}(\underline{x}) = c^2(\lambda_1 - \lambda_2)^2 \frac{c^2 D_p^2}{f + c^2 D_p^2} . \qquad (4.13)$$

We therefore obtain the following analysis of variance table:

TABLE 4

Regression of ξ on $\underset{\sim}{x}$

Source	d.f.	s.s.	F ratio
Regression on $\underset{\sim}{x}$	p	$c^2(\lambda_1 - \lambda_2)^2 \dfrac{c^2D_p^2}{f + c^2D_p^2}$	$\dfrac{n_1+n_2-p-1}{p} \dfrac{c^2D_p^2}{f}$
Error	n_1+n_2-p-1	By subtraction, say SSE	
Total	n_1+n_2-1	$c^2(\lambda_1 - \lambda_2)^2$	

A test of significance of all the regression coefficients is the same as the test of "independence" or "lack of association" of $\underset{\sim}{x}$ and ξ, and from (4.2) we observe that there is no association between $\underset{\sim}{x}$ and ξ if $\underset{\sim}{\beta} = \underset{\sim}{0}$, i.e., if $\underset{\sim}{\mu}(1) = \underset{\sim}{\mu}(2)$. In a standard regression analysis, this is provided by the F ratio in the analysis of variance table above. But let us first examine whether the basic assumptions in a standard regression analysis are true. They are:

(i) The independent variables are fixed.

(ii) The dependent variable has a normal distribution.

(iii) The observations on the dependent variable have a common variance σ^2.

In our regression analysis, the dependent variable ξ does not have a normal distribution. It is not a random variable at all. It is a pseudo or dummy variable with only 2 values, λ_1 and λ_2. On the contrary, the independent variables $\underset{\sim}{x}$ are normally distributed and are not fixed at all, as required. The roles of independent and dependent variables are thus switched. We, therefore, do not have any justification at all to use the F test for the null hypothesis that the true regression coefficients are all null, which is equivalent, as seen earlier, to $\underline{\mu}(1) = \underline{\mu}(2)$. Even so, the F ratio, viz.,

$$\frac{\text{Regression s.s./d.f.}}{\text{Error s.s./d.f.}} = \frac{n_1+n_2-p-1}{p} \frac{c^2D_p^2}{f} \tag{4.14}$$

<u>does</u> <u>have</u> an F distribution with p, n_1+n_2-p-1 d.f., if $\underline{\mu}(1) = \underline{\mu}(2)$, i.e., if $\Delta_p^2 = 0$. This was proved in Chapter 5, Section 4, Equation (4.26). The justification of the test does not come from the usual regression theory but from the distribution of D_p^2. This remarkable result of the validity of the F test, when the "dependent" and "independent" variables switch their roles is due to the "duality" in their relationship, discussed in (A_6) of Section 4, of Chapter 5. Also, notice in this connection that, if R is the multiple correlation coefficient between $\underline{\xi}$ and $\underline{x}$, then, by (1.3.8),

$$R^2 = \frac{\text{Regression s.s.}}{\text{total s.s.}} \text{ in Table 4}$$

$$= \frac{c^2D_p^2/f}{1+c^2D_p^2/f} . \tag{4.15}$$

The distribution of R^2, under the null hypothesis [which is $\underline{\mu}(1) = \underline{\mu}(2)$ here] is the same whether $\underline{\xi}$ is fixed and $\underline{x}$ normal or vice versa. From (4.15) and (5.4.23),

$$\frac{T_p^2}{f} = \frac{c^2 D_p^2}{f} = \frac{R^2}{1 - R^2} . \tag{4.16}$$

Showing the relationship among Hotelling's T^2, Mahalanobis's D^2, and R^2 used by Fisher.

Here we considered a test of significance of all the regression or discriminant function coefficients. We can consider a test of significance of one only or some p-k coefficients only, using regression theory. But we then will come up woth the same tests as we derived earlier, viz., (3.4) or (3.7). This regression analysis, therefore, serves as a useful "mnemonic" device to deal with discriminant analysis. Computer programs written for regression problems can thus be used directly in discrimination problems also.

The reader must have noticed that the arbitrary values λ_1 and λ_2 do not enter anywhere in the F test (or other F tests also), and one can take more convenient values like $\lambda_1 = 1$, $\lambda_2 = 0$ or $\lambda_1 = 1$, $\lambda_2 = -1$ to avoid complications. Fisher used $\lambda_1 = n_2/(n_1+n_2)$ and $\lambda_2 = -n_1/(n_1+n_2)$, so that $\lambda_1 - \lambda_2 = 1$, and $\bar{\xi}$, the mean of all the ξ's, is zero. Different choices of λ_1, λ_2 yield different values of $\underline{b}$ in (4.11), but all such $\underline{b}$'s are proportional to each other and to $\underline{\ell}'\underline{x}$, and this does not matter in discrimination, as we have to standardize the discriminant function by dividing by its standard deviation before using it. This difference between regression and discrimination should be emphasized: while the regression coefficients are unique, the discriminant function coefficients are not, only their ratios are unique.

In standard least squares theory, the variance-covariance matrix of $\underline{b}$ is $(S + c^2\underline{d}\underline{d}')^{-1}\sigma^2$, where σ^2 is the variance of ξ and is estimated by error s.s. and d.f. in the analysis of variance table. This is of course not true here; ξ is not a normally distributed variable. However, if we formally write

$$V(\underline{b}) = (S + c^2\underline{d}\underline{d}')^{-1}\sigma^2 , \tag{4.17}$$

$$\text{Estimate of } \sigma^2 = \text{S.S.E.}/(n_1+n_2-p-1) = \frac{c^2(\lambda_1 - \lambda_2)^2}{f+p-1} \frac{f}{f + c^2D_p^2} , \tag{4.18}$$

and use

$$\frac{b_i}{\{\text{Estimate of } V(b_i)\}^{1/2}} = t, \text{ with } f-p+1 \text{ d.f.} \tag{4.19}$$

to test the significance of b_i, the test will still be valid, as a little algebra will show that the square of this t is identical with the F of (3.4), with $k = p-1$, and $D_k^2 = D^2$ based on all $\underline{x}$ except x_i. Thus we can pretend (4.17) to be true, if we want to test the significance of any element of $\underline{b}$. The actual variance-covariance matrix of $\underline{b}$ (or $\underline{\ell}$, which is proportional to $\underline{b}$) will be derived in the next section.

Table 4 displays the regression and error s.s. in the regression of ξ on $\underline{x}$. However, if we consider the regression of $\underline{x}$ on ξ, the matrix of regression s.s. and s.p. is [from (2.4.24), (4.4), (4.5), (4.6)]

$$\{c^2(\lambda_1 - \lambda_2)\underline{d}\}\ \{c^2(\lambda_1 - \lambda_2)^2\}^{-1}\ \{c^2(\lambda_1 - \lambda_2)\underline{d}'\}$$
$$= c^2\underline{d}\underline{d}', \tag{4.20}$$

and the matrix of error s.s. and s.p. is

$$(S + c^2\underline{d}\underline{d}') - c^2\underline{d}\underline{d}' = S. \tag{4.21}$$

Compare these with the "between populations" and "within populations" s.s. in Table 2, in the analysis of variance for $\underline{\ell}'\underline{x}$. The comparison shows that the "between populations" s.s. can be looked upon as the regression s.s. with regression of $\underline{\ell}'\underline{x}$ on the dummy variable ξ and the "within populations" s.s. as the error s.s. in the regression. This is also true, in general, when there are several populations or groups, but, in that case, one has to consider more than one dummy variable. If the d.f. of the "between populations" s.s. are q, we need q dummy variables like ξ to represent the analysis of variance as a regression analysis. We shall come to this point again in Chapter 9 when we deal with discrimination among several groups.

5. STANDARD ERRORS OF DISCRIMINANT FUNCTION COEFFICIENTS

We have seen in (2.2) that $c\underline{d}$ has the $N_p(c\underline{d}|c\underline{\delta}|\Sigma)$ distribution and S has an independent $W_p(S|f|\Sigma)$ distribution. We shall use these two distributions to derive the expected value and variance-covariance matrix of $\underline{\ell} = fS^{-1}\underline{d}$, the coefficient vector of the sample discriminant function $\underline{\ell}'\underline{x}$.

As S and $\underline{d}$ are independent,

$$E(\underline{\ell}) = f\, E(S^{-1})\, E(\underline{d}) = \frac{f}{f-p-1} \Sigma^{-1} \underline{\delta}\,, \tag{5.1}$$

because, from Lemma 10 of Chapter 4, $E(S^{-1}) = [1/(f-p-1)]\Sigma^{-1}$. Let Γ be a lower triangular matrix such that $\Gamma\Sigma\Gamma' = I_p$. Then, by Theorem 2 of Chapter 2,

$$\underline{u} = c\,\Gamma\underline{d} \text{ has the } N_p(\underline{u}|c\,\Gamma\,\underline{\delta}|I_p) \text{ distribution,} \tag{5.2}$$

and by Lemma 1 of Chapter 3,

$$A = \Gamma S \Gamma' = \left[\begin{array}{c|c} A_{11} & A_{12} \\ \hline A_{21} & A_{22} \end{array}\right] \begin{matrix} p-1 \\ 1 \end{matrix} \tag{5.3}$$
$$\qquad\qquad p-1 \quad 1$$

has the $W_p(A|f|I_p)$ distribution, independent of $\underline{u}$. As in Section 7 of Chapter 4, we partition A^{-1} as

$$A^{-1} = \left[\begin{array}{c|c} P & \underline{h} \\ \hline \underline{h}' & a^{pp} \end{array}\right] \begin{matrix} p-1 \\ 1 \end{matrix}, \tag{5.4}$$
$$\qquad\qquad p-1 \quad 1$$

where

$$a^{pp} = \frac{1}{A_{22 \cdot 1}}, \quad A_{22\cdot 1} = A_{22} - A_{21} A_{11}^{-1} A_{12},$$
$$\underline{h} = -a^{pp} A_{11}^{-1} A_{12}\,. \tag{5.5}$$

We now keep $\underline{u}$ fixed. This does not affect the distribution of A, as A and $\underline{u}$ are independent. Consider a random orthogonal matrix L, such that its last row is $\underline{u}'/(\underline{u}'\underline{u})^{1/2}$, i.e., L is of the form

$$L = \left[\frac{L_1}{\underline{u}'/(\underline{u}'\underline{u})^{1/2}}\right] . \tag{5.6}$$

Since L is orthogonal, $L_1 L_1' = I_{p-1}$ and $L_1 \underline{u} = \underline{0}$. By Lemma 1 of Chapter 3, A and L'AL have the same distribution, and so

$$\begin{aligned} E_1(A^{-1}\underline{u}\underline{u}'A^{-1}) &= E_1\{(L'AL)^{-1}\ \underline{u}\underline{u}'\ (L'AL)^{-1}\} \\ &= L'E_1(A^{-1}L\ \underline{u}\underline{u}'\ L'A^{-1})L , \end{aligned} \tag{5.7}$$

where E_1 stands for the conditional expectation operator when $\underline{u}$ is fixed. From (5.6), $L\underline{u} = [0, 0, \ldots, 0, (\underline{u}'\underline{u})^{1/2}]$, and so (5.7) reduces to [on using (5.4)]

$$E_1(A^{-1}\underline{u}\underline{u}'A^{-1}) = L'E_1\left\{\left[\frac{\underline{h}}{a^{pp}}\right]\left[\frac{\underline{h}}{a^{pp}}\right]'\right\}L \cdot (\underline{u}'\underline{u}). \tag{5.8}$$

Now use $\underline{h} = -a^{pp}\ A_{11}^{-1}\ A_{12}$ from (5.5) and the independence of $a^{pp} = A_{22.1}^{-1}$ and $A_{11}^{-1}\ A_{12}$ established in Section 4 of Chapter 4, to obtain [from (5.8)]

$$E_1(A^{-1}\ \underline{u}\underline{u}'\ A^{-1}) = L'\ E_1\left[\begin{array}{c|c} A_{11}^{-1}A_{12}A_{21}A_{11}^{-1} & -A_{11}^{-1}A_{12} \\ \hline -A_{21}A_{11}^{-1} & 1 \end{array}\right] L \cdot E_1(a^{pp})^2(\underline{u}'\underline{u}). \tag{5.9}$$

Finally, use (4.7.5), (4.7.8), and (4.7.11). This gives

$$E_1(A^{-1}\ \underline{u}\underline{u}'\ A^{-1}) = \frac{(\underline{u}'\underline{u})}{(f-p-1)(f-p-3)}\ L'\left[\begin{array}{c|c} (f-p)^{-1}\ I_{p-1} & 0 \\ \hline 0 & 1 \end{array}\right] L. \tag{5.10}$$

Using (5.6) and the orthogonality of L, the above simplifies to

$$E_1(A^{-1}\ \underline{u}\underline{u}'\ A^{-1}) = \frac{(\underline{u}'\underline{u})I_p + (f-p-1)\underline{u}\underline{u}'}{(f-p)(f-p-1)(f-p-3)} . \tag{5.11}$$

To obtain the unconditional expectation of $A^{-1}\ \underline{u}\underline{u}'\ A^{-1}$, we use

(5.2), from which

$$E(\underline{uu}') = V(\underline{u}) + E(\underline{u})\,E(\underline{u}') = I_p + c^2\,\Gamma\,\underline{\delta\delta}'\,\Gamma', \quad (5.12)$$

and

$$E(\underline{u}'\underline{u}) = \text{tr}\,E(\underline{uu}') = \text{tr}(I_p + c^2\,\Gamma\,\underline{\delta\delta}'\,\Gamma')$$

$$= p + c^2\,\underline{\delta}'\,\Gamma'\Gamma\,\underline{\delta}. \quad (5.13)$$

Hence,

$$E(A^{-1}\,\underline{uu}'\,A^{-1}) = E\{E_1(A^{-1}\,\underline{uu}'\,A^{-1})\}$$

$$= \frac{E(\underline{u}'\underline{u})I_p + (f-p-1)\,E(\underline{uu}')}{(f-p)\,(f-p-1)\,(f-p-3)}$$

$$= \frac{(p + c^2\underline{\delta}'\Gamma'\Gamma\underline{\delta})I_p + (f-p-1)(I_p + c_2\Gamma\underline{\delta\delta}'\Gamma')}{(f-p)\,(f-p-1)\,(f-p-3)}. \quad (5.14)$$

Substitute now $\underline{u} = c\,\Gamma\,\underline{d}$ and $A = \Gamma S \Gamma'$. Also use the fact that $\Sigma^{-1} = \Gamma'\Gamma$ (as $\Gamma\Sigma\Gamma' = I_p$). This gives

$$c^2\,E(\Gamma'^{-1}\,S^{-1}\,\underline{dd}'\,S^{-1}\,\Gamma^{-1})$$

$$= \frac{(p + c^2\underline{\delta}'\Sigma^{-1}\underline{\delta})I_p + (f-p-1)(I_p + c^2\Gamma\underline{\delta\delta}'\Gamma')}{(f-p)\,(f-p-1)\,(f-p-3)} \quad (5.15)$$

or, premultiplying by Γ' and postmultiplying by Γ,

$$E(S^{-1}\,\underline{dd}'\,S^{-1})$$

$$= \frac{p\,\Sigma^{-1} + c^2\Delta_p^2\Sigma^{-1} + (f-p-1)(\Sigma^{-1} + c^2\Sigma^{-1}\underline{\delta\delta}'\Sigma^{-1})}{c^2(f-p)\,(f-p-1)\,(f-p-3)}$$

$$= \frac{(f-1+c^2\Delta_p^2)\Sigma^{-1} + c^2(f-p-1)\Sigma^{-1}\underline{\delta\delta}'\Sigma^{-1}}{c^2(f-p)\,(f-p-1)\,(f-p-3)}. \quad (5.16)$$

From this and (5.1), the variance-covariance matrix of $\underline{\ell}$ is

$$V(\underline{\ell}) = V(f\,S^{-1}\,\underline{d})$$

$$= f^2\,E(S^{-1}\,\underline{dd}'\,S^{-1}) - f^2\,E(S^{-1}\,\underline{d})\,E(S^{-1}\,\underline{d})'$$

$$= \frac{f^2}{(f-p)(f-p-1)(f-p-3)}\left\{\Delta_p^2\Sigma^{-1} + \frac{(f-1)}{c^2}\Sigma^{-1} + \frac{f-p+1}{f-p-1}\Sigma^{-1}\delta\delta'\Sigma^{-1}\right\}. \quad (5.17)$$

This result is due to Das Gupta [12]. He has also proved that, as $f \to \infty$,

$$\sqrt{f}(f\, S^{-1}\, \underline{d} - \Sigma^{-1}\, \underline{\delta}),\ \text{i.e.,}\ \sqrt{f}(\underline{\ell} - \underline{a}) \tag{5.18}$$

is asymptotically normally distributed with zero means and variance-covariance matrix

$$\Sigma^{-1}\, \underline{\delta}\underline{\delta}'\, \Sigma^{-1} + \Delta^2\, \Sigma^{-1} + \Sigma^{-1}\,. \tag{5.19}$$

6. CHANCE OF MISCLASSIFICATION WHEN ANDERSON'S STATISTIC IS USED

From Section 2 we know that when $\underline{\mu}(1)$, $\underline{\mu}(2)$ and Σ are unknown and when their sample estimates are used, the classification procedure is given by (2.4) and is based on Anderson's classification statistic [given by (2.5)]

$$W(\underline{x}) = f\, \underline{d}'\, S^{-1}\, \underline{x} - \frac{f}{2}\, (\bar{\underline{x}} + \bar{\underline{y}})'\, S^{-1}\, \underline{d}\,, \tag{6.1}$$

the quantities f, S, $\underline{d}$, $\bar{\underline{x}}$, $\bar{\underline{y}}$ being defined in (2.1). The two components of the chance of misclassification, when (2.4) is used, are

$$P_1 = \text{Prob}(W(\underline{x}) < 0 \,|\, \underline{x} \in \pi_1;\ \bar{\underline{x}},\ \bar{\underline{y}},\ S),$$

$$P_2 = \text{Prob}(W(\underline{x}) \geq 0 \,|\, x \in \pi_2;\ \bar{\underline{x}},\ \bar{\underline{y}},\ S). \tag{6.2}$$

To obtain these, one needs the distribution of $W(\underline{x})$. This distribution is too complicated, as is shown by the investigations of Bowker [6], Bowker and Sitgreaves [5], and Sitgreaves [24]. Lachenbruch and Mickey [17] have suggested a number of estimates of P_1 and P_2. These are as follows:

(1) Resubstitution method: The sample observations X and Y (of Section 2) were used to obtain $\underline{d}$ and S from which $W(\underline{x})$ is computed, for classifying a future observation $\underline{x}$. The Resubstitution method consists of using the same sample observations X and Y to assess the performance of $W(\underline{x})$. Each of these n_1+n_2 observations in X and Y is substituted in $W(\underline{x})$, and the proportions of misclassified observations from among these are obtained. These proportions are taken as estimates of P_1 and P_2. This method was suggested by Smith [25]. But this technique is often misleading and gives estimates of P_1 and P_2 that are too optimistic, as the same observations are used to compute $W(\underline{x})$ and also to evaluate its performance.

(2) D and DS methods: When the population parameters are known, we have already seen [see (1.21)] that the two components of the chance of misclassification are both equal to $\Phi(-\frac{1}{2}\Delta_p)$. A method of estimating P_1 and P_2 is, therefore, to use sample quantities instead of the population parameters in this itself, as $W(\underline{x})$ is obtained by this method only, from the true discriminant function. This gives $\Phi(-\frac{1}{2}D_p)$, where $D_p^2 = f\,\underline{d}'\,S^{-1}\,\underline{d}$, as an estimate of both P_1 and P_2. This method is known as the D method of estimation of P_1, P_2. From (2.13) we know that D_p^2 overestimates Δ_p^2 and so $\Phi(-\frac{1}{2}D_p)$ is an underestimate. A modification of this method will be to use an unbiased estimate of Δ_p^2. Such an unbiased estimate is given by (2.13), and so both P_1 and

P_2 can be estimated by

$$\Phi\left\{-\frac{1}{2}\left(\frac{f-p-1}{f} D_p^2 - \frac{p}{c^2}\right)^{1/2}\right\} . \tag{6.3}$$

Lachenbruch and Mickey [17] call this the DS method of estimation.

(3) The O method: Okamoto [18] has given an asymptotic expansion for the distribution of $W(\underline{x})$ and from that he obtains the following asymptotic expansion:

$$\begin{aligned}\text{Prob}(W(\underline{x}) < 0 | x \in \pi_1) \\ &= \Phi(-\frac{1}{2}\Delta_p) + \frac{a_1}{n_1} + \frac{a_2}{n_2} + \frac{a_3}{f} + \frac{b_{11}}{n_1^2} + \frac{b_{22}}{n_2^2} + \frac{b_{12}}{n_1 n_2} + \\ &+ \frac{b_{13}}{n_1 f} + \frac{b_{23}}{n_2 f} + \frac{b_{33}}{f^2} + O_3 . \end{aligned} \tag{6.4}$$

He gives a similar expansion for

$$\text{Prob}(W(\underline{x}) \geq 0 | \underline{x} \in \pi_2)$$

and also the expressions for the a's and b's in terms of the parameters of the two populations, π_1 and π_2. He has also tabulated the values of these a's and b's for some special cases. P_1 and P_2 can be estimated from these by substituting D_p for Δ_p in these asymptotic expansions. This method of estimation is the O method, but a better method is to use an unbiased estimator of Δ_p^2 given by (2.13). When this modification is used, we shall call the modified method the OS method.

(4) The U method: In this method, $W(\underline{x})$ is not calculated from all the n_1+n_2 observations X and Y, but one observation is omitted, either from X or from Y, and $W(\underline{x})$ is obtained from the

n_1+n_2-1 observations. Suppose this omitted observation is the rth one from the first sample of size n_1 from the population π_1. In other words, the rth column of X, which we shall denote by $\underline{x}_r$, is deleted. Let X* denote the $p \times (n_1-1)$ matrix of the remaining n_1-1 observations, when $\underline{x}_r$ is deleted from X. Then it is easy to see that

$$\begin{aligned} S_x^* &= X^*\left(I - \frac{1}{n_1-1} E_{n_1-1,\, n_1-1}\right)X^{*\prime} \\ &= X\left(I - \frac{1}{n_1} E_{n_1 n_1}\right)X' - \frac{n_1}{n_1-1} (\underline{x}_r - \underline{\bar{x}})(\underline{x}_r - \underline{\bar{x}})' \\ &= S_x - \frac{n_1}{n_1-1} \underline{U}_r \underline{U}_r' , \end{aligned} \tag{6.5}$$

where

$$\underline{U}_r = \underline{x}_r - \underline{\bar{x}} . \tag{6.6}$$

The pooled matrix of the corrected s.s. and s.p. of all the n_1+n_2-1 observations, excluding $\underline{x}_r$, is, therefore,

$$\begin{aligned} S^* &= S_x^* + S_y \\ &= S - \frac{n_1}{n_1-1} \underline{U}_r \underline{U}_r' \end{aligned} \tag{6.7}$$

and its inverse is, by (1.4.16),

$$S^{*-1} = S^{-1} + \frac{n_1 S^{-1} \underline{U}_r \underline{U}_r' S^{-1}}{n_1-1-n_1 \underline{U}_r' S^{-1} \underline{U}_r} . \tag{6.8}$$

Also observe that the vector of the means of the n_1-1 observations, excluding $\underline{x}_r$, is

$$\underline{\bar{x}}^* = \frac{1}{n_1-1} X^* E_{n_1-1,\, 1} = \frac{n_1}{n_1-1} \underline{\bar{x}} - \frac{1}{n_1-1} \underline{x}_r , \tag{6.9}$$

and so,

$$\underline{d}^* = \bar{\underline{x}}^* - \bar{\underline{y}} = \underline{d} - \frac{1}{n_1-1}\, \underline{U}_r \; . \tag{6.10}$$

Finally, therefore, Anderson's classification statistic based on the n_1+n_2-1 observations X, Y excluding $\underline{x}_r$ is, from (6.1),

$$W_r^*(\underline{x}) = (f-1)\; \underline{d}^{*\prime}\; S^{*^{-1}} \{\underline{x} - \tfrac{1}{2}(\bar{\underline{x}}^* + \bar{\underline{y}})\} \; . \tag{6.11}$$

Since $\underline{x}_r$ is not used in the construction of this statistic, it is independent of $\underline{x}_r$, and we can, therefore, legitimately use it to classify $\underline{x}_r$, which we already know to be from π_1. If we substitute $\underline{x}_r$ for $\underline{x}$ in (6.11), we get

$$W_r^*(\underline{x}_r) = \frac{(f-1)}{2}\left(\underline{d} + \frac{2n_1-1}{n_1-1}\,\underline{U}_r\right)'\left\{S^{-1} + \frac{n_1 S^{-1}\underline{U}_r\underline{U}_r' S^{-1}}{n_1-1-n_1\underline{U}_r' S^{-1}\underline{U}_r}\right\}\left(\underline{d} - \frac{\underline{U}_r}{n_1-1}\right). \tag{6.12}$$

According to our classification rule, $\underline{x}_r$ will be correctly classified as belonging to π_1 if $W_r^*(\underline{x}_r) \geq 0$ and will be misclassified as belonging to π_2 if $W_r^*(\underline{x}_r) < 0$. To estimate P_1, therefore, we calculate $W_r^*(\underline{x}_r)$ for each $r = 1, 2, \ldots, n_1$. Let m_1 of these $W_r^*(\underline{x}_r)$'s be < 0. Then m_1/n_1 is the proportion of misclassified observations out of the n_1 observations of π_1, and m_1/n_1 estimates P_1. To estimate P_2, a similar procedure is to be carried out on the n_2 columns of Y. Omitting each column of Y successively, we construct Anderson's statistic based on the remaining n_1+n_2-1 observations and substitute the deleted observation from Y in it, to classify it. A little algebra will show that, in this case, we need

$$W_r^*(\underline{y}_r) = \frac{(f-1)}{2}\left(-\underline{d} + \frac{2n_2-1}{n_2-1}\underline{\nu}_r\right)'\left\{S^{-1} + \frac{n_2 S^{-1}\underline{\nu}_r\underline{\nu}_r' S^{-1}}{n_2-1-n_2\underline{\nu}_r' S^{-1}\underline{\nu}_r}\right\}\left(\underline{d} + \frac{\underline{\nu}_r}{n_2-1}\right), \tag{6.13}$$

where

$$\underline{y}_r \text{ is the rth column of Y and } \underline{\nu}_r = \underline{y}_r - \bar{\underline{y}} . \tag{6.14}$$

$\underline{y}_r$ will be misclassified as belonging to π_1 if $W_r^*(\underline{y}_r) \geq 0$. If m_2 out of the n_2 observations Y are misclassified, m_2/n_2 will be an estimate of P_2.

This is the U method of estimation. It is similar to the "jacknife" technique (see Quenouille [20], Tukey [27], or Schucany et.al. [23]). Intuitively, it is felt that this method may not be sensitive to the assumption of normality.

(5) The $\bar{U}$ method: Lachenbruch and Mickey propose $\Phi(-\bar{D}_1/S_{D_1})$ as an estimate of P_1 and $\Phi(+\bar{D}_2/S_{D_2})$ for P_2, where

$$\bar{D}_1 = \frac{1}{n_1}\sum_{r=1}^{n_1} W_r^*(\underline{x}_r) , \tag{6.15}$$

$$\bar{D}_2 = \frac{1}{n_2}\sum_{r=1}^{n_2} W_r^*(\underline{y}_r) , \tag{6.16}$$

and

$$S_{D_1}^2 = \text{the sample variance of the } n_1 \text{ quantities } W_r^*(\underline{x}_r), \tag{6.17}$$

$$S_{D_2}^2 = \text{the sample variance of the } n_2 \text{ quantities } W_r^*(\underline{y}_r). \tag{6.18}$$

The reasoning behind this is as follows. Consider the variable

$$u = \{\underline{x} - \tfrac{1}{2}(\underline{\mu}(1) + \underline{\mu}(2))\}'\Sigma^{-1}\underline{\delta} . \tag{6.19}$$

It is easy to see that

$$E(u) = \frac{1}{2}\Delta_p^2, \quad \text{if } \underline{x} \in \pi_1$$
$$= -\frac{1}{2}\Delta_p^2, \quad \text{if } \underline{x} \in \pi_2,$$
$$V(u) = \Delta_p^2 . \tag{6.20}$$

Hence, the chance of misclassification when the true discriminant function is used, viz., $\Phi(-\frac{1}{2}\Delta_p)$, can be written as

$$\Phi\left(\frac{(-1)^k E(u)}{\sqrt{V(u)}}\right), \tag{6.21}$$

where k = 1 if $\underline{x}$ comes from π_1 and k = 2 if $\underline{x}$ comes from π_2. Now u itself is not known. Its estimate, when $\underline{\mu}(1)$, $\underline{\mu}(2)$, Σ are estimated from X and Y, excluding $\underline{x}_r$, and when $\underline{x}$ is replaced by $\underline{x}_r$, is $W_r^*(\underline{x}_r)$. $W_r^*(x_r)$ $(r = 1, 2, \ldots, n_1)$ are thus n_1 observations on $\underline{u}$ (with parameters replaced by estimates) and their (sample) mean $\bar{D}_1$ is thus expected to estimate E(u) and their (sample) variance $S_{D_1}^2$ is expected to estimate V(u) and so, from (6.21), $\Phi(-\bar{D}_1/S_{D_1})$ is expected to estimate the misclassification probability P_1. Similarly, $\Phi(D_2/S_{D_2})$ will estimate P_2. This method of estimation is known as the $\bar{U}$ method.

Lachenbruch and Mickey made a comparative evaluation of all these methods of estimation of P_1 and P_2 on the basis of a series of Monte Carlo experiments. They have come to the conclusion that the Resubstitution method and D method are relatively poor. The O method does fairly well overall, but is poorer than OS, U, and $\bar{U}$ methods. If approximate normality can be assumed, the OS and $\bar{U}$ methods are good. The U method does not take advantage of the assumption of normality. For small sample sizes, the OS method should not be used; the $\bar{U}$ or U method should be used. Cochran [11], who has commented on these investigations, has also reached the conclusion that the OS method ranks first, with the $\bar{U}$ and U methods not far behind.

While dealing with the O or OS methods, it should be borne in mind that Okamoto's [18] expansion (6.4) is for

$$\mathrm{Prob}(W(\underline{x}) < 0 \mid \underline{x} \in \pi_1) . \tag{6.22}$$

This is not P_1. P_1 is the conditional probability:

$$\mathrm{Prob}(W(\underline{x}) < 0 \mid \underline{x} \in \pi_1, \bar{\underline{x}}, \bar{\underline{y}}, S) . \tag{6.23}$$

Thus (6.22) is the expected value of P_1 with respect to $\bar{\underline{x}}$, $\bar{\underline{y}}$, S. Still the OS method gives a very good estimate of P_1. The smaller the value p, the better is the performance of the O or OS method.

7. PENROSE'S SIZE AND SHAPE FACTORS

If the common variance-covariance matrix Σ of two p-variate normal populations π_1 and π_2 has the particular form

$$\Sigma = \begin{bmatrix} 1 & \rho & \rho \cdots & \rho \\ \rho & 1 & \rho \cdots & \rho \\ \vdots & & & \\ \rho & \rho & \rho \cdots & 1 \end{bmatrix} = (1-\rho)I_p + \rho E_{pp} . \tag{7.1}$$

then the discriminant function $\underline{\delta}'\Sigma^{-1}\underline{x}$, where $\underline{\delta}$ is the difference in the vectors of the population means of π_1 and π_2, can be expressed as [on using (1.4.16)]

$$\begin{aligned}\underline{\delta}'\Sigma^{-1}\underline{x} &= \underline{\delta}'\{(1-\rho)I_p + \rho E_{pp}\}^{-1}\underline{x} \\ &= \underline{\delta}'\{(1-\rho)^{-1}I_p - \rho(1-\rho)^{-1}[1+(\rho-1)\rho]^{-1}E_{pp}\}\underline{x} \\ &= \frac{(E_{1p}\underline{\delta})}{p(1-\rho)}\left\{\left[\frac{p\underline{\delta}'}{(E_{1p}\underline{\delta})} - E_{1p}\right] + \left[1 - \frac{p\rho}{1+(p-1)\rho}\right]E_{1p}\right\}\underline{x} \\ &= \frac{(E_{1p}\underline{\delta})}{p(1-\rho)}\left\{\underline{h}'\underline{x} + \frac{1-\rho}{1+(p-1)\rho}E_{1p}\,\underline{x}\right\}, \qquad (7.2)\end{aligned}$$

where

$$\underline{h}'\underline{x} = \left[\frac{p\underline{\delta}'}{(E_{1p}\underline{\delta})} - E_{1p}\right]\underline{x}\,. \qquad (7.3)$$

The discriminant function thus depends on only two factors, $\underline{h}'\underline{x}$ and $E_{1p}\underline{x} = x_1 + x_2 + \ldots + x_p$. Penrose [19] calls $E_{1p}\underline{x}$ the "size" factor, as $x_1 + \ldots + x_p$ measures the "total size", and $\underline{h}'\underline{x}$ the "shape" factor. This terminology is derived from applications of discriminant analysis to biological organs, where Σ is generally of the form (7.1) and $E_{1p}\underline{x}$, $\underline{h}'\underline{x}$ measure the size and shape of an organ. Observe that the size and shape factors are uncorrelated and hence independently distributed. This follows from

$$\mathrm{Cov}(E_{1p}\underline{x}, \underline{h}'\underline{x}) = E_{1p}\,\Sigma\,\underline{h} = 0\,, \qquad (7.4)$$

on using (7.1). Further,

$$\sigma_1^2 = V(E_{1p}\underline{x}) = E_{1p}\,\Sigma\,E_{p1} = p(1 + p\rho - \rho)\,, \qquad (7.5)$$

$$\sigma_2^2 = V(\underline{h}'\underline{x}) = \underline{h}'\,\Sigma\,\underline{h} = p(1-\rho)\left[\frac{p\underline{\delta}'\underline{\delta}}{(E_{1p}\underline{\delta})^2} - 1\right], \qquad (7.6)$$

$$\begin{aligned}\delta_1^* &= \text{difference in the means of } E_{1p}\underline{x} \text{ in } \pi_1 \text{ and } \pi_2 \\ &= E_{1p}\underline{\delta} \qquad (7.7)\end{aligned}$$

$$\delta_2^* = \text{difference in the means of } \underline{h}'\underline{x} \text{ in } \pi_1 \text{ and } \pi_2$$
$$= \underline{h}'\underline{\delta}$$
$$= \frac{p\underline{\delta}'\underline{\delta}}{E_{1p}\underline{\delta}} - E_{1p}\underline{\delta} \tag{7.8}$$

and so, from (7.2), the discriminant function can also be expressed as

$$\underline{h}'\underline{x} + \frac{\delta_1^*}{\delta_2^*} \frac{\sigma_2^2}{\sigma_1^2} E_{1p}\underline{x} \, ,$$

or, by multiplying by a suitable constant, we can even take it to be

$$\frac{\delta_1^*}{\sigma_1^2} E_{1p}\underline{x} + \frac{\delta_2^*}{\sigma_2^2} \underline{h}'\underline{x} \, . \tag{7.9}$$

Discrimination by size and shape factors is thus quicker as it depends on only two factors, their variances and mean differences. However, it is exact only when Σ has the special pattern (7.1). Even if Σ does not have this special pattern, it can be approximated by $(1-\rho)I_p + \rho E_{pp}$, by first standardizing the variates to achieve unit variance for every x_i and then replacing each ρ_{ij} (correlation between x_i and x_j) by ρ, the average correlation among all the variables. The discriminant analysis carried out with this approximate variance-covariance matrix is of course not as efficient as with the true Σ but is certainly economical, and in many biological and anthropological applications, the ρ_{ij}'s do not differ greatly, so that the approximation works well and not much efficiency is lost.

8. DISCRIMINANT ANALYSIS IN THE CASE OF UNEQUAL VARIANCE-COVARIANCE MATRICES

So far, we considered discrimination and classification for two p-variate normal populations with the same variance-covariance matrix. However, if the variance-covariance matrices of the two normal populations π_1 and π_2 are Σ_1 and Σ_2, respectively, and $\Sigma_1 \neq \Sigma_2$, we can still apply the likelihood ratio method (1.7), which is valid for any two populations. But this method does not now lead to a linear discriminant function. It does not give an elegant solution in this case. When $p = 1$, the classification region R_1 of π_1 is an interval. When $p = 2$, the regions are defined by conic sections, and so on. Anderson and Bahadur [2], therefore, consider a "linear" procedure for this problem. They consider a linear discriminant function $\underline{b}'\underline{x}$, with $\underline{b}$ suitably chosen. An observation $\underline{x}$ is classified as from π_1 if $\underline{b}'\underline{x} \leq c$ and as from π_2 if $\underline{b}'\underline{x} > c$, where c is also to be suitably determined. It can be readily seen that the probabilities of misclassification with this procedure are

$$P_1 = \text{Prob}(\underline{b}'\underline{x} > c \mid \underline{x} \in \pi_1) = 1 - \Phi(y_1)$$

$$P_2 = \text{Prob}(\underline{b}'\underline{x} \leq c \mid \underline{x} \in \pi_2) = 1 - \Phi(y_2) , \tag{8.1}$$

where

$$y_1 = \frac{c - \underline{b}'\underline{\mu}(1)}{(\underline{b}'\Sigma_1\underline{b})^{1/2}} , \quad y_2 = \frac{\underline{b}'\underline{\mu}(2) - c}{(\underline{b}'\Sigma_2\underline{b})^{1/2}} , \tag{8.2}$$

and $\underline{\mu}(1)$ and $\underline{\mu}(2)$ are, as before, the means of π_1 and π_2. Observe that P_1, P_2 are small if y_1, y_2 are large. Also, for a given

y_2, we can express y_1 as [using (8.2) and eliminating c]

$$y_1 = \frac{-\underline{b}'\underline{\delta} - y_2(\underline{b}'\Sigma_2\underline{b})^{1/2}}{(\underline{b}'\Sigma_1\underline{b})^{1/2}}, \tag{8.3}$$

where $\underline{\delta} = \underline{\mu}(1) - \underline{\mu}(2)$ as before. Anderson and Bahadur then choose that $\underline{b}$ that maximizes y_1 for a given y_2. By differentiating y_1 with respect to $\underline{b}$, it has been shown that the solution consists of solving the following equations in $\underline{b}$ and a scalar t:

$$[t\,\Sigma_1 + (1-t)\Sigma_2]\underline{b} = -\underline{\delta} \tag{8.4}$$

$$y_2 = (1-t)\,(\underline{b}'\Sigma_2\,\underline{b})^{1/2}\ . \tag{8.5}$$

It is assumed that $y_2 > 0$, i.e., $P_2 < \frac{1}{2}$. The solution of these equations is to be obtained by trial and error. C is then obtained from (8.2) as

$$c = \underline{b}'\underline{\mu}(1) + t\underline{b}'\Sigma_1\underline{b} = \underline{b}'\underline{\mu}(2) - (1-t)\underline{b}'\Sigma_2\underline{b}\ .$$

By substituting (8.5) back into (8.3) and using (8.4), we get

$$y_1 = t(\underline{b}'\Sigma_1\underline{b})^{1/2}\ . \tag{8.6}$$

An alternative method will be to choose that $\underline{b}$ for which the two misclassification probabilities are equal, i.e., $y_1 = y_2$. From (8.5) and (8.6), we get, in that case,

$$0 = y_1^2 - y_2^2 = t^2\ \underline{b}'\Sigma_1\underline{b} - (1-t)^2\ \underline{b}'\Sigma_2\underline{b}$$

$$= \underline{b}'[t^2\Sigma_1 - (1-t)^2\ \Sigma_2]\underline{b}\ . \tag{8.7}$$

Anderson and Bahadur have proved the admissibility of these procedures, and hence this latter discrimination procedure, where $y_1 = y_2$ is chosen, is the minimax procedure. It consists in finding that $\underline{b}$ and t for which (8.4) and (8.7) are satisfied. This is done by trial and error.

It is a well known result in matrix algebra that Σ_1 and Σ_2 can be expressed as

$$\Sigma_1 = N'\Lambda N, \quad \Sigma_2 = N'N, \tag{8.8}$$

where

$$\Lambda = \text{diag}(\lambda_1, \lambda_2, \ldots, \lambda_p), \tag{8.9}$$

and $\lambda_1, \lambda_2, \ldots, \lambda_p$ are the roots of the determinantal equation

$$|\Sigma_1 - \lambda\Sigma_2| = 0 . \tag{8.10}$$

We therefore set

$$\underline{b}^* = N\underline{b} , \tag{8.11}$$

and then (8.7) reduces to

$$\sum_{i=1}^{p} (\lambda_i - \nu)b_i^{*2} = 0 , \tag{8.12}$$

where b_i^* are the elements of $\underline{b}^*$ and $\nu = (1-t)^2/t^2$. If $\lambda_i - \nu$ are all positive or are all negative, (8.12) will not have a solution for $\underline{b}^*$ and hence for $\underline{b}$. Therefore, ν must lie between the minimum and maximum of the roots $\lambda_1, \lambda_2, \ldots, \lambda_p$. This will help in the trial and error procedure for determining t and $\underline{b}$ of the minimax method. This result is due to Banerjee and Marcus [3].

9. DISCRIMINATION IN THE CASE OF ZERO MEAN DIFFERENCES

We now consider discrimination between two p-variate normal populations, with the same mean, which we shall assume to be $\underline{0}$ but unequal variance-covariance matrices Σ_1 and Σ_2. If, in particular, $\Sigma_1 = \sigma_1^2 I_p$ and $\Sigma_2 = \sigma_2^2 I_p$, the logarithm of the likelihood ratio (1.7) reduces to

$$-\frac{1}{2}\left(\frac{1}{\sigma_1^2} - \frac{1}{\sigma_2^2}\right) \underline{x}'\underline{x} + \text{a constant} \tag{9.1}$$

and the discriminant function is the quadratic function

$$z = \underline{x}'\underline{x} \; . \tag{9.2}$$

The regions R_1 and R_2 of Section 1 will then be the inside and outside of a hypersphere in p dimensions.

We now consider another particular case, where

$$\Sigma_1 = (1 - \rho_1)I_p + \rho_1 E_{pp} \tag{9.3}$$

and

$$\Sigma_2 = \sigma^2[(1-\rho_2)I_p + \rho_2 E_{pp}] \, . \tag{9.4}$$

In actual practice, Σ_1 and Σ_2 may not have this form exactly, but as we remarked in the case of Penrose's size and shape factors, Σ_1 and Σ_2 can be approximated at least by (9.3) and (9.4), by standardizing the variables and replacing the true correlations by average correlations ρ_1 and ρ_2 in π_1 and π_2, respectively. The variances of all the x's in π_1 are reduced to 1 by standardization, but at the same time they cannot be reduced to 1 in π_2 also. Hence, σ^2 is introduced in (9.4) as the common standardized variance in π_2. Using (1.4.16) to invert Σ_1 and Σ_2, it is only a matter of algebra to show that the likelihood ratio method of Section 1 leads to

$$aZ_1 - bZ_2 = C \tag{9.5}$$

as the boundary separating the two regions R_1 and R_2 of classification, where

$$Z_1 = \underline{x}'\underline{x}\,, \tag{9.6}$$

$$Z_2 = (E_{1p}\underline{x})^2 = \text{square of the size factor}, \tag{9.7}$$

$$a = \frac{1}{1-\rho_1} - \frac{1}{\sigma^2(1-\rho_2)}\,, \tag{9.8}$$

$$b = \frac{\rho_1}{1-\rho_1}\,\frac{1}{1+(p-1)\rho_1} - \frac{\rho_2}{1-\rho_2}\,\frac{1}{\sigma^2\{1+(p-1)\rho_2\}}\,. \tag{9.9}$$

The constant C in (9.5) is, in general, so chosen that the two components of the chance of misclassification are equal.

In the case of the further assumption $\rho_1 = \rho_2 = \rho$, say, (9.5) reduces to

$$z_1 - \frac{\rho}{1+(p-1)\rho}\, z_2 = c'\,. \tag{9.10}$$

By making a transformation from $\underline{x}$ to $\underline{y}$, i.e.,

$$[y_1, y_2, \ldots, y_p]' = H\underline{x}\,, \tag{9.11}$$

where H is any orthogonal matrix, with its last row as $\frac{1}{\sqrt{p}} E_{1p}$, it can be shown that $y_1, y_2, \ldots, y_p$ are all normal independent variables with zero means and variances given by

$$\begin{aligned} V(y_i) &= (1-\rho), && \text{if } \underline{x} \text{ comes from } \pi_1, \\ &= \sigma^2(1-\rho), && \text{if } \underline{x} \text{ comes from } \pi_2 \quad i = 1,2,\ldots,p-1 \end{aligned} \tag{9.12}$$

and

$$\begin{aligned} V(y_p) &= 1 + (p-1)\rho, && \text{if } \underline{x} \text{ comes from } \pi_1, \\ &= \sigma^2[1+(p-1)\rho], && \text{if } \underline{x} \text{ comes from } \pi_2. \end{aligned} \tag{9.13}$$

Therefore,

$$\frac{y_p^2}{\sigma_i^2[1 + (p-1)\rho]} = \frac{z_2}{p\sigma_i^2[1 + (p-1)\rho]} \tag{9.14}$$

is a χ^2 with 1 d.f., σ_i^2 being 1 if $\underline{x}$ comes from π_1 and σ^2 if $\underline{x}$

comes from π_2. Further, this is independent of

$$\frac{y_1^2 + \dots + y_{p-1}^2}{\sigma_i^2(1-\rho)} = \frac{\underline{x}'\underline{x} - (E_{1p}\underline{x})^2/p}{\sigma_i^2(1-\rho)}$$

$$= \frac{z_1 - z_2/p}{\sigma_i^2(1-\rho)}, \tag{9.15}$$

which is a χ^2 with (p-1) d.f. From these two results, it follows that the discriminant function (9.10), viz.,

$$u = z_1 - \frac{\rho}{(1 + (p-1)\rho} z_2 \tag{9.16}$$

has a $\sigma_i^2(1-\rho) \cdot \chi^2$ distribution with p d.f. If our classification procedure is to assign $\underline{x}$ to π_1 when $u \geq c'$ and to π_2 when $u < c'$, the two components of the chance of misclassification are [see (1.4)]

$$\alpha_1 = \int_{u<c'} f_1(\underline{x})d\underline{x}, \quad \alpha_2 = \int_{u\geq c'} f_2(\underline{x})d\underline{x}. \tag{9.17}$$

From (9.16), it follows that

$$\alpha_1 = \int_0^{c'/(1-\rho)} \chi_p^2(v)dv, \tag{9.18}$$

$$\alpha_2 = \int_{c'/\sigma^2(1-\rho)}^{\infty} \chi_p^2(v)dv, \tag{9.19}$$

where, $\chi_p^2(v)$ is the p.d.f. of a χ^2 distribution with p d.f. The constant c' of (9.10) is thus to be so determined as to have (9.18) and (9.19) equal.

Bartlett and Please [4] suggest that the boundary (9.10) [or (9.5) in the general case] is probably better fitted by eye when only samples from the two populations π_1 and π_2 are available and the parameters are unknown. Bartlett and Please [4] give an interesting and natural example of discriminating between monozygotic and dizygotic pairs of twins (with like sex), using the procedure described in this section. The discriminant function is quadratic in $\underline{x}$ but is linear in z_1 and z_2.

10. DISCRIMINATION IN THE CASE OF QUALITATIVE DATA

So far, we considered discrimination and classification procedures when quantitative measurements on one or more variables were available. We now consider two populations π_1 and π_2 consisting of m categories, such as eye or hair colors or political parties. What is usually available is only the category to which an individual belongs. The problem is then to classify this individual as belonging to π_1 or π_2 on the basis of this qualitative information. Let p_i $(i = 1, 2, \ldots, m)$ be the probability that an individual from π_1 falls in the ith category; $\sum_1^m p_i = 1$. Let p_i' $(i = 1, 2, \ldots, m)$ be the corresponding probabilities for π_2. The likelihood ratio is then p_i/p_i' $(i = 1, 2, \ldots, m)$. The classification procedure based on the likelihood ratio principle of Section 1 is to assign an individual falling in the rth category to π_1 if $p_r \geq p_r'$, and to π_2 otherwise. The two components of the chance misclassification, with this procedure, are

$$\alpha_1 = \sum_{i=1}^{m} \xi_i p_i , \quad \alpha_2 = \sum_{i=1}^{m} \xi_i p_i' , \tag{10.1}$$

where

$$\begin{aligned} \xi_i &= 0 \quad \text{if} \quad p_i \geq p_i' \\ &= 1 \quad \text{if} \quad p_i < p_i' \ (i = 1, 2, \ldots, m). \end{aligned} \tag{10.2}$$

In practice, the p_i, p_i' are rarely known and are estimated by r_i/n, r_i'/n', respectively, where n, n' are the sizes of the samples from π_1 and π_2 and r_i, r_i' are, respectively, the number of individuals out of these samples falling in the ith category.

Cochran and Hopkins [9] discuss an example of classifying a voter in a Presidential election on the basis of his political faith and inclination toward the personality of the Presidential candidate. They have also considered the loss of efficiency when a continuous variable x with a normal distribution is replaced by a qualitative one. Let us assume, for the sake of simplicity, that the variance of x is unity and its mean is δ if the individual belongs to the population π_1 and is 0 if the individual belongs to π_2. The range $-\infty$ to ∞ of x is then subdivided into six parts:

(i) Category A: $-\infty$ to $(\delta/2) - u_2$

(ii) Category a: $(\delta/2) - u_2$ to $(\delta/2) - u_1$

(iii) Category α: $(\delta/2) - u_1$ to $(\delta/2)$

(iv) Category β: $(\delta/2)$ to $(\delta/2) + u_1$

(v) Category b: $(\delta/2) + u_1$ to $(\delta/2) + u_2$

(vi) Category B: $(\delta/2) + u_2$ to ∞ .

u_1 and u_2 are arbitrary. Let $p_1, p_2, \ldots, p_6$ be the probabilities that an individual from π_1 will fall in these six categories A, a, α, β, b, B, respectively. Let $p_1', \ldots, p_6'$ be the corresponding probabilities for an individual from π_2. It is easy to see that

$$p_1 = \int_{-\infty}^{\delta/2 - u_2} \frac{1}{\sqrt{2\pi}} e^{-(x-\delta)^2/2} dx$$

$$= \int_{-\infty}^{-(\delta/2)-u_2} \frac{1}{\sqrt{2\pi}} e^{-z^2/2} dz \tag{10.3}$$

and

$$p_6' = \int_{(\delta/2)+u_2}^{\infty} \frac{1}{\sqrt{2\pi}} e^{-x^2/2} dx . \tag{10.4}$$

Thus $p_1 = p_6'$. Similarly

$$p_1' = p_6, \quad p_2' = p_5, \quad p_3' = p_4, \quad p_4' = p_3, \; p_5' = p_2 . \tag{10.5}$$

Define a variable y, which takes the values

$$w_i = \log p_i - \log p_i' \quad (i = 1, 2, \ldots, 6) . \tag{10.6}$$

Then, if x comes from π_1,

$$\begin{aligned} E(y|x \in \pi_1) &= w_1 p_1 + \ldots + w_6 p_6 \\ &= w_1(p_1 - p_6) + w_2(p_2 - p_5) + w_3(p_3 - p_4) , \end{aligned} \tag{10.7}$$

because of (10.5). However, if x comes from π_2,

$$\begin{aligned} E(y|x \in \pi_2) &= w_1 p_1' + \ldots + w_6 p_6' \\ &= - E(y|x \in \pi_1) , \end{aligned} \tag{10.8}$$

again due to (10.5). In either case,

$$V(y) = w_1^2(p_1 + p_6) + w_2^2(p_2 + p_5) + w_3^2(p_3 + p_4) - \{E(y)\}^2 . \tag{10.9}$$

Let

$$g(t) = \frac{1}{\sqrt{2\pi}} e^{-t^2/2} , \quad I(t) = \int_u^\infty g(t)dt , \tag{10.10}$$

$$I_1 = I(u_1), \quad I_2 = I(u_2), \quad g_1 = g(u_1), \quad g_2 = g(u_2),$$
$$g_o = g(o) . \tag{10.11}$$

Then, for sufficiently large δ,

$$\begin{aligned} p_1' = p_6 &= \int_{-\infty}^{(\delta/2)-u_2} g(x)dx \\ &= I(u_2 - \delta/2) \\ &= \text{approximately } I(u_2) - \frac{\delta}{2}\frac{d}{du_2} I(u_2) \\ &\doteqdot I_2 + \frac{\delta}{2} g_2 . \end{aligned} \tag{10.12}$$

Similarly

$$p_2' = p_5 \doteqdot I_1 - I_2 + \frac{\delta}{2}(g_1 - g_2) \tag{10.13}$$

$$p_3' = p_4 \doteqdot \frac{1}{2} - I_1 + \frac{\delta}{2}(g_o - g_1) \tag{10.14}$$

$$p_4' = p_3 \doteqdot \frac{1}{2} - I_1 - \frac{\delta}{2}(g_o - g_1) \tag{10.15}$$

$$p_5' = p_2 = I_1 - I_2 - \frac{\delta}{2}(g_1 - g_2) \tag{10.16}$$

$$p_6' = p_1 = I_2 - \frac{\delta}{2} g_2 . \tag{10.17}$$

Also,

$$w_1 \doteqdot \log \frac{I_2 - \frac{\delta}{2} g_2}{I_2 + \frac{\delta}{2} g_2} \doteqdot - \frac{\delta g_2}{I_2} = - w_6 , \tag{10.18}$$

$$w_2 \doteqdot - \frac{\delta(g_1 - g_2)}{I_1 - I_2} = - w_4 , \tag{10.19}$$

and

$$w_3 \doteq \frac{\delta(g_o - g_1)}{\frac{1}{2} - I_1} = - w_5 . \tag{10.20}$$

Hence, from (10.7) and (10.8),

$$E(y|x \in \pi_1) \doteq \delta^2\left\{\frac{g_2^2}{I_2} + \frac{(g_1 - g_2)^2}{I_1 - I_2} + \frac{(g_o - g_1)^2}{\frac{1}{2} - I_1}\right\} \tag{10.21}$$

and, from (10.9),

$$V(y) \doteq 2E(y|x_1 \in \pi_1) = 2\ \delta^2\ \psi(u_1, u_2), \tag{10.22}$$

where

$$\psi(u_1, u_2) = \frac{g_2^2}{I_2} + \frac{(g_1 - g_2)^2}{I_1 - I_2} + \frac{(g_o - g_1)^2}{\frac{1}{2} - I_1} . \tag{10.23}$$

If we have only one variable x and are classifying only on the basis of the category in which an individual falls, the classification procedure is, as seen earlier in this section, to assign the individual to π_1 or π_2, according as $p_i \geq p_i'$ or not, if he falls in the ith category. From (10.5), this is the same as classifying on the basis of y and assigning to π_1 if w_i, the value of y is ≥ 0 and to π_2, otherwise. However, if we have a large number of independent normal variables $x_1, x_2, \ldots, x_k$ and for each of them, 6 categories are defined as for a single x above, there will be, in all, 6^k categories. As the variables are assumed to be independent, the probability that an individual falls in the i_1^{th} category of x_1, i_2^{th} of x_2, ..., i_k^{th} of x_k is the product of the k individual probabilities. Then, by taking the logarithm of the likelihood ratio, it is easy to see that the discriminant function will be

$$R = \sum_{r=1}^{k} y_r , \tag{10.24}$$

where y_r is defined corresponding to x_r $(r = 1, 2, \ldots, k)$, in exactly the same way as y was defined for x in (10.5). Therefore, from (10.21),

$$E(R|\pi_1) \doteq \sum_{r=1}^{k} \delta_r^2 \, \psi(u_{1r}, u_{2r}) , \tag{10.25}$$

where δ_r, u_{1r}, u_{2r} are defined in the same way for x_r as δ, u_1, u_2 were defined for x $(r = 1, 2, \ldots, k)$. Also, from (10.22),

$$V(R) = 2 \sum_{r=1}^{k} \delta_r^2 \, \psi(u_{1r}, u_{2r}) . \tag{10.26}$$

When k is large, R is the sum of a large number of independent random variables, and hence by the central limit theorem, under certain conditions its distribution can be approximated by the normal distribution. The rule of classification is

Assign the individual under consideration to

π_1 if $R \geq 0$ and to π_2 otherwise. (10.27)

The probability of misclassifying an individual from π_1 is, therefore,

$$\text{Prob}(R < 0|\pi_1) = \text{Prob}\left(\frac{R - E(R|\pi_1)}{\sqrt{V(R)}} < 0|\pi_1\right) \tag{10.28}$$

and is approximately (for large k, assuming a normal distribution for R)

$$I\left(\left\{\frac{1}{2} \sum_{r=1}^{k} \delta_r^2 \, \psi(u_{1r}, u_{2r})\right\}^{1/2}\right), \tag{10.29}$$

where I(t) is defined in (10.10). This chance of misclassification is minimized if $\sum_{r=1}^{k} \delta_r^2 \, \psi(u_{1r}, u_{2r})$ is maximized with respect to the arbitrary quantities u_{1r}, u_{2r}. But, as $\psi(u_{1r}, u_{2r})$ does not involve δ_r, the values of u_{1r}, u_{2r} that maximize the expression are the same for each r. If they are denoted by u_1^*, u_2^*, the maximum value of $\sum_{r=1}^{k} \delta_r^2 \psi(u_{1r}, u_{2r})$ will be $\psi(u_1^*, u_2^*)\Delta_k^2$, where

$$\Delta_k^2 = \sum_{r=1}^{k} \delta_r^2 = k\,\delta_o^2 \text{ , say,} \tag{10.30}$$

is the square of the distance (Mahalanobis) between π_1 and π_2 based on the <u>independent</u> variables $x_1, x_2, \ldots, x_k$, all of which have unit variances. The chance of misclassification is, therefore, from (10.29),

$$I\left(\{\psi\,(u_1^*, u_2^*)\Delta_k^2/2\}^{1/2}\right). \tag{10.31}$$

This chance of misclassification, associated with the procedure (10.27) can now be compared with the corresponding chance of misclassification [see (1.22)]

$$\Phi(-\tfrac{1}{2}\,\Delta_k) = I(\tfrac{1}{2}\,\Delta_k), \tag{10.32}$$

when the standard discriminant function procedure (1.22) based on $x_1, \ldots, x_k$ is used. We thus see that the effect of replacing each continuous normal variable x_r by a qualitative variable, by defining six categories and classifying an individual on the basis of the categories in which he falls, is to increase the chance of misclassification from (10.32) to (10.31). In other words, the procedure (10.27) effectively reduces the distance from Δ_k to Δ_k', where

$$\Delta_k'^2 = 2\,\psi(u_1^*, u_2^*)\Delta_k^2 , \tag{10.33}$$

because the chance of misclassification (10.31) is $I(\frac{1}{2}\Delta_k')$ and that, in the standard procedure, is $I(\frac{1}{2}\Delta_k)$. Cochran and Hopkins [9] therefore call

$$\frac{\Delta_k'^2}{\Delta_k^2} = 2\,\psi(u_1^*, u_2^*) \tag{10.34}$$

the relative discriminating power of the procedure (10.27). Another way of looking at this is as follows. If we have k identical normal independent variables, each with unit variance and distance δ_o [defined by (10.30)], the total distance is $\Delta_k = (k\,\delta_o^2)^{1/2}$, and the chance of misclassification is, by (10.32), $I(\frac{1}{2}\Delta_k)$. If, however, we use only a fraction f k of these variables, the distance will be reduced from Δ_k to $\Delta_k' = (fk\delta_o^2)^{1/2} = (f)^{1/2}\,\Delta_k$ and the chance of misclassification will be increased from $I(\frac{1}{2}\Delta_k)$ to $I(\frac{1}{2}\Delta_k')$. From (10.34), therefore, it follows that the procedure (10.27) is thus equivalent to choosing only a fraction fk of these variables, where f is given by

$$f = 2\,\psi(u_1^*, u_2^*)\,. \tag{10.35}$$

Cochran and Hopkins [9] have found that $u_1^* = 0.7$, $u_2^* = 1.4$, and $f = 0.942$. If, instead of six categories, one wishes to consider only 5, 4, or a still smaller number, this can be studied by suitably combining the classes A, a, α, β, b, B and redefining u_1, u_2. Cochran and Hopkins give values of u_1, u_2, and f for such cases also.

11. CONCLUDING REMARKS

In this chapter, we did not consider classification procedures when there are several populations. This will be done in Chapter 9. We also did not consider classification procedures when (i) the prior probabilities q_1, q_2 of an individual to belong to π_1 and π_2 are known and (ii) the costs c_1, c_2 of misclassifying an individual from π_1, π_2 are known. The regions of classification R_1 and R_2 of Section 1, derived from the likelihood ratio $f_1(\underline{x})/f_2(\underline{x})$ [of (1.7)], are then modified:

$$R_1: \quad \frac{f_1(\underline{x})c_1q_1}{f_2(\underline{x})c_2q_2} \geq \text{a suitable constant } h,$$

$$R_2: \quad \frac{f_1(\underline{x})c_1q_1}{f_2(\underline{x})c_2q_2} < h .$$

Usually, in practice, the parameters of $f_1(\underline{x})$ and $f_2(\underline{x})$ are estimated from samples from these populations. The effect of this on the discriminant function will depend not only on the accuracy of the estimates but also on whether all the sample observations assumed to be from π_1 (or from π_2) were really so. Some work in this area, about the effect of misclassification in the initial samples themselves, has been done by Lachenbruch [17].

Kendall [15] has suggested a nonparametric discriminating procedure, which he illustrates with the classical data on two species of flowers, *Iris Setosa* and *Iris Versicolor*, analyzed by Fisher [13]. Much remains to be done, to this data at least, about the properties of such a procedure.

Chien-Pai Han [7] extends and combines the analysis of Penrose [19] and Bartlett and Please [4], when the two normal populations π_1 and π_2 with variance-covariance matrices Σ_1 and Σ_2 given by (9.3) and (9.4) have different means also.

REFERENCES

[1] Anderson, T. W. (1958). Introduction to Multivariate Statistical Analysis. John Wiley and Sons, New York.

[2] Anderson, T. W., and Bahadur, R. R. (1962). "Classification into two multivariate normal distributions with different covariance matrices", Ann. Math. Statist., 33, p. 420.

[3] Banerjee, K. S., and Marcus, L. F. (1965). "Bounds in a minimax classification procedure", Biometrika, 52, p. 653.

[4] Bartlett, M. S., and Please, N. W. (1963). "Discrimination in the case of zero mean differences", Biometrika, 50, p. 17.

[5] Bowker, A. H., and Sitgreaves, R. (1959). "An asymptotic expansion for the distribution function of the classification statistic W", Technical Report No. 53, Applied Math. and Stat. Labs., Stanford University.

[6] Bowker, A. H. (1960). "A representation of Hotelling's T^2 and Anderson's classification statistic W in terms of simple statistics", No. 12 in Contributions to Probability and Statistics, Essays in honor of H. Hotelling, Stanford University Press.

[7] Chien-Pai Han (1968). "A note on discrimination in the case of unequal covariance matrices", Biometrika, 55, p. 586.

[8] Cochran, W. G., and Bliss, C. I. (1948). "Discriminant functions with covariance", Ann. Math. Statist., 19, p. 151.

[9] Cochran, W. G., and Hopkins, C. E. (1961). "Some classification problems with multivariate qualitative data", Biometrics, 17, p. 10.

[10] Cochran, W. G. (1964). "Comparison of two methods of handling covariates in discriminatory analysis", Ann. Inst. Stat. Math., 16, p. 43.

[11] Cochran, W. G. (1968). "Commentary on estimation of error rates in discriminant analysis", Technometrics, 10, p. 204.

[12] Das Gupta, S. (1968). "Some aspects of discrimination function coefficients", Sankhyā, A, 30, p. 387.

[13] Fisher, R. A. (1936). "The use of multiple measurements in taxonomic problems", Ann. Eugen., 7, p. 179.

[14] Hodges, Joseph L. (1950). "Discriminatory Analysis", (survey of discriminatory analysis), Project No. 21-49-004, Report No. 1, USAF School of Aviation Medicine, Randolph Field, Texas.

[15] Kendall, M. G. (1966). "Discrimination and classification", Multivariate Analysis, (edited by Krishnaiah, P. R.), Academic Press, New York.

[16] Lachenbruch, P. A. (1966). "Discriminant analysis when the initial samples are misclassified", Technometrics, 8, p. 657.

[17] Lachenbruch, P. A., and Mickey, M. R. (1968). "Estimation of error rates in discriminant analysis", Technometrics, 10, p. 1.

[18] Okamoto, M. (1963). "An asymptotic expansion for the distribution of the linear discriminant function", Ann. Math. Statist., 34, p. 1286.

[19] Penrose, L. S. (1947). "Some notes on discrimination", Ann. Eugen., 13, p. 228.

[20] Quenouille, M. (1956). "Notes on bias in estimation", Biometrika, 43, p. 353.

[21] Rao, C. Radhakrishna (1965). Linear Statistical Inference and Its Applications. John Wiley and Sons, New York.

[22] Rao, C. Radhakrishna (1966). "Covariance adjustment and related problems in multivariate analysis", p. 87. Multivariate Analysis I, (edited by Krishnaiah, P. R.), Academic Press, New York.

[23] Schucany, W. R., Gray, H. L., and Owen, D. B. (1971). "On bias reduction in estimation", J. Am. Stat. Assoc., 66, p. 524.

[24] Sitgreaves, R. (1958). "Some results on the distribution of W-classification statistics", USAF School of Aviation Medicine, Report 58-3.

[25] Smith, C. A. B. (1947). "Some examples of discrimination", Ann. Eugen., 13, p. 272.

[26] Subrahmaniam, Kathleen, and Subrahmaniam, Kocherlakota, (1971). "On the distribution of $(D^2_{p+q} - D^2_q)$ statistic: percentage points and the power of the test", University of Manitoba (Canada), Department of Statistics, Technical Report No. 7.

[27] Tukey, J. W. (1958). "Bias and confidence in not quite large samples", (Abstract), Ann. Math. Statist., 20, p. 614.

CHAPTER 7

CANONICAL VARIABLES AND CANONICAL CORRELATIONS

1. SINGULAR DECOMPOSITION OF A MATRIX [11]

We shall first prove the following theorem:

Theorem 1: If A is a $p \times q$ matrix ($p \leq q$), of rank k, it can be expressed as

$$A = L \Delta M' ,$$

where L is a $p \times p$ orthogonal matrix, M is a $q \times q$ orthogonal matrix, and Δ is the $p \times q$ matrix

$$[\text{diag}(\delta_1, \delta_2, \ldots, \delta_k, 0, \ldots, 0 | \underset{p \times (q-p)}{0}] .$$

$\delta_1^2, \delta_2^2, \ldots, \delta_k^2$ being the nonzero eigenvalues of $\mathbf{AA'}$ or $\mathbf{A'A}$.

Proof: Since $\delta_1^2, \ldots, \delta_k^2$ are the nonzero eigenvalues of $\mathbf{AA'}$, there exists an orthogonal matrix L of order p, such that

$$\mathbf{AA'} = L \,\text{diag}(\delta_1^2, \ldots, \delta_k^2, 0, \ldots, 0)L'. \tag{1.1}$$

Let

$$L = [\underline{\ell}_1 | \underline{\ell}_2 | \ldots | \underline{\ell}_p] = [\underset{k}{L_1} | \underset{p-k}{L_2}]p . \tag{1.2}$$

Then, from (1.1), we have

$$\mathbf{AA'}L_1 = L_1 \Delta_1^2 , \tag{1.3}$$

where

$$\Delta_1 = \text{diag}(\delta_1, \delta_2, \ldots, \delta_k). \tag{1.4}$$

From (1.3), premultiplying by A' and postmultiplying by Δ_1^{-1}, we get

$$A'AM_1 = M_1\Delta_1^2, \tag{1.5}$$

where

$$M_1 = A'L_1\Delta_1^{-1}. \tag{1.6}$$

Observe that M_1 is of order $q \times k$ and satisfies

$$M_1'M_1 = \Delta_1^{-1}L_1'AA'L_1\Delta_1^{-1} = I_k, \tag{1.7}$$

because of (1.3). Since A has q columns and only k of them are linearly independent, we can always find a $q \times (q-k)$ matrix M_2 such that

$$A\,M_2 = 0, \tag{1.8}$$

and

$$M_2'M_2 = I_{q-k}. \tag{1.9}$$

Observe that

$$M_1'M_2 = \Delta_1^{-1}\,L_1'\,A\,M_2 = 0 \tag{1.10}$$

[because of (1.8)], and so

$$M = [M_1 | M_2] \tag{1.11}$$

is a $q \times q$ orthogonal matrix. This follows from (1.7), (1.9), and (1.10).

Finally, from (1.3), by postmultiplying by Δ_1^{-1} and using (1.6), we have

$$A\,M_1 = L_1\Delta_1.$$

Postmultiplying both sides by M_1', noting that $M_1M_1' = I - M_2M_2'$, and using $AM_2 = 0$, we get

$$A = L_1 \Delta_1 M_1' \ . \tag{1.12}$$

This can also be written as

$$A = L\,\Delta\,M' \ , \tag{1.13}$$

where

$$\Delta = [\mathrm{diag}(\delta_1, \ldots, \delta_k, 0, \ldots, 0) \mid \underset{p \times (q-p)}{0}\ ] . \tag{1.14}$$

This proves the theorem. If

$$M = [\underline{m}_1 | \underline{m}_2 | \ \cdots \ | \underline{m}_q] \ , \tag{1.15}$$

(1.12) or (1.13) are equivalent to

$$A = \delta_1\, \underline{\ell}_1 \underline{m}_1' + \delta_2\, \underline{\ell}_2 \underline{m}_2' + \cdots + \delta_k\, \underline{\ell}_k \underline{m}_k' \ . \tag{1.16}$$

From (1.13), it also follows that $AA' = L\Delta\Delta'L'$ and $A'A = M\Delta'\Delta M'$; as L and M are orthogonal, this also shows that the columns of L are eigenvectors of AA' and those of M are eigenvectors of $A'A$. The first k columns correspond to $\delta_1^2, \ldots, \delta_k^2$, the remaining to the zero eigenvalues.

2. CANONICAL VARIABLES AND CANONICAL CORRELATIONS

Let $\underline{x}$ and $\underline{y}$ be two column vectors of p and q random variables, respectively, with $p \leq q$. Let the variance-covariance matrix of all these $p+q$ random variables be

$$V\begin{bmatrix} \underline{x} \\ \hline \underline{y} \end{bmatrix} = \underset{\quad p \qquad\quad q}{\begin{bmatrix} \Sigma_{11} & \Sigma_{12} \\ \Sigma_{21} & \Sigma_{22} \end{bmatrix}}\begin{matrix} p \\ q \end{matrix} \ . \tag{2.1}$$

We now prove the following theorem:

Theorem 2: There exists a transformation

$$\underline{u} = L\,\underline{x} \ , \quad \underline{v} = M\,\underline{y} \ ,$$

from $\underline{x}$, $\underline{y}$ to $\underline{u}$, $\underline{v}$, such that

$$V\left[\begin{array}{c} \underline{u} \\ \hline \underline{v} \end{array}\right] = \left[\begin{array}{c|c} I_p & P \\ \hline P' & I_q \end{array}\right] ,$$

where

$$P = [\operatorname{diag}(\rho_1, \rho_2, \ldots, \rho_r, 0, \ldots, 0) \mid \underset{q-p}{0}] ,$$

r being the rank of Σ_{12}. Σ_{11} and Σ_{22} are nonsingular.

Proof: Define the $p \times q$ matrix C as

$$C = \Sigma_{11}^{-1/2} \Sigma_{12} \Sigma_{22}^{-1/2} , \tag{2.2}$$

where $\Sigma_{11}^{1/2}$, $\Sigma_{22}^{1/2}$ are matrix square roots of Σ_{11} and Σ_{22}, i.e., $(\Sigma_{11}^{1/2})^2 = \Sigma_{11}$, $(\Sigma_{22}^{1/2})^2 = \Sigma_{22}$. As Σ_{11}, Σ_{22} are positive definite, it has been shown that $\Sigma_{11}^{1/2}$, $\Sigma_{22}^{1/2}$ exist and are symmetric and positive definite. Since the rank of a matrix is unaltered by pre- or postmultiplication by a nonsingular matrix, the rank of C is the same as that of Σ_{12}, namely, r. Hence, by Theorem 1 of this Section (the Singular Decomposition Theorem), there exist two orthogonal matrices A and B or order p and q, respectively, such that

$$C = A P B' , \tag{2.3}$$

where P is already defined in the statement of the theorem and $\rho_1^2, \rho_2^2, \ldots, \rho_r^2$ are the nonzero eigenvalues of

$$CC' = \Sigma_{11}^{-1/2} \Sigma_{12} \Sigma_{22}^{-1} \Sigma_{21} \Sigma_{11}^{-1/2} \tag{2.4}$$

or

$$C'C = \Sigma_{22}^{-1/2} \Sigma_{21} \Sigma_{11}^{-1} \Sigma_{12} \Sigma_{22}^{1/2} , \tag{2.5}$$

i.e., ρ_i^2 $(i = 1, 2, \ldots, r)$ are the nonzero roots of the determinantal equation $|CC' - \rho^2 I| = 0$ or $|C'C - \rho^2 I| = 0$, in

ρ^2. After a little manipulation, these equations can be written in the form

$$|-\rho^2 \Sigma_{11} + \Sigma_{12} \Sigma_{22}^{-1} \Sigma_{21}| = 0 \tag{2.6}$$

or

$$|-\rho^2 \Sigma_{22} + \Sigma_{21} \Sigma_{11}^{-1} \Sigma_{12}| = 0 . \tag{2.7}$$

Define

$$L = A' \Sigma_{11}^{-1/2} , \quad M = B' \Sigma_{22}^{-1/2} , \tag{2.8}$$

which is the same as

$$A = \Sigma_{11}^{1/2} L' , \quad B = \Sigma_{22}^{1/2} M' . \tag{2.9}$$

Since A and B are orthogonal, $A'A = I$ and $B'B = I$. Therefore, we get

$$L \Sigma_{11} L' = I_p , \quad M \Sigma_{22} M' = I_q . \tag{2.10}$$

Also, from (2.8) and (2.3),

$$\begin{aligned} L \Sigma_{12} M' &= A' \Sigma_{11}^{-1/2} \Sigma_{12} \Sigma_{22}^{-1/2} B \\ &= A' C B \\ &= A' A P B' B \\ &= P \end{aligned} \tag{2.11}$$

as $A'A = I$, $B'B = I$.

Therefore, the variance-covariance matrix of the transformed variables $\underline{u} = L\, \underline{x}$, $\underline{v} = M\, \underline{y}$ is

$$V \left[\begin{array}{c} \underline{u} \\ \hline \underline{v} \end{array} \right] = \left[\begin{array}{c|c} L \Sigma_{11} L' & L \Sigma_{12} M' \\ \hline M \Sigma_{21} L' & M \Sigma_{22} M' \end{array} \right] = \left[\begin{array}{c|c} I_p & P \\ \hline P' & I_q \end{array} \right] , \tag{2.12}$$

because of (2.10) and (2.11). This proves the theorem. $\underline{u}$ are linear combinations of $x_1, \ldots, x_p$, the elements of $\underline{x}$ and $\underline{v}$ are linear combinations of $y_1, \ldots, y_q$, the elements of $\underline{y}$. $\underline{u}$ are called the canonical variables of the x-space and v are called canonical variables of the y-space. The columns of L' and M' are called canonical vectors. If $\rho_1^2, \rho_2^2, \ldots, \rho_p^2$ are so arranged that

$$\rho_1^2 \geq \rho_2^2 \geq \cdots \geq \rho_r^2 .$$

u_1 is called the first canonical variable of the x-space, and v_1 is the first canonical variable of the y-space. u_2 will be the second, and so on. The entire relationship between the p variables x and the q variables y is expressed only in terms of r parameters $\rho_1^2, \rho_2^2, \ldots, \rho_r^2$. Hence the name canonical variables. Since, from (2.12), we observe that the variances of each u_i, v_i is 1, the covariances are correlations. From (2.12) we find that

$$\begin{aligned} \text{corr}\,(u_i, v_j) &= \rho_i \quad \text{only when } i = j \\ &= 0 \quad \text{otherwise.} \end{aligned} \tag{2.13}$$

Also observe from (2.12) that

$$\text{corr}\,(u_i, u_j) = 0, \quad \text{corr}\,(v_i, v_j) = 0, \quad \text{when } i \neq j. \tag{2.14}$$

In other words, the first canonical variable of the x-space is correlated only with the first canonical variable of the y-space, the second with the second, and so on. These correlations are $\rho_1, \rho_2, \ldots, \rho_r$. Hence $\rho_1, \ldots, \rho_r$, the positive square roots of $\rho_1^2, \ldots, \rho_r^2$, the roots of (2.6) or (2.7), are known as the canonical correlations. This relationship between u and v can be depicted as

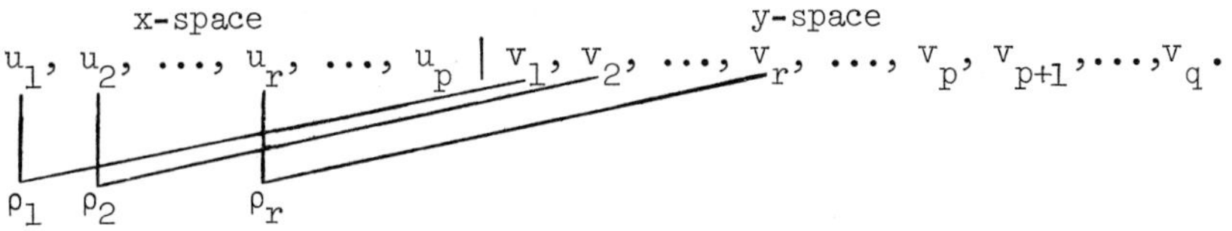

From (2.3), premultiplying by A' and using A'A = I, we get

$$A'C = P\,B'.$$

Expressing A and B in terms of L and M by (2.9), and using (2.2), the above relation yields

$$L\,\Sigma_{12}\,\Sigma_{22}^{-1} = P\,M. \tag{2.15}$$

This is the same as

$$\begin{aligned}\Sigma_{22}^{-1}\,\Sigma_{21}\,\underline{\ell}_i &= \rho_i\underline{m}_i \quad (i = 1, 2, \ldots, r)\\ &= 0 \qquad (i = r+1, \ldots, p).\end{aligned} \tag{2.16}$$

Similarly, from (2.3), we have CB = AP and this leads to

$$\Sigma_{11}^{-1}\,\Sigma_{12}\,M' = L'P, \tag{2.17}$$

which is the same as

$$\begin{aligned}\Sigma_{11}^{-1}\,\Sigma_{12}\,\underline{m}_i &= \rho_i\underline{\ell}_i \quad (i = 1, 2, \ldots, r)\\ &= 0 \qquad (i = r+1, \ldots, q).\end{aligned} \tag{2.18}$$

$\Sigma_{11}^{-1}\,\Sigma_{12}$ is the matrix of regression coefficients in the regression of the vector $\underline{y}$ on the vector $\underline{x}$ [see (2.4.11)]. Equation (2.18) therefore shows that $\underline{\ell}_i$ is the "projection of $\underline{m}_i$ on the $\underline{x}$-space. Similarly (2.16) shows that $\underline{m}_i$ is the "projection" of $\underline{\ell}_i$ on the $\underline{y}$-space. Equations (2.16) and (2.18) together can be expressed in the form

$$\left[\begin{array}{c|c} -\rho_i\,\Sigma_{11} & \Sigma_{12} \\ \hline \Sigma_{21} & -\rho_i\,\Sigma_{22}\end{array}\right]\left[\begin{array}{c}\underline{\ell}_i \\ \hline \underline{m}_i\end{array}\right] = 0 \quad (i = 1, 2, \ldots, r), \tag{2.19}$$

$$\Sigma_{21}\,\underline{\ell}_i = 0, \quad (i = r+1, \ldots, p),$$

$$\Sigma_{12}\,\underline{m}_i = 0, \quad (i = r+1, \ldots, q).$$

Again, eliminating $\underline{m}_i$ between (2.16) and (2.18), we get

$$[-\rho_i^2\,\Sigma_{11} + \Sigma_{12}\,\Sigma_{22}^{-1}\,\Sigma_{21}]\underline{\ell}_i = \underline{0} \quad (i = 1, 2, \ldots, r). \tag{2.20}$$

Similarly, by eliminating $\underline{\ell}_i$, we obtain

$$[-\rho_i^2\,\Sigma_{22} + \Sigma_{21}\,\Sigma_{11}^{-1}\,\Sigma_{12}]\underline{m}_i = \underline{0} \quad (i = 1, 2, \ldots, r). \tag{2.21}$$

These could also have been obtained from (2.3) by writing it as $CC'A = APP'$ or $C'CB = BP'P$ and then using (2.8). From (2.20), by premultiplying by $\underline{\ell}_i'$, we obtain

$$\frac{\underline{\ell}_i'(\Sigma_{12}\,\Sigma_{22}^{-1}\,\Sigma_{21})\underline{\ell}_i}{\underline{\ell}_i'\,\Sigma_{11}\,\underline{\ell}_i} = \rho_i^2 \quad (i = 1, 2, \ldots, r). \tag{2.22}$$

From (1.3.8), it is easy to see that ρ_i is thus the multiple correlation coefficient between $u_i = \underline{\ell}_i'\underline{x}$ and $y_1, y_2, \ldots, y_q$. Similarly from (2.21), we can show that ρ_i is the multiple correlation coefficient between $v_i = \underline{m}_i'\underline{y}$ and $x_1, x_2, \ldots, x_p$.

Again, from (2.2) and (2.3)

$$CC' = APP'A'$$

$$\Sigma_{11}^{-1/2}\,\Sigma_{12}\,\Sigma_{22}^{-1}\,\Sigma_{21}\,\Sigma_{11}^{-1/2} = APP'A',$$

$$\Sigma_{12}\,\Sigma_{22}^{-1}\,\Sigma_{21} = (\Sigma_{11}^{1/2}\,A)(PP')(\Sigma_{11}^{1/2}\,A)'.$$

This, from (2.8), can be written as (A being orthogonal)

$$\text{(i)} \quad \Sigma_{12}\Sigma_{22}^{-1}\Sigma_{21} = (L^{-1})\,\mathrm{diag}(\rho_1^2, \rho_2^2, \ldots, \rho_r^2, 0, \ldots, 0)(L^{-1})'$$

$$\text{(ii)} \quad \Sigma_{11} = (L^{-1})(L^{-1})'. \tag{2.23}$$

Observe, from (2.12), that the determinant of the variance-covariance matrix of the canonical variables $\underline{u}$, $\underline{v}$ is (from 1.3.14)

$$\left|\begin{array}{c|c} I_p & P \\ \hline P' & I_q \end{array}\right| = |I_p|\ |I_q - P'I_p^{-1}P| = |I - P'P| = \prod_{i=1}^{r} (1 - \rho_i^2)\ . \tag{2.24}$$

The canonical variables and correlations can be obtained and interpreted in a different manner, as follows. Consider the problem of finding a linear combination $\underline{\ell}'\underline{x}$ of $\underline{x}$ and a linear combination $\underline{m}'\underline{y}$ of $\underline{y}$ such that

$$V(\underline{\ell}'\underline{x}) = 1,\ V(\underline{m}'\underline{y}) = 1,\ \text{corr}(\underline{\ell}'\underline{x},\ \underline{m}'\underline{y}) \text{ is maximum.}$$

We have, therefore, to maximize

$$\phi = \underline{\ell}'\Sigma_{12}\underline{m} - (\lambda_1/2)(\underline{\ell}'\Sigma_{11}\underline{\ell} - 1) - (\lambda_2/2)(\underline{m}'\Sigma_{22}\underline{m} - 1)$$

with respect to $\underline{\ell}$, $\underline{m}$. $\lambda_1/2$, $\lambda_2/2$ are Lagrangian multipliers. By equating the partial derivatives of ϕ with respect to $\underline{\ell}$, $\underline{m}$ to zero, and a little algebra, we find that the $\underline{\ell}$, $\underline{m}$ satisfy equation (2.19), with $i = 1$, showing that $u_1 = \underline{\ell}_1'\underline{x}$, $v_1 = \underline{m}_1'\underline{y}$ are the required linear combinations, yielding ρ_1 as the maximum correlation. If we now try to find a second pair of variables u_2, v_2, such that u_2 is uncorrelated with u_1, v_2 is uncorrelated with v_1, $V(u_2) = V(v_2) = 1$, and subject to these conditions, maximize the correlation between u_2, v_2, we come up again with the second pair of canonical variables as our solution. We can proceed in this manner until we get the last, namely, the rth pair of canonical variables.

3. SAMPLE CANONICAL CORRELATIONS AND CANONICAL VARIABLES

If we have a sample of N observations on each of the $p+q$ variables $\underline{x}$ and $\underline{y}$, and if we denote these sample observations on $\underline{x}$ and $\underline{y}$ by two matrices X and Y of order $p \times N$ and $q \times N$, respectively, the matrix of corrected s.s. and s.p. of observations will be

$$S = \left[\begin{array}{c|c} S_{11} & S_{12} \\ \hline S_{21} & S_{22} \end{array}\right]\begin{matrix} p \\ q \end{matrix} , \qquad (3.1)$$
$$\qquad\quad p \qquad q$$

where

$$S_{11} = X(I - \frac{1}{N} E_{NN})X' ,$$

$$S_{12} = S_{21}' = X(I - \frac{1}{N} E_{NN})Y' ,$$

$$S_{22} = Y(I - \frac{1}{N} E_{NN})Y' . \qquad (3.2)$$

Under the assumption of a joint normal distribution for $\underline{x}$ and $\underline{y}$, $(1/N)S$ will be the maximum likelihood estimate of the variance-covariance matrix (2.1) of $\underline{x}$, $\underline{y}$. The sample canonical variables and the sample canonical correlations are then defined in exactly the same manner as the true (or population) canonical variables and canonical correlations, by replacing Σ_{11}, Σ_{12}, Σ_{22} by their maximum likelihood estimate, viz., $(1/N)S_{11}$, $(1/N)S_{12}$, $(1/N)S_{22}$, respectively, in (2.16), (2.18), (2.19), (2.20), (2.21), (2.22), and (2.23). Thus the squares of the sample canonical correlation coefficients r_i^2 $(i = 1, 2, \ldots, p)$ will be the roots of the determinantal equation in r^2,

$$|-r^2 S_{11} + S_{12} S_{22}^{-1} S_{21}| = 0 . \qquad (3.3)$$

Usually r_1^2 is taken to be the largest root, r_2^2 as the next largest, and so on. r_p^2 will be the smallest root. When we deal with sample canonical correlations, it should be remembered (see Section 4, of Chapter 3) that S is a positive definite matrix with probability 1. Thus, the rank of S_{12} will be the smaller of p, q, which, by our assumption is p. The equation (3.3) will, therefore, have all the p roots nonzero. However, if we consider the equation corresponding to (2.7), viz.,

$$|-r^2 S_{22} + S_{21} S_{11}^{-1} S_{12}| = 0 , \qquad (3.4)$$

it will have the same nonzero roots r_i^2 $(i = 1, 2, \ldots, p)$ and in addition will have $(q - p)$ roots which are zero. [If $p > q$, there will be only q nonzero roots of (3.3) and $p - q$ roots will be zero. But all the q roots of (3.4) will be nonzero.]

The sample canonical variables will be $\underline{g}_i'\underline{x}$ and $\underline{h}_i'\underline{y}$, where the vectors $\underline{g}_i$, $\underline{h}_i$ satisfy (assuming $p < q$)

$$S_{22}^{-1} S_{21} \underline{g}_i = r_i \underline{h}_i, \quad (i = 1, 2, \ldots, p) , \qquad (3.5)$$

$$\begin{aligned} S_{11}^{-1} S_{12} \underline{h}_i &= r_i \underline{g}_i, \quad (i = 1, 2, \ldots, p) , \\ &= 0 \qquad (i = p+1, \ldots, q) , \end{aligned} \qquad (3.6)$$

$$\left[\begin{array}{c|c} -r_i S_{11} & S_{12} \\ \hline S_{21} & -r_i S_{22} \end{array}\right] \begin{bmatrix} \underline{g}_i \\ \underline{h}_i \end{bmatrix} = \underline{0} \quad (i = 1, 2, \ldots, p), \qquad (3.7)$$

$$S_{12} \underline{h}_i = 0 \quad (i = p+1, \ldots, q) ,$$

$$[-r_i^2 S_{11} + S_{12} S_{22}^{-1} S_{21}]\underline{g}_i = 0 \quad (i = 1, 2, \ldots, p), \qquad (3.8)$$

$$[-r_i^2 S_{22} + S_{21} S_{11}^{-1} S_{12}]\underline{h}_i = 0 \quad (i = 1, 2, \ldots, p), \qquad (3.9)$$

r_i = the sample multiple correlation coefficient between $\underline{g}_i'\underline{x}$ and $\underline{y}$, or $\underline{h}_i'\underline{y}$ and $\underline{x}$; (3.10)

$$\text{(i)} \quad S_{12}\, S_{22}^{-1}\, S_{21} = D F D' ,$$
$$\text{(ii)} \quad S_{11} = DD' , \tag{3.11}$$

where

$$D^{-1} = G = [\underline{g}_1 | \underline{g}_2 | \cdots | \underline{g}_p] , \tag{3.12}$$

$$F = \text{diag}(r_1^2, r_2^2, \ldots, r_p^2) . \tag{3.13}$$

Note that the ith column of D^{-1} is the coefficient vector $\underline{g}_i$ of the ith canonical variable $\underline{g}_i'\underline{x}$ of the sample.

Using (1.3.14) and (3.11), we observe that

$$\begin{aligned} \frac{|S|}{|S_{11}|\,|S_{22}|} &= \frac{|S_{22}|\,|S_{11} - S_{12}\, S_{22}^{-1}\, S_{21}|}{|S_{11}|\,|S_{22}|} \\ &= \frac{|DD' - DFD'|}{|DD'|} \\ &= |I - F| \\ &= \prod_{i=1}^{p} (1 - r_i^2) . \end{aligned} \tag{3.14}$$

This statistic is known as Wilks's Λ and is used to measure the "association" or relationship between the two vectors $\underline{x}$ and $\underline{y}$. We shall come to this again, in a later chapter.

We shall now derive the distribution of the sample canonical correlations and the sample canonical vectors. But, for this, we need certain preliminary results about an orthogonal matrix. We therefore consider these first in the next section.

4. REPRESENTATION OF AN ORTHOGONAL MATRIX IN TERMS OF ROTATION ANGLES

If $A = [a_{ij}]$ is a $p \times p$ orthogonal matrix, all its p^2 elements are not functionally independent, because of the $p(p+1)/2$ restrictions

$$\sum_{j=1}^{p} a_{ij}^2 = 1, \quad \sum_{j=1}^{p} a_{ij}a_{kj} = 0, \quad (i \neq k), \quad (i, k = 1, 2, \ldots, p).$$

Only $p^2 - p(p+1)/2 = p(p-1)/2$ elements are thus functionally independent; the remaining $p(p+1)/2$ elements can, at least theoretically, be expressed in terms of the independent elements. Thus, when $p = 2$, the matrix A can be written as

$$A = \begin{bmatrix} a_{11} & -\sqrt{1 - a_{11}^2} \\ +\sqrt{1 - a_{11}^2} & a_{11} \end{bmatrix}$$

in terms of a single element, a_{11}, as a_{12}, a_{21}, a_{22} can be expressed in terms of a_{11}. A better way of representation will be to let $a_{11} = \cos\theta$, so that

$$A = \begin{bmatrix} \cos\theta & -\sin\theta \\ \sin\theta & \cos\theta \end{bmatrix}.$$

θ is called the rotation angle, as a transformation using A is equivalent to rotating the axes, in a two-dimensional space, through an angle θ. In the general case, a $p \times p$ orthogonal matrix A can be represented in terms of $p(p-1)/2$ rotational angles θ_{ij} $(i = 1, 2, \ldots, p-1;\ j = i, i+1, \ldots, p-1)$ in the following way. Define

$$R_j(\theta) = \left[\begin{array}{c|c|c} I_{j-1} & 0 & 0 \\ \hline 0 & \begin{matrix}\text{Cos } \theta, & -\text{Sin } \theta \\ \text{Sin } \theta, & \text{Cos } \theta\end{matrix} & 0 \\ \hline 0 & 0 & I_{p-j-1} \end{array}\right]. \tag{4.1}$$

Using the relation (4.1), Tumura [32] has shown that a p x p orthogonal matrix A can, in general, be represented as

$$A = \prod_{i=1}^{p-1} \prod_{j=i}^{p-1} \{R_j(\theta_{ij})\} \tag{4.2}$$

in terms of the $p(p-1)/2$ rotational angles θ_{ij} ($j \geq i$, $i = 1, 2, \ldots, p-1$). Let K be any symmetric positive definite matrix of order p and eigenvalues $k_1 > k_2 > \ldots > k_p$. There exists, then, an orthogonal matrix A such that

$$K = A \operatorname{diag}(k_1, k_2, \ldots, k_p)A' . \tag{4.3}$$

Representing A in terms of the rotational angles θ_{ij} as in (4.2), we can consider a transformation from K to A and k_i ($i = 1, \ldots, p$). Strictly speaking, it is a transformation from the $p(p+1)/2$ distinct elements of K to k_i ($i = 1, \ldots, p$) and θ_{ij} ($j \geq i$, $i = 1, 2, \ldots, p-1$). Denote the Jacobian of this transformation by $J(K \to A, k_i)$. Tumura [32] has proved the following theorem.

<u>Theorem 2</u>: The Jacobian of transformation $J(K \to A, k_i)$ is the absolute value of

$$\prod_{i=1}^{p} \prod_{j>i} (k_i - k_j) \cdot \prod_{i=1}^{p-2} \prod_{j=1}^{p-2} \sin^{p-j-1} \theta_{ij} .$$

The transformation from K to A, k_i is not one to one. While the θ's range in the intervals

$$0 \leq \theta_{ij} \leq \pi \ (j \leq p-2); \quad 0 \leq \theta_{i,p-1} \leq 2\pi, \tag{4.4}$$

K will cover the region $2^{(p-1)}$ times. Whenever we integrate over the θ_{ij}'s, it is, therefore, necessary to divide the result by 2^{p-1}. In certain cases, the integrand is an even function of $\cos \theta_{ij}$ for all i, j. In such situations, we can replace (4.4) by

$$0 \leq \theta_{ij} \leq \pi/2, \ (j \leq p-2); \ 0 \leq \theta_{i,p-1} \leq \pi, \tag{4.5}$$

and then the divisor will be unnecessary. Using the well-known result

$$\int_0^{\pi} \sin^a \theta \, d\theta = 2 \int_0^{\pi/2} \sin^a \theta = \frac{\sqrt{\pi}\, \Gamma\left(\frac{a+1}{2}\right)}{\Gamma\left(\frac{a+2}{2}\right)},$$

we can easily establish that

$$\int J(K \to A, k_i) \prod_{i=1}^{p-1} \prod_{j=i}^{p-1} d\theta_{ij} = \frac{\prod_{i=1}^{p} \prod_{j>i} (k_i - k_j) \cdot \pi^{p(p+1)/4}}{\prod_{i=1}^{p} \Gamma\left(\frac{p+1-i}{2}\right)} \tag{4.6}$$

where the range of integration of the θ_{ij}'s is (4.4). Therefore, the result on the right-hand side of (4.6) is written after dividing by 2^{p-1}. Alternatively, we can take the range to be (4.5) and need not divide by 2^{p-1}.

5. DISTRIBUTION OF THE SAMPLE CANONICAL CORRELATION COEFFICIENTS

Let X be a $p \times N$ matrix denoting the sample observations on a p-component vector $\underline{x}$. Similarly, let Y be the $q \times N$ matrix of the corresponding observations on a q-component vector $\underline{y}$. The matrices of the corrected s.s. and s.p. of these N observations are

$$S_{11} = XJX', \quad S_{12} = S_{21}' = XJY', \quad S_{22} = YJY', \tag{5.1}$$

where J denotes the $N \times N$ idempotent matrix $I - N^{-1} E_{NN}$. The sample canonical correlations between $\underline{x}$ and $\underline{y}$ are r_i ($i = 1, 2, \ldots, k$), where $r_1^2 > r_2^2 > \ldots > r_k^2$ are the nonzero roots of the equation in r^2

$$|-r^2 S_{11} + S_{12} S_{22}^{-1} S_{21}| = 0, \tag{5.2}$$

where, as in Section 3, $k = p$ if $p \leq q$ and $k = q$ if $p > q$. We now make the following assumptions:

(i) Y is a fixed matrix.

(ii) Each column of X has a p-variate nonsingular normal distribution, with the same variance-covariance matrix Σ. All the columns of X are independently distributed.

(iii) $E(X) = \underline{\alpha} E_{1N} + \beta Y$, where $\underline{\alpha}$ is a $p \times 1$ vector, and β is a $p \times q$ matrix.

(iv) $N > p+q+1$. (5.3)

The distribution of $r_1^2, r_2^2, \ldots, r_k^2$, under these assumptions and the assumption $\beta = 0$, is known as the null distribution of the sample canonical correlations for a fixed $\underline{y}$. If $\beta \neq 0$, the distribution is known as the non-null distribution. We propose to derive only the null distribution.

Observe that

$$S_{12}\, S_{22}^{-1}\, S_{21} = XJY'\, S_{22}^{-1}\, YJX'$$
$$= XLX' , \qquad (5.4)$$

where

$$L = JY'\, S_{22}^{-1}\, YJ . \qquad (5.5)$$

Similarly,

$$S_{11\cdot 2} = S_{11} - S_{12}\, S_{22}^{-1}\, S_{21}$$
$$= XJX' - XLX'$$
$$= XMX' , \qquad (5.6)$$

where,

$$M = J - L . \qquad (5.7)$$

Notice that $L^2 = L$ and $M^2 = M$. In other words, L and M are symmetric $N \times N$ idempotent matrices. Therefore,

$$\text{rank } L = \text{tr } L = \text{tr } JY'S_{22}^{-1}YJ$$
$$= \text{tr } YJ^2Y'S_{22}^{-1}$$
$$= \text{tr } YJY'S_{22}^{-1}$$
$$= \text{tr } S_{22}S_{22}^{-1}$$
$$= \text{tr } I_q = q , \qquad (5.8)$$

as $J^2 = J$, and as the trace of the product of a number of matrices remains unaltered, if the matrices are permuted cyclically. Similarly,

$$\text{rank } M = \text{tr}(J - L) = N-1-q = n-q, \quad \text{where } n = N-1. \qquad (5.9)$$

Again observe that

$$L\, E_{N1} = 0, \quad M\, E_{N1} = 0, \quad LM = 0 , \qquad (5.10)$$

$$L + M + \frac{1}{N} E_{NN} = I_N \; . \tag{5.11}$$

The matrices L, M, $\frac{1}{N} E_{NN}$, therefore, satisfy the conditions (2.5.4) of James' theorem. Thus, there exists an $N \times N$ orthogonal matrix

$$P = \underset{\quad q \qquad n-q \qquad 1}{[P_1 \mid P_2 \mid P_3]\, N} \tag{5.12}$$

such that

$$L = P\,\mathrm{diag}(\underbrace{1, 1, \ldots, 1}_{q \text{ times}}, 0, \ldots, 0)P'$$

$$= P_1 P_1' \; , \tag{5.13}$$

$$M = P\,\mathrm{diag}(\underbrace{0, \ldots, 0}_{q}, \underbrace{1, \ldots, 1}_{n-q}, 0)P'$$

$$= P_2 P_2' \; . \tag{5.14}$$

As P is orthogonal, $P_1'P_1 = I_q$, $P_2'P_2 = I_{n-q}$; and so, from (5.13) and (5.14), we obtain

$$L\,P_1 = P_1 \; , \quad M\,P_2 = P_2 \; . \tag{5.15}$$

Also, from (5.7) and (5.5), we find that

$$YM = Y(J - L)$$

$$= YJ - YJY'S_{22}^{-1}YJ$$

$$= YJ - S_{22}S_{22}^{-1}YJ$$

$$= YJ - YJ = 0 \; . \tag{5.16}$$

Define a $p \times N$ matrix Z by

$$Z = \underset{\quad q \qquad n-q \qquad 1}{[Z_1 \mid Z_2 \mid Z_3]\, N} = XP \; . \tag{5.17}$$

On account of (5.12), this is the same as

$$Z_1 = XP_1 , \quad Z_2 = XP_2 , \quad Z_3 = XP_3 \; . \tag{5.18}$$

Observe that

$$\begin{aligned} E(Z_1) &= E(X)P_1 \\ &= (\underline{\alpha}\, E_{1N} + \beta Y)P_1 \\ &= \underline{\alpha}\, E_{1N}\, LP_1 + \beta YP_1 \\ &= \beta YP_1 , \end{aligned} \tag{5.19}$$

as $P_1 = LP_1$ and $E_{1N}L = 0$, due to (5.15) and (5.10). Similarly,

$$\begin{aligned} E(Z_2) &= E(X)P_2 \\ &= (\underline{\alpha}\, E_{1N} + \beta Y)P_2 \\ &= \alpha\, E_{1n}MP_2 + \beta YMP_2 \\ &= 0 , \end{aligned} \tag{5.20}$$

as $P_2 = MP_2$ [see (5.15)] and $YM = 0$ [see (5.16)]. It should be emphasized here that $E(Z_2) = 0$, whatever $\underline{\alpha}$, β may be. But $E(Z_1) = 0$ only when $\beta = 0$.

The distribution of X, from our assumptions (5.3), is [see (2.2.4)]

$$\frac{1}{(2\pi)^{NP/2}|\Sigma|^{N/2}} \exp\left\{-\frac{1}{2}\operatorname{tr}\Sigma^{-1}(X - E(X))(X - E(X))'\right\}dX . \tag{5.21}$$

Transform from X to Z by (5.17). The Jacobian of transformation (see Deemer and Olkin [7]) is unity, as P is orthogonal. Noting that

$$\begin{aligned} (X - E(X))(X - E(X))' &= (X - E(X))PP'(X - E(X))' \\ &= (Z - E(Z))(Z - E(Z))' \\ &= (Z_1 - \beta YP_1)(Z_1 - \beta YP_1)' + Z_2Z_2' \\ &\quad + (Z_3 - E(Z_3))(Z_3 - E(Z_3))' , \end{aligned}$$

we find the distribution of Z, i.e., of Z_1, Z_2, Z_3, to be

$$\frac{1}{(2\pi)^{Np/2}\ |\Sigma|^{N/2}} \exp\Big\{-\frac{1}{2}\operatorname{tr}\Sigma^{-1}(Z_1-\beta YP_1)(Z_1-\beta YP_1)' - \frac{1}{2}\operatorname{tr}\Sigma^{-1}Z_2Z_2' - \frac{1}{2}\operatorname{tr}\Sigma^{-1}(Z_3-E(Z_3))(Z_3 - E(Z_3))'\Big\}\, dZ_1dZ_2dZ_3 . \qquad (5.22)$$

Integrating out Z_3, the distribution of Z_1 and Z_2 is (using $n = N-1$)

$$\frac{1}{(2\pi)^{pq/2}\ |\Sigma|^{q/2}} \exp\Big\{-\frac{1}{2}\operatorname{tr}\ \Sigma^{-1}(Z_1-\beta YP_1)(Z_1-\beta YP_1)'\Big\}\, dZ_1$$

$$\times \frac{1}{(2\pi)^{(n-q)p/2}|\Sigma|^{(n-q)/2}} \exp\Big\{-\frac{1}{2}\operatorname{tr}\Sigma^{-1}Z_2Z_2'\Big\}\, dZ_2 . \qquad (5.23)$$

Because of (5.2), (5.4), (5.6), (5.13), (5.14) and (5.18), r_i^2's are the roots of the determinantal equation

$$\begin{aligned} 0 &= |-r^2S_{11} + S_{12}S_{22}^{-1}S_{21}| \\ &= |-r^2(S_{11\cdot 2} + S_{12}S_{22}^{-1}S_{21}) + S_{12}S_{22}^{-1}S_{21}| \\ &= |-r^2(XMX' + XLX') + XLX'| \\ &= |-r^2(XP_2P_2'X' + XP_1P_1'X') + XP_1P_1'X'| \\ &= |-r^2(Z_2Z_2' + Z_1Z_1') + Z_1Z_1'| . \end{aligned} \qquad (5.24)$$

The problem of deriving the distribution of the r_i^2's thus reduces to finding the distribution of the nonzero roots of

$$|-r^2(Z_2Z_2' + Z_1Z_1') + Z_1Z_1'| = 0 \qquad (5.25)$$

when Z_1, Z_2 have the distribution (5.23). We shall derive only the null distribution, i.e., the distribution when $\beta = 0$. When $\beta = 0$, we observe from (5.3) that $\underline{x}$ and $\underline{y}$ have no relationship or association at all.

The distribution of Z_2, given by (5.23), shows that [compare with (2.2.4)] the n-q columns of Z_2 are independently and identically distributed as a p-variate nonsingular normal distribution with zero means and variance-covariance matrix Σ. In other words Z_2 is a random sample of size n-q from a p-variate normal distribution $N_p(\underline{z}|\underline{0}|\Sigma)$. Therefore

$$A = Z_2 Z_2' \tag{5.26}$$

is a Wishart matrix based on n-q d.f. Since we have assumed $N > p+q+1$, the d.f. $n-q > p$, and so A has the $W_p(A|n-q|\Sigma)$ distribution. The joint distribution of A and Z_1, when $\beta = 0$, therefore comes out as

$$W_p(A|n-q|\Sigma) \frac{1}{(2\pi)^{pq/2}|\Sigma|^{q/2}} \exp\left\{-\frac{1}{2}\operatorname{tr} \Sigma^{-1} Z_1 Z_1'\right\} dA dZ_1 . \tag{5.27}$$

This can also be written as

$$\frac{K(p, n-q)}{(2\pi)^{pq/2}|\Sigma|^{n/2}} |A|^{(n-q-p-1)/2} \exp\left\{-\frac{1}{2}\operatorname{tr} \Sigma^{-1}(A + Z_1 Z_1')\right\} dA dZ_1, \tag{5.28}$$

where the constant K(p, n-q) is defined by (3.2.20). Now transform from A and Z_1 to G and U, where

$$G = A + Z_1 Z_1' , \tag{5.29}$$

$$U = G^{-1/2} Z_1 . \tag{5.30}$$

Note that U is a $p \times q$ matrix, and $G^{-1/2}$ is a matrix square root of G^{-1}. The Jacobian of the transformation from A to G is unity and that of Z_1 to U is (see Deemer and Olkin [7]) $|G|^{q/2}$. Also

$$
\begin{aligned}
|A| &= |G - Z_1 Z_1'| \\
&= |G - G^{1/2}\, UU'\, G^{1/2}| \\
&= |G^{1/2}|\; |I_p - UU'|\; |G^{1/2}| \\
&= |I_p - UU'| \, . \qquad (5.31)
\end{aligned}
$$

The joint distribution of G and U is, therefore,

$$
\frac{K(p,\, n-q)}{(2\pi)^{pq/2}|\Sigma|^{n/2}}\; |G|^{(n-p-1)/2}\, e^{-(1/2)\operatorname{tr}\Sigma^{-1}G} |I-UU'|^{(n-q-p-1)/2} dG dU. \qquad (5.32)
$$

This can also be expressed as

$$
W_p(G|n|\Sigma)dG \cdot \frac{K(p,\, n-q)}{K(p,n)(2\pi)^{pq/2}}\; |I - UU'|^{(n-q-p-1)/2}\; dU\,, \qquad (5.33)
$$

showing that G and U are independently distributed. G has the Wishart distribution $W_p(G|n|\Sigma)$ and U has the distribution

$$
\frac{K(p,\, n-q)}{K(p,n)(2\pi)^{pq/2}}\; |I - UU'|^{(n-q-p-1)/2}\; dU\,. \qquad (5.34)
$$

From (5.25), (5.26), (5.29), and (5.30), r_i^2's are seen to be the roots of the equation

$$
\begin{aligned}
0 &= |-r^2(Z_2Z_2' + Z_1Z_1') + Z_1Z_1'| \\
&= |-r^2(A + Z_1Z_1') + Z_1Z_1'| \\
&= |-r^2 G + G^{1/2}\, UU'\, G^{1/2}| \\
&= |G|\; |-r^2 I + UU'| \, . \qquad (5.35)
\end{aligned}
$$

Our problem is thus reduced to finding the distribution of the nonzero roots of $|UU' - r^2 I| = 0$, when U has the distribution (5.34). We shall consider the two cases $p \le q$ and $p > q$ separately.

Case 1: $p \leq q$. Transform from the pq variables in U to the $p(p+1)/2$ distinct variables in the symmetric $p \times p$ matrix $H = UU'$ and the remaining $pq - \frac{1}{2}p(p+1)$ variables. Then integrate out these latter variables. To obtain this integral, we use Lemma 6 of Chapter 3, Section 5. An essential condition for the application of this lemma is that the order of H be smaller than the number of columns of U, i.e., $p \leq q$. This is satisfied here. So the distribution of H is, by the application of the lemma,

$$\frac{K(p, n-q)}{K(p,n)\ (2\pi)^{pq/2}} \int_{UU'=H} |I - UU'|^{(n-q-p-1)/2}\, dU$$

$$= \frac{K(p, n-q)\, K(p,q)}{K(p,n)} |H|^{(q-p-1)/2} |I - H|^{(n-q-p-1)/2}\, dH\,,$$

$$H > 0,\ I - H > 0\,. \quad (5.36)$$

From (5.35), $r_1^2 > r_2^2 > \ldots > r_p^2$ are the nonzero roots of $|H - r^2 I| = 0$, i.e., they are the eigenvalues of H. There exists, therefore, an orthogonal matrix $K(\theta_{ij})$, represented in terms of its rotational angles θ_{ij}, such that

$$H = K(\theta_{ij})\ \mathrm{diag}(r_1^2, r_2^2, \ldots, r_p^2)\, K'(\theta_{ij})\,. \quad (5.37)$$

Transform from H to $r_i^2 (i = 1, 2, \ldots, p)$ and the θ_{ij}'s. The Jacobian of this transformation is given by Theorem 2 of Section 4. Then integrate out the θ_{ij}'s. Use (4.6) for this integration. The distribution of r_i^2's, therefore, finally comes out as

$$\frac{K(p,q)\, K(p, n-q)\ \pi^{p(p+1)/4}}{K(p,n) \prod_{i=1}^{p} \Gamma\left(\frac{p+1-i}{2}\right)} \prod_{i=1}^{p} \left\{(r_i^2)^{(q-p-1)/2} (1-r_i^2)^{(n-q-p-1)/2}\right.$$

$$\left. \cdot\ dr_i^2\right\} \prod_{i=1}^{p} \prod_{j>i} (r_i^2 - r_j^2)\,. \quad (5.38)$$

Substituting for K(p,q), K(p,n) and K(p, n-q), the distribution of $r_1^2 > \dots > r_p^2$ simplifies to

$$C(n,p,q)\ \ell_{n,p,q}(r_1^2, \dots, r_p^2) \prod_{i=1}^{p} dr_i^2 \tag{5.39}$$

where

$$C(n,p,q) = \pi^{p/2} \prod_{i=1}^{p} \left\{ \frac{\Gamma\left(\frac{n+1-i}{2}\right)}{\Gamma\left(\frac{p+1-i}{2}\right)\Gamma\left(\frac{q+1-i}{2}\right)\Gamma\left(\frac{n-q+1-i}{2}\right)} \right\} \tag{5.40}$$

and

$$\ell_{n,p,q}(r_1^2, \dots, r_p^2) = \prod_{i=1}^{p} \left\{ (r_i^2)^{(q-p-1)/2} (1 - r_i^2)^{(n-q-p-1)/2} \right\}$$

$$\times \prod_{i=1}^{p} \prod_{j>i} (r_i^2 - r_j^2) . \tag{5.41}$$

<u>Case 2: p > q</u>. In the distribution of U given by (5.34), we now transform from the pq variables in U to the q(q+1)/2 distinct variables of the symmetric $q \times q$ matrix $H^* = U'U$ and the remaining $pq - q(q+1)/2$ variables and then integrate out these latter variables. For this, we again use Lemma 6 of Chapter 3. This is possible because q, the order of H*, is less than p, the number of columns of U'. We also notice that [by (1.4.5)]

$$|I_p - UU'| = |I_q - U'U| .$$

Hence, from (5.34) and Lemma 6 of Chapter 3, the distribution of H* is

$$\frac{K(p,\ n-q)}{K(p,n)\ (2\pi)^{pq/2}} \int_{U'U=H^*} |I_q - U'U|^{(n-q-p-1)/2}\ dU'$$

$$= \frac{K(p,\ n-q)\ K(q,p)}{K(p,n)} |H^*|^{(p-q-1)/2} |I - H^*|^{(n-q-p-1)/2}\ dH^* ,$$

$$H^* > 0,\ I - H^* > 0 . \tag{5.42}$$

We have already seen that the r_i^2's are the nonzero roots of

$$|UU' - r^2 I_p| = 0 .$$

But because of (1.4.5) and because $p > q$, we can write the above equation as

$$(-r^2)^p \; |I_p - UU'| = 0$$

or $$(-r^2)^p \; |I_q - U'U| = 0$$

or $$(-r^2)^{p-q} |U'U - r^2 I_q| = 0,$$

or $$(-r^2)^{p-q} \; |H^* - r^2 I_q| = 0.$$

This shows that only q roots are nonzero, and they are all the eigenvalues of H*.

We therefore use the same procedure used in Case 1 ($p \leq q$); transform from H* to its eigenvalues $r_1^2 > r_2^2 > \ldots > r_q^2$ and the associated orthogonal matrix; then integrate out the rotational angles in terms of which orthogonal matrix is represented. The final result, as is obvious by inspection and anology with Case 1, is

$$C(n, q, p) \; \ell_{n,q,p}(r_1^2, \ldots, r_q^2) \prod_{i=1}^{q} dr_i^2 , \tag{5.43}$$

where the functions C and ℓ are already defined in (5.40) and (5.41). The distribution of r_i^2's, when $p > q$, is thus obtained simply by <u>interchanging p and q</u>, in the distribution of the r_i^2's, when $p \leq q$.

The distribution of the r_i^2's was independently obtained by Fisher [9], Hsu [14], and Roy [27]. It is therefore referred to in the literature as the Fisher-Hsu-Roy distribution.

We shall now consider the case when $\underline{y}$, and hence Y, is not fixed but is itself a stochastic variable. In that case, we can keep Y fixed first and consider the conditional distribution (null) of the r_i^2's. This is obviously (5.39) or (5.43), depending on whether $p \leq q$ or $p > q$. But this conditional distribution does not involve $\underline{y}$ or Y at all. The distribution of the r_i^2's is thus independent of $\underline{y}$ and (5.39) or (5.43) represents the distribution of the r_i^2's even when $\underline{y}$ is random and has <u>any</u> distribution. This holds only in the null case. It should be remembered here that the r_i^2's are independently distributed of

$$G = A + Z_1Z_1' = Z_2Z_2' + Z_1Z_1' = S_{11} = XJX'$$

also. This follows from (5.33), where we proved that G and U are independently distributed.

The non-null case, i.e., the case when β is not null, is more complicated. Constantine [6] has derived the distribution of the r_i^2's in this more general case. But its derivation requires the knowledge of noncentral Wishart distribution and zonal polynomials. It is beyond the scope of this book. However, we would like to point it out that, when $\beta \neq 0$, $E(Z_1) = \beta Y P_1 \neq 0$, and the distribution of Z_1Z_1' will involve the eigenvalues of the noncentrality matrix $\Sigma^{-1} E(Z_1) E(Z_1') = \Sigma^{-1}\beta Y P_1 P_1' Y' \beta' = \Sigma^{-1}\beta Y L Y' \beta'$ $= \Sigma^{-1}\beta S_{22} \beta'$, as $YLY' = S_{22}$, the matrix of the corrected s.s. and s.p. of the observations on $\underline{y}$. The conditional distribution of the r_i^2's, therefore, involves the roots of the equation

$$|\Sigma^{-1}\beta S_{22}\beta' - \delta^2 I| = 0 . \qquad (5.44)$$

The distribution, therefore, involves $\underline{y}$ in the form S_{22}. If we desire the marginal distribution of the roots, we must multiply the conditional p.d.f. by the p.d.f. of S_{22}, and integrate it out. This can be achieved in the case where $\underline{x}$ and $\underline{y}$ have a joint (p+q)-variate normal distribution with the variance-covariance matrix

$$\begin{bmatrix} \Sigma_{11} & \Sigma_{12} \\ \Sigma_{21} & \Sigma_{22} \end{bmatrix} \begin{matrix} p \\ q \end{matrix} .$$
$$\quad p \qquad q$$

In this case, the conditional distribution of $\underline{x}$ when $\underline{y}$ is fixed has mean of the form

$$\underline{\alpha} + \beta\, \underline{y} ,$$

where $\beta = \Sigma_{12}\Sigma_{22}^{-1}$ and the variance-covariance matrix is

$$\Sigma_{11\cdot 2} = \Sigma_{11} - \Sigma_{12}\Sigma_{22}^{-1}\Sigma_{21} .$$

Therefore, from (5.44), we find that the conditional distribution of the r_i^2's involves the roots of

$$|\Sigma_{11\cdot 2}^{-1}\beta S_{22}\beta' - \delta^2 I| = 0,$$

i.e , of

$$|\Sigma_{11\cdot 2}^{-1}\Sigma_{12}\Sigma_{22}^{-1} S_{22}\Sigma_{22}^{-1}\Sigma_{21} - \delta^2 I| = 0.$$

Also S_{22} will have a $W_q(S_{22}|n|\Sigma_{22})$ distribution, and so the unconditional or marginal distribution of the r_i^2's will involve the roots of (replace S_{22} by Σ_{22})

$$|\Sigma_{11.2}^{-1}\Sigma_{12}\Sigma_{22}^{-1}\Sigma_{22}\Sigma_{22}^{-1}\Sigma_{22} - \delta^2 I| = 0,$$

i.e., of $|\Sigma_{11.2}^{-1}\Sigma_{12}\Sigma_{22}^{-1}\Sigma_{21} - \delta^2 I| = 0,$

i.e., of $|-\rho^2\Sigma_{11} + \Sigma_{12}\Sigma_{22}^{-1}\Sigma_{21}| = 0$, where $\rho^2 = \delta^2/(1+\delta^2)$.

But these roots are the true or population canonical correlation coefficients (squared) between $\underline{x}$ and $\underline{y}$. In the null case, $\beta = 0$, i.e., $\Sigma_{12} = 0$, or $\underline{x}$ and $\underline{y}$ are independent and all the true canonical correlations are null.

From all this derivation and discussion, it is evident that the distribution of the r_i^2's is the same, whether $\underline{x}$ is normal and $\underline{y}$ is fixed or vice versa, in the null case. This is one more exhibition of the duality of relationship, already referred to in Chapter 5, Section 4 and Chapter 6, Section 4.

6. DISTRIBUTION OF THE CANONICAL VECTORS

We shall consider only the null case and also assume $p \leq q$. From (3.11), we know that

$$S_{12}S_{22}^{-1}S_{21} = DFD', \quad S_{11} = DD', \tag{6.1}$$

where F is the diagonal matrix with $r_1^2, \ldots, r_p^2$ along the diagonal and the columns of D^{-1} are the canonical vectors of the sample canonical variables. From (5.2), (5.4), (5.6), (5.13), (5.14), and (5.18), we know that

$$S_{11} = S_{11.2} + S_{12}S_{22}^{-1}S_{21} = Z_1Z_1' + Z_2Z_2',$$

$$S_{12}S_{22}^{-1}S_{21} = Z_1Z_1', \tag{6.2}$$

where the joint distribution of Z_1 and Z_2, in the null case, is given by (5.27). The columns of Z_1 and Z_2 are therefore independently and identically distributed as $N_p(\underline{z}|0|\Sigma)$. Hence $A = Z_2 Z_2'$ and $B = Z_1 Z_1'$ are both Wishart matrices based on n-q and q d.f., respectively. Since $p < n-q$ and $p \leq q$, A and B have independent $W_p(A|n-q|\Sigma)$, $W_p(B|q|\Sigma)$ distributions, respectively. From (6.1) and (6.2), we find

$$\begin{aligned} A + B &= DD' , \\ B &= DFD' . \end{aligned} \tag{6.3}$$

Transform from A and B to D and F by (6.3). It has been proved (see Deemer and Olkin [7]) that the Jacobian of this transformation is

$$2^p |D|^{p+2} \prod_{i=1}^{p} \prod_{j>i} (r_i^2 - r_j^2) . \tag{6.4}$$

Making the transformation (6.3), in the joint distribution of A and B, stated above, and expressing A and B in terms of D and F by (6.3), we find the joint distribution of D and F to be

$$\frac{2^p K(p,q)K(p, n-q)}{|\Sigma|^{n/2}} |F|^{(q-p-1)/2} |I-F|^{(n-q-p-1)/2}$$

$$\times \prod_{i=1}^{p} \prod_{j>i} (r_i^2 - r_j^2) dF \; |D|^{n-p} \exp\left\{-\frac{1}{2} \operatorname{tr} \Sigma^{-1} DD'\right\} dD . \tag{6.5}$$

Since $|I-F| = \prod_1^p (1 - r_i^2)$ and $|F| = \prod_1^p r_i^2$, the above can be written as

$$C(n,p,q) \; \ell_{n,p,q}(r_1^2, \ldots, r_p^2) \prod_1^p dr_i^2 .$$

$$\times \; d(n,p) \frac{|D|^{n-p}}{|\Sigma|^{n/2}} e^{-(1/2) \operatorname{tr} \Sigma^{-1} DD'} dD \tag{6.6}$$

where $C(n,p,q)$, $\ell_{n,p,q}$ are defined in (5.39) to (5.41) and

$$d(n,p) = \prod_{i=1}^{p} \left\{ \frac{\Gamma\left(\frac{p+1-i}{2}\right)}{\Gamma\left(\frac{n+1-i}{2}\right)} \right\} \cdot 2^{-np/2} \pi^{-p^2/2} . \qquad (6.7)$$

Equation (6.6) shows that $r_1^2, \ldots, r_p^2$ and D are independently distributed. Since $C(n,p,q)\ \ell_{n,p,q}\ (r_1^2, \ldots, r_p^2)$ is the density of the r_i^2's, it is obvious that

$$d(n,p)\ |\Sigma|^{-n/2}\ |D|^{n-p} \exp\left(-\frac{1}{2}\operatorname{tr}\Sigma^{-1}DD'\right) d(D) \qquad (6.8)$$

is the distribution of D.

To find the distribution of the sample canonical vectors, we must find the distribution of D^{-1}, because the columns of D^{-1} are the canonical vectors, as seen in (3.11). The Jacobian of transformation from D to D^{-1} is (Deemer and Olkin [7]) $|D^{-1}|^{-2p}$. The distribution of the canonical vectors is therefore given by

$$d(n,p)\ |\Sigma|^{-n/2}\ |D^{-1}|^{-(n+p)} \exp\left(-\frac{1}{2}\operatorname{tr}\Sigma^{-1}DD'\right) d(D^{-1}) . \qquad (6.9)$$

When $\Sigma = I$, Tumura [32] has shown that every unit vector $\underline{d}r/(\underline{d}_r'\underline{d}_r)^{1/2}$, where $\underline{d}_r$ is the rth column of D, is distributed uniformly on a p-dimensional sphere of unit radius. He proves this by making a polar transformation from $\underline{d}_r$. He also shows that $\underline{d}_r'\underline{d}_r$ are independently distributed as χ^2 variables with n d.f. When $\Sigma \neq I$, we have to consider $d^* = \Sigma^{-1/2}D$, and then the columns of D* and their squared lengths will have these properties.

When $\Sigma = I$, apart from the factor $|D|^{n-p}$, (6.8) is the distribution of normal independent variables. But because of $|D|^{n-p}$, they are not independent. Bartlett [2] calls $|D|^{n-p}$ the "linkage" factor among the elements of D. $\underline{d}_r'\underline{d}_r$ would have a χ^2 distribution with p d.f., if $|D|^{n-p}$ were not there. But as we have just seen, it has a χ^2 distribution with n d.f. The effect of the linkage factor is thus to elevate the d.f. from p to n.

When $p > q$, the transformation (6.3) cannot be made, as the transformation matrix D is not uniquely determined. The distribution of D, in that case, depends on the way in which D is constructed and cannot be derived in general.

7. SOME MISCELLANEOUS RESULTS ABOUT CANONICAL CORRELATIONS

(1) When $p = 1$, equation (2.6) reduces to

$$-\rho^2\Sigma_{11} + \Sigma_{12}\Sigma_{22}^{-1}\Sigma_{21} = 0 . \tag{7.1}$$

Its solution is

$$\rho^2 = \Sigma_{12}\Sigma_{22}^{-1}\Sigma_{21}/\Sigma_{11} , \tag{7.2}$$

showing that there is only one canonical correlation, and by (1.3.8) it is the multiple correlation coefficient between x and $\underline{y}$. Similarly, when $q = 1$, there is only one (nonzero) canonical correlation, and it is the multiple correlation coefficient between $\underline{x}$ and $\underline{y}$. It is needless to add that, when both p and q are equal to 1, the canonical correlation reduces to the ordinary correlation coefficient.

(2) Canonical correlations are invariant for a nonsingular linear transformation of $\underline{x}$ and a nonsingular linear transformation of $\underline{y}$. In other words, the canonical correlations between $\underline{x}$ and $\underline{y}$ are the same as those between $P\underline{x}$ and $Q\underline{y}$, where P and Q are any two nonsingular matrices of order p and q, respectively. This follows from the fact that the variance-covariance matrix of $P\underline{x}$ and $Q\underline{y}$ is [from (2.1)]

$$\left[\begin{array}{c|c} P\,\Sigma_{11}\,P' & P\,\Sigma_{12}\,Q' \\ \hline Q\,\Sigma_{21}\,P' & Q\,\Sigma_{22}\,Q' \end{array}\right], \tag{7.3}$$

and hence, by (2.6), the canonical correlations between $P\underline{x}$ and $Q\underline{y}$ are the roots of the determinantal equation

$$|-\rho^2(P\,\Sigma_{11}\,P') + (P\,\Sigma_{12}\,Q')(Q\,\Sigma_{22}Q')^{-1}\,(Q\,\Sigma_{21}\,P')| = 0 . \tag{7.4}$$

which reduces to

$$|P|\;|-\rho^2\,\Sigma_{11} + \Sigma_{12}\Sigma_{22}^{-1}\Sigma_{21}|\;|P'| = 0 ,$$

$$|-\rho^2\Sigma_{11} + \Sigma_{12}\Sigma_{22}^{-1}\Sigma_{21}| = 0 ,$$

as P is nonsingular. This is the same as (2.6) and hence the result.

(3) As we have seen in Sections 5 and 6, the distribution theory of canonical correlations and canonical vectors is complicated even in the null case and is of questionable usefulness. The important case from the practical point of view is the non-null case, and the distribution theory associated with it is almost untractable. Lawley [19] has, therefore, employed

a laborious algebraic method of expanding the matrix $S_{11}^{-1}S_{12}S_{22}^{-1}S_{21}$ to obtain some useful results. These results are valid only for sufficiently large n, the d.f. on which S_{11} is based. The following is a summary:

It is assumed that only $\rho_1, \rho_2, \ldots, \rho_k$ are nonzero, and the remaining population canonical correlations are all null, i.e., Σ_{12} is of rank k. It is also assumed that $p \leq q$ and that the sample canonical correlations are $r_1 > r_2 > \ldots > r_p$. ρ_i are assumed to be neither very near one another, nor small.

$$2\rho_s E(r_s - \rho_s) = \frac{(1-\rho_s^2)}{n}\left\{p+q-2-\rho_s^2 + 2(1-\rho_s^2)\sum_{\substack{i=1\\i\neq s}}^{q}\frac{\rho_i^2}{\rho_s^2-\rho_i^2}\right\} + O(n^{-2}), \tag{7.5}$$

$$V(r_s) = \frac{1}{n}(1-\rho_s^2)^2 + O(n^{-2}). \tag{7.6}$$

Lawley gives the third and fourth cumulants of r_s also.

The correlation coefficient between r_i and r_j $(i < j \leq k)$ is

$$\frac{2\rho_i\rho_j(1-\rho_i^2)(1-\rho_j^2)}{n(\rho_i^2-\rho_j^2)^2} + O(n^{-2}). \tag{7.7}$$

This breaks down if either ρ_i^2, ρ_j^2 or the difference between them is small.

When $k = 1$, i.e., ρ_1 is the only nonzero population canonical correlation coefficient, or when $\rho_1, \ldots, \rho_{k-1}$ are near unity and ρ_k is moderate, Lawley shows (writing ρ in place of ρ_k) that

$$E(r_k) \doteq \rho + \frac{1}{2n}(1-\rho^2)(m\rho^{-1} - \rho), \tag{7.8}$$

$$V(r_k) \doteq \frac{1}{n}(1-\rho^2)^2 + \frac{1}{2n^2}(1-\rho^2)^2 \left\{11\ \rho^2 - 2(m-k+1) - m\rho^{-2}\right\}. \tag{7.9}$$

where $m = p+q-2k$.

Fisher's Z transformation (4.1.16) does not work well in the canonical correlation case. Lawley gives the moments of z in this case, but it fails to stabilize the variance to any marked extent. In practice, one does not know the ρ's, and so it is necessary to replace ρ_i by r_i in the preceding formulas, to estimate the moments of the ρ_i's.

The investigations carried out by Roy [28, 29], Pillai [22, 23, 24, 25, 26], Khatri [17], Marriott [20], and Chambers [5] on individual roots are worth mentioning. However, the important statistics connected with these canonical correlations are Wilks's $\Lambda = \pi(1-r_i^2)$ and its factors. Their usefulness and the distributional theory associated with them will, therefore, form the subject matter of the next chapter.

8. CANONICAL ANALYSIS

Predicting or forecasting the values of one or more variables from other variables is an important problem in almost every branch of science. This leads to the relationship between two groups of variables, in general, as it is this relationship which must be exploited to yield fruitful predictions. The theory of canonical variables and canonical correlations is, therefore, important in statistics, as it summarizes the relationship in terms of only a few variables and parameters. Most of the practical problems arising in statistics can be translated, in some form or the other, as the problems of measurement of association between two vector variates $\underline{x}$ and $\underline{y}$. Hotelling [13] was the first to tackle the classical problem of identifying and measuring relations between two sets of variables, and his work still stands as an important reference in multivariate analysis. He [12] utilized this canonical analysis in Psychology to study the relationship between a set of mental test variables and a set of physical measurement variables. He could thereby determine and characterize the number and nature of the independent relations of mind and body by extracting them from the multiplicity of correlations in the system (a very good illustration of the use of canonical analysis in educational research is given by Barnett and Lewis [1]). Canonical analysis, however, is by no means useful only in psychological or educational research. It is equally applicable in Econometrics, Anthropometry, Botany (see Pearse and Holland [21]), and several other branches of science. Tintner [30, 31], Waugh [33], and Kendall [15] give many applications of this technique to economic data. Bartlett [3] gives an interesting application of the theory of canonical correlations to estimate supply and demand relations from time series. We will have occasion to refer to this in more detail again.

Canonical analysis is a generalization of the technique of regression of one variable on another, to the regression of one vector on another vector. It therefore brings together multivariate analysis of variance and covariance and design of experiments, discriminant analysis (especially in the case of several groups), structural equations in Econometrics, and analysis of contingency tables, in a unified manner. Each of the above topics can be shown to lead to the relation between two vector variables $\underline{x}$ and $\underline{y}$. In certain cases, both $\underline{x}$ and $\underline{y}$ are random; in certain cases, only one of them is a vector of random variables, the other being either a vector of "dummy" or "pseudo" variables (which take only two values like 0 or 1) or nonrandom variables like time; yet, in certain other situations, both $\underline{x}$ and $\underline{y}$ are "dummy". In multivariate analysis of variance and covariance (as well as in design of experiments) $\underline{x}$ is random, $\underline{y}$ is dummy, and we consider the regression of $\underline{x}$ on $\underline{y}$. In discriminant analysis, the situation is reversed; we consider the regression of $\underline{y}$, the dummy variables, on $\underline{x}$, which is random. The canonical variables of the $\underline{x}$-space, corresponding to the nonzero canonical correlations, turn out to be the discriminators, while the canonical variables of the $\underline{y}$-space provide optimum scores to be assigned to the different populations, yielding a quantitative assessment to the relative positions of the means of the populations. The number of nonzero canonical correlations gives the adequate number of discriminant functions. When $\underline{y}$ consists of nonrandom variables like t or some exogeneous variables, canonical variables of the $\underline{x}$-space corresponding to the zero canonical correlations yield structural equations, if $\underline{x}$ is a vector of economic variables. In the analysis of contingency tables, one can define dummy variables to correspond to each category, and then the canonical variables provide optimum scores to be attached to the different classes or categories. Fisher [10] and Bartlett [4] discuss an example of this type, where they represented certain blood reactions denoted by symbols of the type ?, +, (+), - on a linear scale by using canonical analysis, where both $\underline{x}$ and $\underline{y}$ are dummy. (In this connection, Klatzky and Hodge's [18] paper, "A canonical correlation analysis of occupational mobility", where they assign weights to the categories of occupational mobility tables, is also worth mentioning.) In psychological and educational analysis involving relationship of two batteries of tests, usually $\underline{x}$ and $\underline{y}$ are both random. We shall be considering the details of all these situations in subsequent chapters. A brief reference is made here only to indicate the wide range of applicability of canonical analysis to statistical problems in different fields.

If canonical analysis had no other use, it could at least be used as a descriptive and exploratory tool. It summarizes the complex relationships and provides a useful method of reducing the dimensionality of a problem, in a many-variable situation by considering only the first few important canonical variables.

Bartlett refers to canonical analysis as "external" analysis because it is concerned with the relation of $\underline{x}$ with an external vector $\underline{y}$. As opposed to this, "Principal Components" analysis, which we will discuss in Chapter 11 is "internal" analysis, as it is concerned with the internal relationship and variability of $\underline{x}$ itself. Some authors like Kendall and Stuart [16] regard canonical analysis to be more difficult and less useful than principal components analysis, but this author has a different opinion because of the wider applicability and richness of canonical analysis. Further, the canonical analysis method is less vague, and distributional theory and tests of significance associated with it are more precise and rest on a better foundation than their counterparts in principal components analysis (and factor analysis, which is also intimately related to principal components analysis). In short, it is a systematic technique which will help in understanding, as far as possible, a whole interacting complex system. It has certainly been found to be useful in quantitative biology, though in social sciences its usefulness is rather limited.

No doubt, canonical analysis requires considerably more computational work, but with the easy access to electronic computers these days and availability of ready-made library programs, this can hardly be called an obstacle.

It must be pointed out here that the canonical variables constructed by linearly combining the original variables are, almost always, artificial, in the sense that no physical meaning can, in general, be given to them. However, they are of interest for their own sake, as they help to deepen the understanding of the original variables and may even suggest new measures. Like any other tool, canonical analysis also must be used judiciously; an indiscriminate use may lead to pitfalls, especially when a "joint" analysis of several variables is not called for at all and only a series of separate univariate analysis is more meaningful.

Canonical correlations can be interpreted geometrically as the cosines of angles between two flats. This, as well as several other geometrical aspects of the theory of canonical correlations and its applications, have been discussed very well by Dempster [8].

REFERENCES

[1] Barnett, V. D., and Lewis, T. (1963). "A study of the relation between G. C. E. and degree results", J. Roy. Stat. Soc., A, 126, p. 187.

[2] Bartlett, M. S. (1947). "The general canonical correlation distribution", Ann. Math. Statist., 18, p. 1.

[3] Bartlett, M. S. (1948). "A note on the statistical estimation of demand and supply relations from time series", Econometrica, 16, p. 323.

[4] Bartlett, M. S. (1951). "The goodness of fit of a single hypothetical discriminant function in the case of several groups", Ann. Eugen., 16, p. 199.

[5] Chambers, J. M. (1967). "On methods of asymptotic approximation for multivariate distributions", Biometrika, 54, p. 367.

[6] Constantine, A. G. (1963). "Some noncentral distribution problems in multivariate analysis", Ann. Math. Statist., 34, p. 1270.

[7] Deemer, W. L., and Olkin, I. (1951). "The jacobians of certain matrix transformations useful in multivariate analysis", Biometrika, 38, p. 345.

[8] Dempster, A. P. (1969). Elements of Continuous Multivariate Analysis. Addison-Wesley Publishing Company, Massachusetts.

[9] Fisher, R. A. (1939). "The sampling distribution of some statistics obtained from nonlinear equations", Ann. Eugen., 9, p. 238.

[10] Fisher, R. A. (1946). Statistical Methods for Research Workers. Oliver and Boyd, Edinburgh.

[11] Good, I. J. (1969). "Some applications of the singular decomposition of a matrix", Technometrics, 11, p. 823.

[12] Hotelling, H. (1935). "The most predictable criterion", J. Educat. Psych., 26, p. 139.

[13] Hotelling, H. (1936). "Relation between two sets of variates", Biometrika, 28, p. 321.

[14] Hsu, P. L. (1939). "On the distribution of the roots of certain determinantal equations", Ann. Eugen., 9, p. 250.

[15] Kendall, M. G. (1961). A Course in Multivariate Analysis. Charles Griffin and Company, London.

[16] Kendall, M. G., and Stuart, A. (1968). The Advanced Theory of Statistics, Vol. 3, Charles Griffin and Company, London.

[17] Khatri, C. G., and Pillai, K. C. S. "Some results on the noncentral multivariate beta distribution and moments of traces of two matrices", Ann. Math. Statist., 36, p. 1511.

[18] Klatzky, S. R., and Hodge, R. W. (1971). "A canonical analysis of occupational mobility", J. Am. Stat. Assoc., 66, p. 16.

[19] Lawley, D. N. (1959). "Tests of significance in canonical analysis", Biometrika, 46, p. 59.

[20] Marriott, F. H. C. (1952). "Tests of significance in canonical analysis", Biometrika, 39, p. 58.

[21] Pearce, S. C., and Holland, D. A. (1960). "Some applications of multivariate methods in botany", Applied Statistics, 9, p. 1.

[22] Pillai, K. C. S. (1956). "On the distribution of the largest or the smallest root of a matrix in multivariate analysis", Biometrika, 43, p. 122.

[23] Pillai, K. C. S. (1957). "Concise tables for statisticians", The Statistical Center, University of the Phillipines.

[24] Pillai, K. C. S., and Mijares, T. H. (1959). "On the moments of the trace of matrix and approximations to its distribution", Ann. Math. Statist., 30, p. 1135.

[25] Pillai, K. C. S., and Bantegui, C. G. (1959). "On the distribution of the largest of six roots of a matrix in multivariate analysis", Biometrika, 46, p. 237.

[26] Pillai, K. C. S. (1964). "On the moments of elementary symmetric functions of the roots of two matrices", Ann. Math. Statist., 35, p. 1704.

[27] Roy, S. N. (1939). "p-statistics or some generalizations in analysis of variance appropriate to multivariate problems", Sankhyā, 4, p. 381.

[28] Roy, S. N. (1945). "The individual sampling distribution of the maximum, the minimum and any intermediate the p-statistics on the null hypothesis", Sankhyā, 7, p. 133.

[29] Roy, S. N. (1957). Some Aspects of Multivariate Analysis. John Wiley and Sons, New York.

[30] Tintner, G. (1946). "Some applications of multivariate analysis in economic data", J. Am. Stat. Assoc., 41, p. 472.

[31] Tintner, G. (1952). Econometrics. John Wiley and Sons, New York.

[32] Tumura, Y. (1965). "The distribution of latent roots and vectors", Tokyo Rica University, Mathematics, 1, p. 1.

[33] Waugh, F. V. (1942). "Regressions between sets of variables", Econometrica, 10, p. 290.

CHAPTER 8

WILKS'S Λ CRITERION AND ITS APPLICATIONS

1. WILKS'S Λ CRITERION

In univariate statistical analysis there are situations where one constructs, from the sample observations, a statistic u which has a $\chi^2\sigma^2$ distribution with f_2 d.f. (i.e., u/σ^2 has a χ^2 distribution) and an independent statistic v, which has, in general, a noncentral $\chi^2\sigma^2$ distribution with f_1 d.f. From these two then, the criterion $F = f_2 v/f_1 u$ is built up and is used to test the hypothesis that the noncentrality parameter in the distribution of v is zero. A large number of tests of significance occurring in regression, analysis of variance and covariance, design of experiments, and in general analysis of observations from a linear statistical model are constructed in this manner. The null hypothesis is then rejected, at a level of significance α, if the observed value of F exceeds the 100 α% point of the distribution of F. The nuisance parameter σ^2 does not enter the distribution of F, and this is achieved by taking the ratio of v to u, in building F.

Wilks's Λ criterion plays the same role in multivariate analysis that F plays in univariate analysis. In many multivariate problems, (see Kshirsagar [36] for an expository paper on Wilks's Λ), it is possible to construct two matrices A and B such that A has the $W_p(A|n-q|\Sigma)$ distribution and B has an independent $W_p(B|q|\Sigma)$ distribution if and only if a certain hypothesis H_o is true. Otherwise B has a noncentral Wishart distribution. Σ is usually unknown and is a nuisance parameter, like σ^2. Since A and B are matrices, we cannot think of a ratio A/B as in univariate analysis. Wilks [66] therefore, proposed the criterion

$$\Lambda = \frac{|A|}{|A + B|} \tag{1.1}$$

to test the validity of H_o, and is known as Wilks's Λ criterion. In order to be able to use this criterion to test H_o, it is necessary to know its distribution and its percentage points. We shall, therefore, consider these first and later turn to several applications of Wilks's Λ in different fields.

Sometimes q, the d.f. on which B is based, is smaller than p, the number of variables, and then B cannot have a Wishart density like (3.3.8). We have already remarked (see Section 6 of Chapter 3) that, in such cases, B is said to have a pseudo-Wishart distribution, i.e., B is distributed as $\sum_{r=1}^{q} \underline{Z}_r \underline{Z}_r'$, where $\underline{Z}_r$ (r = 1, 2, ..., q) are independent and have a p-variate normal distribution with variance-covariance matrix Σ. The means of $\underline{Z}_r$ are null if and only if H_o is true. We, therefore, formally define Wilks's Λ as below:

Definition: If a matrix A has the $W_p(A|n-q|\Sigma)$ distribution and is independent of $\underline{Z}_r$ $(r = 1, 2, \ldots, q)$, which themselves are independently distributed as $N_p(\underline{Z}_r|\underline{0}|\Sigma)$, Wilks's Λ criterion is defined as

$$\Lambda(n,p,q) = \frac{|A|}{|A + B|}, \tag{1.2}$$

where

$$B = \sum_{r=1}^{q} \underline{Z}_r \underline{Z}_r' .$$

This statistic is used to test any hypothesis which is equivalent to

$$H_o: \; E(\underline{Z}_r) = \underline{0} \quad (r = 1, 2, \ldots, q).$$

n,p,q are said to be the parameters of Wilks's Λ and are, respectively the d.f. of A + B, the order of A and B, and the d.f. of B.

In univariate analysis, u (refereed to earlier) corresponds to what is known as the error s.s. and v to the hypothesis s.s. Correspondingly A here is called the error matrix of s.s. and s.p. and B is called the hypothesis matrix of s.s. and s.p. The distribution of $|A|/|A + B|$ is known as the Wilks's $\Lambda(n,p,q)$ distribution. When the hypothesis H_o is not true, the distribution of $|A|/|A + B|$ will depend on $E(\underline{Z}_r)$ and is known as the non-null distribution of Wilks's Λ.

The matrices A, B and their d.f.'s can all be formally presented in a multivariate analysis of variance table (Table 1).

TABLE 1

Source	d.f.	Matrix of s.s. and s.p.
Hypothesis H_o	q	B
Error	n-q	A
Total	n	A+B

Wilks's Λ is then the ratio of the determinants of the error matrix to the total matrix (error + hypothesis). n is the d.f. of this total matrix, p is the number of variables (or order of A, B), and q is the d.f. of the hypothesis matrix. In strict accordance with univariate analysis of variance, the criterion should have been $|B|/|A|$. But $|A|/|A + B|$ is more suitable on account of its tractability, and further it is related to the likelihood ratio criterion.

Wilks's Λ is not the only criterion that can be constructed from A and B. There are other criteria too, and we shall desribe them briefly later. Our emphasis in this book will, however, be on Wilks's Λ only.

2. THE DISTRIBUTION OF WILKS'S Λ

Case 1: We shall first consider the case when $p \leq q$. In this case, B has the $W_p(B|q|\Sigma)$ distribution when H_o is true, and the joint distribution of A and B will be

$$W_p(A|n-q|\Sigma)\; W_p(B|q|\Sigma)\; dA\, dB\,. \tag{2.1}$$

Let C be a l.t. matrix such that

$$A + B = CC' . \tag{2.2}$$

C is not unique in the sense that we can change the signs of all the elements in any column of C without altering (2.2). For definiteness, therefore, we choose C in such a way that the diagonal elements are positive. From A and B, now transform to A + B and the symmetric matrix L defined by

$$A = C L C' . \tag{2.3}$$

The Jacobian of this transformation is (Deemer and Olkin [15]) $|C|^{p+1} = |A+B|^{(p+1)/2}$. The joint distribution of A + B and L is therefore, given by

$$W_p(A+B|n|\Sigma)\ d(A+B) \cdot B_p(L|\tfrac{n-q}{2}, \tfrac{q}{2})\ dL ,$$

showing that A + B and L are independently distributed and that the distribution of L is

$$B_p(L|\tfrac{n-q}{2}, \tfrac{q}{2})\ dL = \begin{cases} \dfrac{1}{B_p\left(\frac{n-q}{2}, \frac{q}{2}\right)} |L|^{(n-q-p-1)/2}\ |I - L|^{(q-p-1)/2}\ dL, & \text{if } L > 0,\ I - L > 0 \\ 0 \quad \text{otherwise.} & \end{cases} \tag{2.4}$$

Here L > 0 stands for L positive definite. Also

$$B_p(\tfrac{f_1}{2}, \tfrac{f_2}{2}) = \pi^{p(p-1)/4} \prod_{i=1}^{p} \left\{ \frac{\Gamma\left(\frac{f_1+f_2+1-i}{2}\right)}{\Gamma\left(\frac{f_1+1-i}{2}\right) \Gamma\left(\frac{f_2+1-i}{2}\right)} \right\} . \tag{2.5}$$

The distribution of L is known as the matrix variate beta distribution because of its similarity to the univariate beta distribution $B(Z|\ell, m)$, defined in (4.2.25).

We now make a further transformation

$$L = TT' , \tag{2.6}$$

where

$$T = \begin{bmatrix} t_{11} & 0 \\ \underline{t} & T_2 \end{bmatrix} \begin{matrix} 1 \\ p-1 \end{matrix} \tag{2.7}$$
$$\quad\; 1 \quad p-1$$

is l.t. and t_{ij} $(i > j)$ are its nonzero elements. T_2 also is evidently l.t. and of order p-1. The Jacobian of transformation $J(L \to T)$ is $2^p \prod_{i=1}^{p} (t_{ii}^{p+1-i})$ (Deemer and Olkin [15]) and so, from (2.4), the distribution of T is (as $|L| = \prod_1^p t_{ii}^2$).

$$\frac{2^p}{B_p\left(\frac{n-q}{2}, \frac{q}{2}\right)} \prod_{i=1}^{p} (t_{ii}^{n-q-i}) \; |I - TT'|^{(q-p-1)/2} \; dT$$

$$= V_p(T|n-q, q) \; dT, \quad \text{say.} \tag{2.8}$$

Because of (1.3.14) and (1.4.5), we find that

$$|I - TT'| = \left|\begin{matrix} 1 - t_{11}^2 & -t_{11}\underline{t}' \\ -t_{11}\underline{t} & I - \underline{t}\underline{t}' - T_2T_2' \end{matrix}\right| \tag{2.9}$$

$$= (1 - t_{11}^2) \; |I - T_2T_2' - (1 - t_{11}^2)^{-1} \; \underline{t}\underline{t}'|$$

$$= (1 - t_{11}^2) \; |I - T_2T_2'| \; |I - (1-t_{11}^2)^{-1}(I-T_2T_2')^{-1}\underline{t}\underline{t}'|$$

$$= (1-t_{11}^2)|I - T_2T_2'| \cdot \{1 - (1-t_{11}^2)^{-1}\underline{t}'(I-T_2T_2')^{-1}\underline{t}\}.$$

The distribution of T given by (2.8) is the same as the joint distribution of t_{11}, $\underline{t}$, and T_2 of (2.7). Make a transformation from $\underline{t}$ to $\underline{w}$ defined by

$$\underline{w} = (1 - t_{11}^2)^{-1/2} \; (I - T_2T_2')^{-1/2} \; \underline{t} \; . \tag{2.10}$$

The Jacobian of the transformation is

$$J(\underline{t} \to \underline{w}) = (1 - t_{11}^2)^{(p-1)/2} \; |I - T_2T_2'|^{1/2} \; .$$

Hence, from (2.8) and (2.9), the joint distribution of t_{11}, $\underline{w}$, and T_2 is

$$B\left(t_{11}^2 \mid \frac{n-q}{2}, \frac{q}{2}\right) d(t_{11}^2) \cdot V_{p-1}(T_2 | n-q-1, q) dT_2 \cdot k(\underline{w}) d\underline{w}, \tag{2.11}$$

where $k(\underline{w})$ is some function of $\underline{w}$ only. This shows that

$$t_{11}^2 \text{ has the } B\left(t_{11}^2 \mid \frac{n-q}{2}, \frac{q}{2}\right) \text{ distribution} \tag{2.12}$$

and is independent of T_2; the distribution of T_2 is exactly the same as T in (2.8), with n changed to n-1 and p to p-1. It is, therefore, evident that, if we now partition T_2 like T in (2.7), and proceed exactly similarly, we shall find that t_{22}^2 has the $B(t_{22}^2 | (n-q-1)/2, q/2)$ distribution independent of t_{11}^2. Continuing this process, we finally obtain the result:

$$t_{ii}^2 (i = 1, 2, \ldots, p) \text{ are independently distributed as } B(t_{ii}^2 | (n-q-i+1)/2, q/2) dt_{ii}^2 . \tag{2.13}$$

From (2.2), (2.3), and (2.6), we notice that

$$\begin{aligned} \Lambda = \frac{|A|}{|A + B|} &= \frac{|CLC'|}{|CC'|} \\ &= \frac{|C| \; |L| \; |C'|}{|C| \; |C'|} \\ &= |L| \\ &= |TT'| = |T|^2 \\ &= \prod_{i=1}^{p} t_{ii}^2 . \end{aligned} \tag{2.14}$$

From the distribution of the t_{ii}^2 given in (2.13) and from (2.4), we have the following two important theorems:

Theorem 1: If the distribution of a symmetric matrix L is $B_p(L|(n-q)/2, q/2)dL$, the variables $t_{11}^2 = |L_1|$, $t_{22}^2 = |L_2|/|L_1|$, ..., $t_{pp}^2 = |L_p|/|L_{p-1}|$ (where L_i is the matrix of the first i rows and columns of L and $|L_o| \equiv 1$) are independently distributed, the distribution of t_{ii}^2 being $B(t_{ii}^2|(n-q-i+1)/2, q/2)dt_{ii}^2$ (i = 1, 2, ..., p).

Since T is l.t., $|L_i| = t_{11}^2 \cdot t_{22}^2 \cdot \ldots \cdot t_{ii}^2$ and so $t_{ii}^2 = |L_i|/|L_{i-1}|$. The above theorem then follows from (2.13).

Theorem 2: Wilks's $\Lambda(n,p,q)$ distribution when $p \leq q$ is the distribution of $\prod_{i=1}^{p} t_{ii}^2$, where t_{11}^2, ..., t_{pp}^2 are independently distributed as $B(t_{ii}^2|(n-q+1-i)/2, q/2)dt_{ii}^2$, (i = 1, 2, ..., p).

Several authors (see for example Consul [10] or Pillai [52]) have attempted to find an explicit expression for the p.d.f. of $\Lambda(n,p,q)$ by writing $\log \Lambda = \sum_{1}^{p} \log t_{ii}^2$ and then considering the convolution of the independent variables $\log t_{ii}^2$. But the p.d.f. is complicated and of little use. We shall, therefore, not pursue this. Instead we shall utilize Theorem 2 itself later on. But first we consider the other case, p > q.

Case 2 (p > q): As p > q, B has not the Wishart density, and so we are required to consider the joint distribution of A and the q normal vectors $\underline{Z}_r$ given in (1.2). If

$$Z = [\underline{Z}_1|\underline{Z}_2| \cdots |\underline{Z}_q],$$

the joint distribution of A and Z is

$$\frac{K(p,\ n-q)}{(2\pi)^{pq/2}\ |\Sigma|^{n/2}}\ |\mathbf{A}|^{(n-q-p-1)/2}\ \exp\left\{-\frac{1}{2}\operatorname{tr}\Sigma^{-1}(A+ZZ')\right\} dAdZ\ , \tag{2.15}$$

where K(p,n-q) is defined in (2.2.20). Now compare (2.15) with (7.5.28). Both are the same, except that we use Z here and had Z_1 in (7.5.28). We therefore proceed in exactly the same way as we did in Chapter 7, Section 5 (Case 2, $p > q$), to derive (7.5.42) from (7.5.28), i.e., we make the transformations

$$G = \mathbf{A} + ZZ',\quad U = G^{-1/2}Z,\quad H^* = U'U\ . \tag{2.16}$$

Corresponding to (7.5.29), and (7.5.30) and from (7.5.42) obtain the distribution of the $q \times q$ matrix H^* to be

$$B_q\left(H^* \middle| \frac{p}{2},\ \frac{n-p}{2}\right) dH^*\ . \tag{2.17}$$

Hence the distribution of $L^* = I - H^*$ is

$$B_q\left(L^* \middle| \frac{n-p}{2},\ \frac{p}{2}\right) dL^*\ . \tag{2.18}$$

So, by Theorem 1,

$$t_{ii}^{*2} = |L_i^*|/|L_{i-1}^*| \text{ has the } B\left(t_{ii}^{*2} \middle| \frac{n-p-i+1}{2},\ \frac{p}{2}\right) dt_{ii}^{*2}\ , \tag{2.19}$$

distribution and the t_{ii}^{*2} $(i = 1, 2, \ldots, q)$ are all independent. Now observe that [from (2.16) and (1.2) and using (1.4.5)]

$$\begin{aligned}\Lambda &= \frac{|\mathbf{A}|}{|\mathbf{A}+B|} = \frac{|\mathbf{A}|}{|\mathbf{A}+ZZ'|} = \frac{|G - ZZ'|}{|G|} = |I - G^{-1/2}ZZ'G^{-1/2}| \\ &= |I - UU'| = |I - U'U| = |I - H^*| = |L^*| \\ &= \prod_{i=1}^{q} \frac{|L_i^*|}{|L_{i-1}^*|} = \prod_{i=1}^{q} t_{ii}^{*2} \quad (\text{as } |L_0^*| = 1).\end{aligned} \tag{2.20}$$

We therefore have the following theorem:

Theorem 3: Wilks's $\Lambda(n,p,q)$ distribution, when $p > q$ is the distribution of $\prod_{i=1}^{q} t^{*2}_{ii}$, where t^{*2}_{ii} $(i = 1, 2, \ldots, q)$ are independently distributed as $B(t^{*2}_{ii} | (n-p+1-i)/2, p/2)dt^{*2}_{ii}$.

A comparison of Theorems 2 and 3 shows that the distribution of $\Lambda(n,p,q)$ when $p > q$ can be obtained from that of $\Lambda(n,p,q)$ when $p \leq q$ only by interchanging p and q. Since

$$E(t^2_{ii})^h = \frac{\Gamma\left(\frac{n+1-i}{2}\right)}{\Gamma\left(\frac{n+1-i}{2}+h\right)} \frac{\Gamma\left(\frac{n-q+1-i}{2}+h\right)}{\Gamma\left(\frac{n-q+1-i}{2}\right)} \tag{2.21}$$

when t^2_{ii} has the $B(t^2_{ii} | (n-q+1-i)/2, q/2)$ distribution, it follows from Theorem 2 that, when $p < q$,

$$E(\Lambda^h) = \prod_{i=1}^{p} \left\{\frac{\Gamma\left(\frac{n+1-i}{2}\right)}{\Gamma\left(\frac{n+1-i}{2}+h\right)} \frac{\Gamma\left(\frac{n-q+1-i}{2}+h\right)}{\Gamma\left(\frac{n-q+1-i}{2}\right)}\right\}. \tag{2.22}$$

This result gives the moments of Wilks's Λ. Since each t^2_{ii} varies in the range (0, 1), Wilks's Λ, which is equal to $\prod_{1}^{p} t^2_{ii}$, also varies in the range (0, 1), which is a finite range. In general, the moments of a distribution do not determine the distribution uniquely. (This is unlike the characteristic function, which always determines a distribution uniquely.) However, under certain conditions, the moments do determine a distribution uniquely. A finite range is a sufficient condition for this purpose. Hence the moments (2.22) of Λ determine the Wilks's $\Lambda(n,p,q)$ distribution uniquely. Now observe that (2.22) can also be written as

$$E(\Lambda^h) = \frac{\Gamma\left(\frac{n}{2}\right)\Gamma\left(\frac{n-1}{2}\right)\cdots\Gamma\left(\frac{n-p+1}{2}\right)\Gamma\left(\frac{n-p}{2}\right)\cdots\Gamma\left(\frac{n-q+1}{2}\right)}{\Gamma\left(\frac{n}{2}+h\right)\Gamma\left(\frac{n-1}{2}+h\right)\cdots\Gamma\left(\frac{n-p+1}{2}+h\right)\Gamma\left(\frac{n-p}{2}+h\right)\cdots\Gamma\left(\frac{n-q+1}{2}+h\right)}$$

$$\times\frac{\Gamma\left(\frac{n-p}{2}+h\right)\cdots\Gamma\left(\frac{n-q+1}{2}+h\right)\Gamma\left(\frac{n-q}{2}+h\right)\cdots\Gamma\left(\frac{n-q-p+1}{2}+h\right)}{\Gamma\left(\frac{n-p}{2}\right)\cdots\Gamma\left(\frac{n-q+1}{2}\right)\Gamma\left(\frac{n-q}{2}\right)\cdots\Gamma\left(\frac{n-q-p+1}{2}\right)}$$

$$= \prod_{i=1}^{q}\left\{\frac{\Gamma\left(\frac{n+1-i}{2}\right)\Gamma\left(\frac{n-p+1-i}{2}+h\right)}{\Gamma\left(\frac{n+1-i}{2}+h\right)\Gamma\left(\frac{n-p+1-i}{2}\right)}\right\}$$

$$= \prod_{i=1}^{q} E(t_{ii}^{*2})^h, \tag{2.23}$$

where t_{ii}^{*2} have the $B(t_{ii}^{*2}|(n-p+1-i)/2,\ p/2)$ distribution. This result, therefore, proves that the moments of Wilks's $\Lambda(n,p,q)$ are unaltered by interchanging p and q. We have so far proved that:

(a) The distribution of $\Lambda(n,p,q)$ when $p > q$ is obtainable from that of $\Lambda(n,p,q)$ when $p \le q$ by interchanging p and q.

(b) The moments of Wilks's Λ determine the distribution uniquely.

(c) The moments of Wilks's Λ remain unaltered if p and q are interchanged.

Putting all these together, it is easy to see that

Theorem 4: $\Lambda(n,p,q)$ and $\Lambda(n,q,p)$ have exactly the same distribution, and it is also the distribution of $\prod_1^p t_{ii}^2$ if

$p \leq q$ and of $\prod_{1}^{q} t_{ii}^{*2}$ if $p > q$, where t_{ii}^2 are independent $B(t_{ii}^2 | (n-q+1-i)/2,\ q/2)$ variables and t_{ii}^{*2} are independent $B(t_{ii}^{*2} | (n-p+1-i)/2,\ p/2)$ variables.

From this theorem, by putting p = 1, 2 we find, in particular, that

$$\frac{1 - \Lambda(n,1,q)}{\Lambda(n,1,q)} \left(\frac{n-q}{q}\right) \text{ has the } F_{q,n-q} \text{ distribution} \tag{2.24}$$

and

$$\frac{1 - \sqrt{\Lambda(n,2,q)}}{\sqrt{\Lambda(n,2,q)}} \left(\frac{n-q-1}{q}\right) \text{ has the } F_{2q,\ 2(n-q-1)} \text{ distribution.} \tag{2.25}$$

Similarly, when q = 1, 2, we find that [by interchanging p and q in (2.24) and (2.25)]

$$\frac{1 - \Lambda(n,p,1)}{\Lambda(n,p,1)} \left(\frac{n-p}{p}\right) \text{ has the } F_{p,\ n-p} \text{ distribution} \tag{2.26}$$

and

$$\frac{1 - \sqrt{\Lambda(n,p,2)}}{\sqrt{\Lambda(n,p,2)}} \left(\frac{n-p-1}{p}\right) \text{ has the } F_{2p,\ 2(n-p-1)} \text{ distribution} \tag{2.27}$$

The general case of $\Lambda(n,p,q)$ for any values of p and q is considered in the next section.

3. BARTLETT'S APPROXIMATION TO WILKS'S Λ

In (2.22) above, put $h = -m\sqrt{-1}\,t$. We obtain

$$\prod_{i=1}^{p} \left\{ \frac{\Gamma\left(\frac{n+1-i}{2}\right)}{\Gamma\left(\frac{n+1-i}{2} - m\sqrt{-1}\,t\right)} \cdot \frac{\Gamma\left(\frac{n-q+1-i}{2} - m\sqrt{-1}\,t\right)}{\Gamma\left(\frac{n-q+1-i}{2}\right)} \right\}$$

$$= E(\Lambda^{-m\sqrt{-1}\,t})$$

$$= E\{\exp[\sqrt{-1}\,t(-m \log_e \Lambda)]\}$$

= the characteristic function of $-m \log_e \Lambda$, with t as the argument. (3.1)

This result will be valid for all real t only when the gamma functions in (3.1) exist. If so, we can expand the gamma functions by using

$$\log \Gamma(x+h) = \log \sqrt{2\pi} + (x+h-\tfrac{1}{2}) \log x - x - \sum_{r=1}^{k} (-1)^r \frac{B_{r+1}(h)}{r(r+1)x^r} + R_{k+1}(x), \quad (3.2)$$

where $R_{k+1}(x) = O(x^{-k-1})$ and $B_r(h)$ are Bernoulli polynomials. By this method, by expanding the characteristic function of $-m \log_e \Lambda$ up to terms of order $O(1/n^2)$ and disposing of the arbitrary constant m suitably, Bartlett [5] derived the result that

$$-\{n - \tfrac{1}{2}(p+q+1)\} \log_e \Lambda \text{ is approximately distributed as a } \chi^2 \text{ with pq d.f.} \quad (3.3)$$

This approximation to the true distribution of Wilks's Λ is sufficiently accurate, for all practical purposes, if n is moderately large. In fact, if $p^2 + q^2 \leq \frac{1}{3}m$ [with $m = n - (p+q+1)/2$] it is accurate to three decimal places. It is, therefore, a very important result from the practical point of view, as it is not necessary to consider the percentage points of the distribution of Λ. The χ^2 table is sufficient, provided n is not too small. One has simply to calculate $-\log_e \Lambda$, multiply it by <u>Bartlett's correction factor</u> $m = n - (p+q+1)/2$, and compare the resulting value with the $100\alpha\%$ point of the χ^2 distribution with pq d.f., to test the hypothesis H_o. Bartlett later extended this technique of correction factors to various tests in multivariate analysis and opened up an easy access to multivariate tests, without any additional statistical tables. Approximation to Wilks's Λ here is so good that it is hardly an approximation; it is almost an exact result.

Box [8] expanded the characteristic function to terms of order n^{-6} and gave the following extension of Bartlett's result, when more accuracy is needed:

$$\begin{aligned}\text{Prob}(-m \log_e \Lambda \leq u) = C_{pq}(u) &+ \frac{r_2}{m^2}\{C_{pq+4}(u) - C_{pq}(u)\} \\ &+ \frac{1}{m^4}[r_4\{C_{pq+8}(u) - C_{pq}(u)\} \\ &\quad - r_2^2\{C_{pq+4}(u) - C_{pq}(u)\}] \\ &+ O(n^{-6}), \qquad (3.4)\end{aligned}$$

where $C_f(u)$ is the cumulative distribution function of a χ_f^2 variable, i.e.,

$$C_f(u) = \int_0^u \chi_f^2(v)dv \qquad (3.5)$$

and

$$\begin{aligned} r_2 &= pq(p^2 + q^2 - 5)/48 \\ r_4 &= \frac{r_2^2}{2} + \frac{pq}{1920}[3p^4 + 3q^4 + 10p^2q^2 - 50(p^2 + q^2) + 159]. \end{aligned} \qquad (3.6)$$

Rao [55] has also given a different approximation. His approximation is

$$\frac{ms + 2\lambda}{2r}\,\frac{1 - \Lambda^{1/s}}{\Lambda^{1/s}} \text{ has the } F_{2r,\ ms+2\lambda} \text{ distribution,} \qquad (3.7)$$

where $m = n - (p+q+1)/2$ as before, and

$$r = pq/2, \quad \lambda = -(pq-2)/4, \quad s = (p^2q^2-4)^{1/2}/(p^2+q^2-5)^{1/2}.$$

Recently, however, Schatzoff [60] and Pillai and Gupta [53] have prepared tables that give the exact percentage points of the distribution of Wilks's Λ. Schatzoff prepared a computer program to obtain the convolution of the independent variables $\log_e t_{ii}^2$, recursively, to yield the distribution of $\log_e \Lambda = \sum_1^p \log_e t_{ii}^2$. These tables of Schatzoff and Pillai and Gupta are included in "Tables for Multivariate Analysis", edited by Pearson [42]. These tables do not give the percentage points of the distribution of Wilks's Λ directly. The tables give the values of a conversion factor

$$C_\alpha(p,q,M) \ , \tag{3.8}$$

depending on p, q, M = n-p-q+1 and α, the level of significance. The following formula is then to be used to obtain the exact $100\,\alpha\%$ point of the distribution of $W = -m \log_e \Lambda(n,p,q)$:

$$W_\alpha(n,p,q) = C_\alpha(p,q,M)\chi_{pq}^2(\alpha) \ , \tag{3.9}$$

where $W_\alpha(n,p,q)$ is the $100\,\alpha\%$ point for $-m \log_e \Lambda(n,p,q)$ and $\chi_{pq}^2(\alpha)$ is the $100\,\alpha\%$ point of a χ^2 distribution with pq d.f. Pearson's table gives $C_\alpha(p,q,M)$ for $\alpha = 0.01$ and 0.05, but the original tables consider $\alpha = 0.10$, 0.05, 0.025, 0.01, 0.005. Since $\Lambda(n,p,q)$ and $\Lambda(n,q,p)$ have the same distribution, $C_\alpha(p,q,M)$ and $C_\alpha(q,p,M)$ are the same.

The null hypothesis H_0 is to be rejected at a level of significance α if the observed value of $W = -m \log_e \Lambda(n,p,q)$ exceeds $W_\alpha(n,p,q)$. The conversion factor $C_\alpha(p,q,M)$ is always greater than unity. Therefore, if $W < \chi^2_{pq}(\alpha)$, it is certainly less than $C_\alpha(p,q,M)\chi^2_{pq}(\alpha) = W_\alpha(n,p,q)$, and H_o will not be rejected. In this case, therefore, it is not necessary to refer to the tables of C, and Bartlett's approximate test based on $\chi^2_{pq}(\alpha)$ is sufficient. But if $W > \chi^2_{pq}(\alpha)$, it could still be $< W_\alpha(n,p,q)$. Bartlett's test would then reject H_o, while in fact, there is no evidence against H_o according to the exact test. So, for greater accuracy, $C_\alpha(p,q,M)$ should be referred to whenever $W > \chi^2_{pq}(\alpha)$. Thus, for example, when $p = 7$, $q = 8$, and $n = 17$, we find $M = 3$, $pq = 56$; and for $\alpha = 0.05$, the 5% point of the χ^2 distribution with 56 d.f. is 83·5134. From Schatzoff's tables for the conversion factor, we obtain $C_{0.05}(7.8.3) = 1{\cdot}210$. The exact 5% point of $W = -m \log_e \Lambda$ is, by (3.9),

$$W_\alpha(n,p,q) = W_{.05}(17.7.8) = (1{\cdot}210)(83{\cdot}5134)$$
$$= 101{\cdot}0512 \; .$$

Hence, it could be immediately concluded that there is no evidence against H_o if the observed value of W is smaller than 83·5134. But if it exceeds 83·5134 and does not exceed 101·0512, Bartlett's test will reject H_o, when actually there is no evidence against H_o. If, however, W exceeds 101·0512, H_o must be rejected at the 5% level of significance.

4. INDEPENDENCE OF TWO VECTORS

Let $\underline{x}$ and $\underline{y}$ be two vectors of p and q components, respectively, having a joint (p+q)-variate nonsingular normal distribution with means $\underline{\mu}$ and $\underline{\nu}$ respectively, and variance-covariance matrix

$$\begin{bmatrix} \Sigma_{11} & \Sigma_{12} \\ \Sigma_{21} & \Sigma_{22} \end{bmatrix} \begin{matrix} p \\ q \end{matrix} . \tag{4.1}$$
$$\begin{matrix} p & \quad q \end{matrix}$$

The regression of $\underline{x}$ on $\underline{y}$ is (see Section 4 of Chapter 2)

$$E(\underline{x}|\underline{y}) = \underline{\mu} + \beta(\underline{y} - \underline{\nu}) , \tag{4.2}$$

where

$$\beta = \Sigma_{12}\Sigma_{22}^{-1} \text{ is the matrix of regression coefficients.} \tag{4.3}$$

The residual variance-covariance matrix is

$$V(\underline{x}|\underline{y}) = \Sigma_{11} - \Sigma_{12}\Sigma_{22}^{-1}\Sigma_{21} = \Sigma_{11\cdot 2} . \tag{4.4}$$

Similarly, the regression of $\underline{y}$ on $\underline{x}$ is

$$E(\underline{y}|\underline{x}) = \underline{\nu} + \beta^*(\underline{x} - \underline{\mu}) , \tag{4.5}$$

where $\beta^* = \Sigma_{21}\Sigma_{11}^{-1}$ is the matrix of regression coefficients. The population canonical correlations between $\underline{x}$ and $\underline{y}$ are ρ_i (i = 1, 2, ..., k; k = smaller of p,q), where ρ_i^2 are the roots of (see Chapter 7)

$$|-\rho^2\Sigma_{11} + \Sigma_{12}\Sigma_{22}^{-1}\Sigma_{21}| = 0 \quad \text{or} \quad |-\rho^2\Sigma_{22} + \Sigma_{21}\Sigma_{11}^{-1}\Sigma_{12}| = 0 . \tag{4.6}$$

We shall consider the hypothesis H_o of independence of $\underline{x}$ and $\underline{y}$. The joint p.d.f. of $\underline{x}$ and $\underline{y}$ factorizes as $N_p(\underline{x}|\underline{\mu}|\Sigma_{11})\ N_q(\underline{y}|\underline{\nu}|\Sigma_{22})$ if and only if Σ_{12} is a null matrix. When Σ_{12} is null, both β and β^*, given by (4.3) and (4.5), are null, as also are all the ρ_i. H_o can therefore be expressed in any of the following alternative forms:

(a) $\Sigma_{12} = 0$, (b) $\beta = 0$, (c) $\beta^* = 0$, (d) $\rho_1^2 = \rho_2^2 = \dots = \rho_k^2 = 0$.

We shall derive a test for H_o on the basis of a random sample of $N = n+1$ observations on $\underline{x}$ and $\underline{y}$. If X and Y denote, respectively, the $p \times N$ and $q \times N$ matrices of these sample observations, the matrices of the corrected s.s. and s.p. of these observations are

$$C_{xx} = XJX', \quad C_{yy} = YJY', \quad C_{xy} = XJY' = C'_{yx} \tag{4.7}$$

where $J = I - N^{-1}E_{NN}$. Previously we had used S_{11}, S_{22}, and S_{12} instead of C_{xx}, C_{yy}, and C_{xy}, respectively, but later, we will have occasion to subdivide $\underline{x}$ and $\underline{y}$ themselves into two parts and then the subscripts 1, 2 will create confusion. So we are switching to this new notation due to Bartlett [4]. We shall use

$$C_{xx\cdot y} = C_{xx} - C_{xy}C_{yy}^{-1}C_{yx} \tag{4.8}$$

instead of $S_{11\cdot 2}$. The matrices of sample regression coefficients of $\underline{x}$ on $\underline{y}$ and $\underline{y}$ on $\underline{x}$ are, respectively,

$$\hat{\beta} = C_{xy}C_{yy}^{-1} \quad \text{and} \quad \hat{\beta}^* = C_{yx}C_{xx}^{-1}, \tag{4.9}$$

respectively. These correspond to the true regression coefficients β and β^*, respectively. From (4.2), we obtain

$$\begin{aligned} E(X|\underline{y}) &= \underline{\mu}\, E_{1N} + \beta(Y - \underline{\nu}\, E_{1N}) \\ &= \underline{\alpha}\, E_{1N} + \beta Y, \end{aligned} \tag{4.10}$$

where $\underline{\alpha} = \underline{\mu} - \beta\underline{\nu}$. Compare this with (iii) of (7.5.3). We find that all the assumptions of (7.5.3) are satisfied if $N > p+q+1$.

We can, therefore, use the results already proved in Section 5 of Chapter 7. We made an orthogonal transformation from X to

$$Z = [Z_1 | Z_2 | Z_3] P \tag{4.11}$$

there and showed [see (7.5.27)] that

$$\begin{aligned} A = Z_2 Z_2' &= C_{xx} - C_{xy} C_{yy}^{-1} C_{yx} \\ &= C_{xx \cdot y} \end{aligned} \tag{4.12}$$

has the $W_p(A | n-q | \Sigma_{11 \cdot 2})$ distribution and is independent of Z_1, and hence of

$$\begin{aligned} B = Z_1 Z_1' &= C_{xy} C_{yy}^{-1} C_{yx} \\ &= \hat{\boldsymbol{\beta}} \, C_{yy} \, \hat{\boldsymbol{\beta}}' \ , \end{aligned} \tag{4.13}$$

$$E(Z_1 | \underline{y}) = \boldsymbol{\beta} \, Y \, P_1 \ , \tag{4.14}$$

and the q columns of Z_1 are independently distributed, each having a p-variate nonsingular normal distribution, with the same variance-covariance matrix $\Sigma_{11 \cdot 2}$. The hypothesis H_o is therefore equivalent to

$$E(Z_1 | \underline{y}) = 0 \ , \tag{4.15}$$

and by (1.2), a criterion for testing H_o will be provided by

$$\begin{aligned} \Lambda(n,p,q) &= \frac{|A|}{|A + B|} \\ &= \frac{|C_{xx \cdot y}|}{|C_{xx \cdot y} + C_{xy} C_{yy}^{-1} C_{yx}|} \\ &= \frac{|C_{xx \cdot y}|}{|C_{xx}|} \ . \end{aligned} \tag{4.16}$$

If H_o is true, $\Lambda(n,p,q)$ will have the $\Lambda(n,p,q)$ distribution, discussed in Sections 2 and 3 of this chapter. H_o will be rejected at a significance level α, if $-m \log_e \Lambda$ exceeds $W_\alpha(n,p,q)$ defined in (3.9); m is $n-(p+q+1)/2$. Since the sample canonical correlations r_i satisfy [see (7.3.3)]

$$|-r_i^2 C_{xx} + C_{xy}C_{yy}^{-1}C_{yx}| = 0 , \tag{4.17}$$

it is easy to see that [use (4.8)]

$$|(1-r_i^2)\, C_{xx} - C_{xx\cdot y}| = 0 , \tag{4.18}$$

and hence Wilks's Λ given by (4.16) is also equal to

$$\prod_1^k (1-r_i^2) \, . \tag{4.19}$$

If, instead of considering the regression of $\underline{x}$ on $\underline{y}$, we had considered the regression of $\underline{y}$ on $\underline{x}$, we would have come up with

$$\Lambda(n,q,p) = \frac{|C_{yy\cdot x}|}{|C_{yy}|} \tag{4.20}$$

as the criterion, but this is the same as $\Lambda(n,p,q)$ of (4.16), as each is equal to $\prod_1^k (1-r_i^2)$ of (4.19). This could also be established from the identity [derivable from (1.3.14)]

$$\left|\begin{array}{c|c} C_{xx} & C_{xy} \\ \hline C_{yx} & C_{yy} \end{array}\right| = |C_{xx}|\;|C_{yy\cdot x}|$$

$$= |C_{yy}|\;|C_{xx\cdot y}| \, . \tag{4.21}$$

The matrices $A = C_{xx\cdot y}$ and $B = C_{xy}C_{yy}^{-1}C_{yx} = \hat{\beta}\, C_{yy}\, \hat{\beta}'$ are known, respectively, as the error s.s. and s.p. matrix and regression s.s. and s.p. matrix, in the regression of $\underline{x}$ on $\underline{y}$. They are presented in Table 2.

TABLE 2

Regression Analysis of $\underline{x}$ on $\underline{y}$

Source	d.f.	s.s. and s.p. matrix
Regression ($\underline{x}$ on $\underline{y}$)	q	$B = C_{xy}C_{yy}^{-1}C_{yx} = \hat{\beta}\, C_{yy}\, \hat{\beta}'$
Error	n-q	$A = C_{xx\cdot y} = C_{xx} - C_{xy}C_{yy}^{-1}C_{yx}$
Total	n	$A + B = C_{xx}$

Wilks's Λ to test H_o is the ratio of the determinants of the error and the total s.s. and s.p. matrices. The corresponding table for the regression of $\underline{y}$ on $\underline{x}$ can be obtained simply by interchanging x and y and p and q. As has been seen, it yields the same Wilks's Λ. Since A has a Wishart distribution (use Lemma 10 of Chapter 3),

$$E(A) = (n-q)\Sigma_{11\cdot 2}\ , \tag{4.22}$$

and since

$$E(Z_1|\underline{y}) = \beta\, Y P_1\ , \quad V\ (\text{any column of } Z_1|\underline{y}) = \Sigma_{11\cdot 2},$$

it follows that

$$E\{(Z_1 - \beta\, Y P_1)(Z_1 - \beta\, Y P_1)'|\underline{y}\} = q\ \Sigma_{11\cdot 2}\ . \tag{4.23}$$

From this, we obtain

$$E(B|\underline{y}) = E(Z_1 Z_1'|\underline{y}) = q\ \Sigma_{11\cdot 2} + \beta\, Y P_1 P_1' Y' \beta'\ . \tag{4.24}$$

But because of (7.5.13) and (7.5.5),

$$Y\ P_1 P_1' Y' = YJY'\ (YJY')^{-1}\ YJY' = C_{yy}$$

and so

$$E(B|\underline{y}) = q\ \Sigma_{11\cdot 2} + \beta\ C_{yy}\ \beta'\ , \tag{4.25}$$

and

$$E(B) = q\,\Sigma_{11\cdot2} + \boldsymbol{\beta}\,E(C_{yy})\boldsymbol{\beta}'$$
$$= q\,\Sigma_{11\cdot2} + n\,\boldsymbol{\beta}\,\Sigma_{22}\,\boldsymbol{\beta}'$$
$$= q\,\Sigma_{11\cdot2} + n\,\Sigma_{12}\Sigma_{22}^{-1}\Sigma_{21}\,. \tag{4.26}$$

When H_o is true, (4.22) and (4.25) or (4.26) reduce to

$$E(A) = (n-q)\Sigma_{11}, \quad E(B) = q\,\Sigma_{11}\,. \tag{4.27}$$

The distribution of Wilks's Λ, when H_o is not true, involves the eigenvalues of the matrix

$$\Omega = \Sigma_{11\cdot2}^{-1}\Sigma_{12}\Sigma_{22}^{-1}\Sigma_{21}\,, \tag{4.28}$$

known as the noncentrality matrix. From (7.5.45), we already know that the eigenvalues δ_i^2 of Ω are related to the true canonical correlations between $\underline{x}$ and $\underline{y}$ by $\rho_i^2 = \delta_i^2/(1 + \delta_i^2)$. An asymptotic expansion of the non-null distribution of Λ, i.e., the distribution of Λ when H_o is not true, is given by Sugiura and Fujikoshi [63], and Yoong-Sin Lee [68], and this will be useful in obtaining the power of this test for H_o.

If the null hypothesis H_o is accepted, it means that $\underline{x}$ and $\underline{y}$ have no relationship at all and neither is useful for predicting the other. No further analysis is necessary. But if H_o is rejected, showing thereby the existence of some relationship between $\underline{x}$ and $\underline{y}$, it becomes necessary to analyze the data further, in order to bring out the nature and form of this relation. We give below a number of hypotheses and derive tests for them for this purpose. These tests will throw more light on the relationship between $\underline{x}$ and $\underline{y}$. All these tests and hypotheses may not be appropriate for every situation. Which of the following tests and hypotheses are meaningful in any particular situation will depend on the physical nature of the variables in $\underline{x}$ and $\underline{y}$.

5. TEST FOR A SUBHYPOTHESIS ASSOCIATED WITH β

Suppose that the variables in $\underline{y}$ are divided into two groups $\underline{u}$ and $\underline{v}$:

$$\underline{y} = \left[\begin{array}{c} \underline{u} \\ \hline \underline{v} \end{array}\right] \begin{array}{l} q_1 \\ q_2 \end{array}, \tag{5.1}$$

where $q_1 + q_2 = q$. Correspondingly, β, the matrix of regression coefficients attached to $\underline{y}$ in (4.2), and Y, the matrix of sample observations on $\underline{y}$, are also partitioned as

$$\beta = p\left[\beta_u \,|\, \beta_v\right] \quad \begin{array}{ll} q_1 & q_2 \end{array} \tag{5.2}$$

and

$$Y = \begin{array}{l} q_1 \\ q_2 \end{array}\left[\begin{array}{c} U \\ \hline V \end{array}\right]. \tag{5.3}$$

We shall set up the hypothesis

$$H: \quad \beta_v = 0. \tag{5.4}$$

Since β_v is a submatrix of β and since rejection of H_o has shown that the whole matrix β is not null, it will be interesting in some cases to test whether at least a submatrix β_v out of β is null. If so, the group of variables $\underline{v}$ does not add any further to the relationship of $\underline{x}$ and $\underline{y}$, once $\underline{u}$ is already considered.

We use the obvious notation C_{uu}, C_{uv}, C_{vv}, C_{xu}, etc. Thus, for example, $C_{uu} = UJU'$, $C_{uv} = UJV'$, and $C_{xu} = XJU'$. Also note that

$$C_{yy} = \begin{bmatrix} C_{uu} & C_{uv} \\ C_{vu} & C_{vv} \end{bmatrix} \begin{matrix} q_1 \\ q_2 \end{matrix} \; ; \quad C_{xy} = [C_{xu} | C_{xv}] . \qquad (5.5)$$
$$\qquad\qquad q_1 \quad q_2$$

The regression equation (4.2) can be written because of (5.2) and (5.3), as

$$E(\underline{x} | \underline{u}, \underline{v}) = \underline{\mu} + \boldsymbol{\beta}_u(\underline{u} - E(\underline{u})) + \boldsymbol{\beta}_v(\underline{v} - E(\underline{v})) \qquad (5.6)$$

and reduces to only

$$\underline{\mu} + \boldsymbol{\beta}_u(\underline{u} - E(\underline{u})) \qquad (5.7)$$

when H is true. A test for H is obtained by splitting the matrix B of table 2 into two parts, one corresponding to the regression of $\underline{x}$ on $\underline{u}$ alone [see (5.7)] and the remainder. Just as $B = C_{xy}C_{yy}^{-1}C_{yx}$ is the regression s.s. and s.p. matrix, in the regression of $\underline{x}$ on $\underline{y}$, the matrix

$$B_u = C_{xu}C_{uu}^{-1}C_{ux} \quad (\text{d.f. } q_1) \qquad (5.8)$$

will be the regression s.s. and s.p. matrix in the regression of $\underline{x}$ and $\underline{u}$ only. The analysis of variance table for this situation is therefore given by Table 3.

TABLE 3

Source	d.f.	s.s. and s.p. matrix
Regression of $\underline{x}$ on $\underline{u}$	q_1	$C_{xu}C_{uu}^{-1}C_{ux} = B_u$
Remainder	q_2	$B - B_u$
Regression of $\underline{x}$ on $\underline{y}$	q	$C_{xy}C_{yy}^{-1}C_{yx} = B$
Error	$n-q$	$C_{xx\cdot y} = A$
Total	n	C_{xx}

We then construct the criterion

$$\Lambda_H = \Lambda(n-q+q_2, p, q_2) = \frac{|A|}{|A+B-B_u|} \tag{5.9}$$

for testing the hypothesis H. If H is true, Λ_H has the Wilks's $\Lambda(n-q+q_2, p, q_2)$ distribution and can therefore be tested by examining whether

$$-\{(n-q+q_2) - \tfrac{1}{2}(p+q_2+1)\} \log_e \Lambda_H \tag{5.10}$$

exceeds $W_\alpha(n-q+q_2, p, q_2)$ [as defined in (3.9)]. If it exceeds $W_\alpha(n-q+q_2, p, q_2)$, H is rejected at a significance level α. To prove this, we use [see (2.4.5) and (5.5)]

$$C_{yy}^{-1} = \left[\begin{array}{c|c} C_{uu}^{-1} + C_{uu}^{-1}C_{uv}C_{vv\cdot u}^{-1}C_{vu}C_{uu}^{-1} & -C_{uu}^{-1}C_{uv}C_{vv\cdot u}^{-1} \\ \hline -C_{vv\cdot u}^{-1}C_{vu}C_{uu}^{-1} & C_{vv\cdot u}^{-1} \end{array}\right] \tag{5.11}$$

and then, from (5.5) and (5.11), we find that

$$
\begin{aligned}
B &= C_{xy}C_{yy}^{-1}C_{yx} \\
&= [C_{xu}|C_{xv}]\ C_{yy}^{-1}[C_{xu}|C_{xv}]' \\
&= C_{xu}C_{uu}^{-1}C_{ux} + (C_{xu}C_{uu}^{-1}C_{uv} - C_{xu})C_{vv\cdot u}^{-1}(C_{vu}C_{uu}^{-1}C_{ux} - C_{vx}) ,
\end{aligned} \tag{5.12}
$$

and hence [see (5.8)]

$$
B - B_u = \hat{\beta}_u\ C_{vv\cdot u}\ \hat{\beta}_v , \tag{5.13}
$$

where

$$
\hat{\beta}_v = (C_{xv} - C_{xu}C_{uu}^{-1}C_{uv})C_{vv\cdot u}^{-1} . \tag{5.14}
$$

Note that, because of (5.5) and (5.11),

$$
\begin{aligned}
\hat{\beta} &= C_{xy}C_{yy}^{-1} \\
&= [C_{xu}|C_{xv}]C_{yy}^{-1} \\
&= [\underset{q_1}{\hat{\beta}_u}|\underset{q_2}{\hat{\beta}_v}]p ,
\end{aligned} \tag{5.15}
$$

where $\hat{\beta}_v$ is given in (5.14). We now refer to Theorem 2 of Section 4, Chapter 4. We have proved there that the elements of $\hat{\beta}$ have a pq-variate normal distribution with mean β and variance-covariance matrix $\Sigma_{11\cdot 2} \otimes C_{yy}^{-1}$ when $\underset{\sim}{y}$ is fixed. From this, it follows (Theorem 2 of Chapter 2) that the conditional distribution of $\hat{\beta}_v$ (the matrix of the last q_2 columns of $\hat{\beta}$), when $\underset{\sim}{y}$ is fixed, is a pq_2-variate normal, with mean β_v and variance-covariance matrix $\Sigma_{11\cdot 2} \otimes C_{vv\cdot u}^{-1}$. [Note, from (5.11), that $C_{vv\cdot u}^{-1}$ is the submatrix of C_{yy}^{-1} formed from the last q_2 rows and columns.] If we let

$$G = \hat{\boldsymbol{\beta}}_v C_{vv\cdot u}^{1/2} , \tag{5.16}$$

it is easy to see that the conditional distribution of G, when $\underline{y}$ is fixed, is $N_{pq_2}(G|\boldsymbol{\beta}_v C_{vv\cdot u}^{1/2}|\Sigma_{11\cdot 2} \otimes I_{q_2})$. In other words, the q_2 columns $\underline{g}_j$ ($j = 1, 2, \ldots, q_2$) of G are independently distributed, as a p-variate normal distribution, with the same variance-covariance matrix $\Sigma_{11\cdot 2}$. Further, when the hypothesis H [given by (5.4)] is true, E(G) = 0, and the distribution of G does not involve $\underline{y}$. From Theorem 1 of Chapter 4, we know that $\hat{\boldsymbol{\beta}}$ and hence $\hat{\boldsymbol{\beta}}_v$ is independent of $A = C_{xx\cdot y}$. Hence A is independent of

$$\begin{aligned} B - B_u &= \hat{\boldsymbol{\beta}}_v C_{vv\cdot u} \hat{\boldsymbol{\beta}}_v' \\ &= GG' \\ &= \sum_{j=1}^{q_2} \underline{g}_j \underline{g}_j' , \end{aligned} \tag{5.17}$$

where, under the hypothesis H, the g_j are independently distributed as $N_p(\underline{g}_j|\underline{0}|\Sigma_{11\cdot 2})$. So, finally, from (1.2), we find that

$$\begin{aligned} \Lambda(n-q+q_2, p, q_2) &= \frac{|A|}{|A + \Sigma \underline{g}_j \underline{g}_j'|} = \frac{|A|}{|A + \hat{\boldsymbol{\beta}}_v C_{vv\cdot u} \hat{\boldsymbol{\beta}}_v'|} \\ &= \frac{|A|}{|B + B - B_u|} \end{aligned} \tag{5.18}$$

has the Wilks's $\Lambda(n-q+q_2, p, q_2)$ distribution when H is true. We have thus established the test.

The criterion (5.9) can be derived from the likelihood ratio principle also. The likelihood of the observations X on $\underline{x}$, under the model (4.10) and (4.4), is

$$L(\Omega_1) = \frac{1}{(2\pi)^{Np/2}|\Sigma_{11\cdot 2}|^{N/2}} \exp\left\{-\frac{1}{2}\mathrm{tr}\Sigma_{11\cdot 2}^{-1}(X-\underline{\alpha}E_{1N}-\beta Y)(X-\underline{\alpha}E_{1N}-\beta Y)'\right\} \tag{5.19}$$

where Ω_1 is the space of all the parameters $\Sigma_{11\cdot 2}$, $\underline{\alpha}$, $\beta = [\beta_u | \beta_v]$ such that $\Sigma_{11\cdot 2} > 0$, $-\infty < \underline{\alpha} < \infty$, $-\infty < \beta < \infty$. By maximizing $L(\Omega_1)$ with respect to these parameters, it can be shown that the maximum likelihood estimates are

$$\hat{\underline{\alpha}} = \frac{1}{N}(X\ E_{N1} - \hat{\beta}\, Y\, E_{N1}) \tag{5.20}$$

$$\hat{\beta} = C_{xy}C_{yy}^{-1} \tag{5.21}$$

$$\hat{\Sigma}_{11\cdot 2} = C_{xx\cdot y} = A\,. \tag{5.22}$$

The maximum value of $L(\Omega_1)$, when these estimates are substituted in (5.19), turns out to be

$$L(\Omega_1 \max) = \frac{e^{-Np/2}}{(2\pi)^{Np/2}|A|^{N/2}}\,. \tag{5.23}$$

When the hypothesis H: $\beta_v = 0$ is true, the likelihood of X reduces to

$$L(\Omega_2) = \frac{1}{(2\pi)^{Np/2}|\Sigma_{11\cdot 2}|^{N/2}} \exp\left\{-\frac{1}{2}\mathrm{tr}\Sigma_{11\cdot 2}^{-1}(X-\underline{\alpha}E_{1N}-\beta_u U)(X-\underline{\alpha}E_{1N}-\beta_u U)'\right\} \tag{5.24}$$

where Ω_2 is the subspace of Ω_1, of the parameters $\Sigma_{11\cdot 2}$, $\underline{\alpha}$, β_u, such that $\Sigma_{11\cdot 2} > 0$, $-\infty < \underline{\alpha} < \infty$, $-\infty < \beta_u < \infty$. As before, maximizing $L(\Omega_2)$ with respect to these parameters, the new maximum likelihood estimates will be found to be

$$\underline{\alpha}^* = \frac{1}{N}(X\ E_{N1} - \beta_u^*\ U\ E_{N1})\,, \tag{5.25}$$

$$\beta_u^* = C_{xu}C_{uu}^{-1}\,, \tag{5.26}$$

$$\Sigma^{*}_{11\cdot 2} = C_{xx\cdot u}$$

$$= C_{xx\cdot y} + (C_{xy}C_{yy}^{-1}C_{yx} - C_{xu}C_{uu}^{-1}C_{ux})$$

$$= A + (B - B_u), \text{ on account of (5.8).} \qquad (5.27)$$

Substituting these in (5.24), we find

$$L(\Omega_2 \text{ max.}) = \frac{e^{-Np/2}}{(2\pi)^{Np/2}\,|A + B - B_u|^{N/2}} . \qquad (5.28)$$

The likelihood ratio statistic for testing H is, therefore,

$$\ell = \frac{L(\Omega_2 \text{ max.})}{L(\Omega_1 \text{ max.})}$$

$$= \frac{|A|^{N/2}}{|A + B - B_u|^{N/2}} , \qquad (5.29)$$

and is thus the (N/2)th power of our statistic Λ_H proposed in (5.9).

If the hypothesis to be tested is not H: $\beta_v = 0$, but H': $\beta_v = \Delta$, a specified $p \times q_2$ matrix, the criterion Λ_H must be modified to [see (5.19)]

$$\Lambda_{H'} = \frac{|A|}{|A + (\hat{\beta}_v - \Delta)\, C_{vv\cdot u}(\hat{\beta}_v - \Delta)'|} , \qquad (5.30)$$

as not G but $G_o = (\hat{\beta}_v - \Delta)C_{vv\cdot u}^{1/2}$ has the normal distribution with zero mean under H'.

Consider now the particular case $p = 1$, i.e., we consider the regression of a single variable x on q variables $y_1, \ldots, y_q$ and test the hypothesis H, that the regression coefficients attached to $y_{q_1+1}, y_{q_1+2}, \ldots, y_q$ are all zero. Putting $p = 1$ in

(5.9), the test statistic is

$$\lambda_H = \frac{A}{A + (B - B_u)}, \tag{5.31}$$

where

$$A = C_{xx\cdot y}, \quad B = C_{xy}C_{yy}^{-1}C_{yx}, \quad B_u = C_{xu}C_{uu}^{-1}C_{ux} \tag{5.32}$$

are no longer matrices but only scalars. This λ_H has the $\Lambda(n-q+q_2, 1, q_2)$ distribution when H is true. But this is the same as saying that (see Theorem 2 and put $p = 1$)

$$\lambda_H \text{ has the } B\left(\lambda_H \mid \frac{n-q}{2}, \frac{q_2}{2}\right) \text{ distribution.} \tag{5.33}$$

If so, it can be readily seen that [from (2.24)]

$$\frac{n-q}{q_2} \frac{1 - \lambda_H}{\lambda_H} = F_{q_2,\, n-q} \,. \tag{5.34}$$

The F test can thus be used to test H, when $p = 1$. From (1.3.8), one can show that

$$B = C_{xx}R^2_{x(\underline{y})}, \quad A = C_{xx}(1 - R^2_{x(\underline{y})}), \quad B_u = C_{xx}R^2_{x(\underline{u})}, \tag{5.35}$$

where $R_{x(\underline{y})}$ is the sample multiple correlation coefficient between x and $\underline{y}$, and similarly $R_{x(\underline{u})}$ is the sample multiple correlation coefficient between x and $\underline{u}$, i.e., $y_1, \ldots, y_{q_1}$ only. Therefore,

$$\lambda_H = \frac{1 - R^2_{x(\underline{y})}}{1 - R^2_{x(\underline{u})}} \,. \tag{5.36}$$

If we put $p = 1$ in the distributions of A, B, B_u derived earlier, we shall find that

$$C_{xx}(1 - R^2_{x(\underline{y})}) \text{ is a } \chi^2\sigma^2 \text{ with } n-q \text{ d.f.}, \tag{5.37}$$

$$C_{xx}(R^2_{x(\underline{y})} - R^2_{x(\underline{u})}) \text{ is a } \chi^2\sigma^2 \text{ with } q_2 \text{ d.f. only if H is true.} \quad (5.38)$$

Further, these two distributions are independent. Here σ^2 is the population variance of x. The test (5.34) follows from these distributions also.

If $\rho_{x(\underline{y})}$ and $\rho_{x(\underline{u})}$ are the population or true multiple correlation coefficients of x with $\underline{y}$ and $\underline{u}$, respectively, it can be shown by a little algebra that, when H is true, $\rho_{x(\underline{y})} = \rho_{x(\underline{u})}$, and conversely, when $\rho_{x(\underline{y})} = \rho_{x(\underline{u})}$, the true regression coefficients attached to the last q_2 variables in $\underline{y}$, in the regression of x on $\underline{y}$ are all zero.

We have, therefore, the following theorem:

<u>Theorem 5</u>: If

$$\underline{y} = \left[\begin{array}{c} \underline{u} \\ \hline \underline{v} \end{array}\right] \begin{array}{l} q_1 \\ q_2 = q - q_1 \end{array}$$

and

$$\rho^2_{x(\underline{y})} = \rho^2_{x(\underline{u})},$$

the statistic

$$\frac{n-q}{q_2} \frac{R^2_{x(\underline{y})} - R^2_{x(\underline{u})}}{1 - R^2_{x(\underline{y})}}$$

has the $F_{q_2,\, n-q}$ distribution, under the assumption of a normal distribution for x, (n+1) being the size of the sample from this distribution.

6. FACTORS OF WILKS'S Λ

Consider the following sequence of hypotheses:

H_1: x_1 has no relationship with the vector $\underline{y}$;

H_2: x_2, when the relationship due to x_1 is eliminated, has no relation with $\underline{y}$;

H_3: x_3, when the relationship due to x_1 and x_2 is eliminated, has no relationship with $\underline{y}$;

and so on, the last hypothesis being

H_p: x_p has no relation with $\underline{y}$, when the relationship due to $x_1, \ldots, x_{p-1}$ is eliminated.

Consider all these hypotheses together and compare with the null hypothesis H_o, of no relationship between the vectors $\underline{x}$ and $\underline{y}$, discussed in Section 4. It is easy to see that, if H_o is true, each of $H_1, \ldots, H_p$ must be true, and conversely, if every one of $H_1, \ldots, H_p$ is true, H_o must be true. H_o is the intersection of the hypotheses H_i ($i = 1, 2, \ldots, p$), and this can be symbolically expressed by

$$H_o = \bigcap_{i=1}^{p} H_i , \tag{6.1}$$

H_i are thus the components of H_o. The hypothesis H_i can be mathematically expressed as

H_i: the regression coefficients attached to $\underline{y}$, in the regression of x_i on $x_1, x_2, \ldots, x_{i-1}$ and $\underline{y}$ are all zero (6.2)

or as [see (5.40)]

$$H: \rho^2_{x_i}(x_1, \ldots, x_{i-1}, \underline{y}) = \rho^2_{x_i}(x_1, \ldots, x_{i-1}),$$

where $\rho_{x_i}(x_1, \ldots, x_{i-1}, \underline{y})$ is the multiple correlation coefficient (true) of x_i with $x_1, \ldots, x_{i-1}$ and $\underline{y}$, with a similar definition of $\rho_{x_i}(x_1, \ldots, x_{i-1})$. From Theorem 5, the test criterion for H_i is

$$\lambda_i = \frac{1 - R^2_{x_i}(x_1, \ldots, x_{i-1}, \underline{y})}{1 - R^2_{x_i}(x_1, \ldots, x_{i-1})}, \tag{6.3}$$

where R denotes a sample multiple correlation coefficient. When H_i is true, from (5.33), we know that λ_i has the $B(\lambda_i | (n-q-i+1)/2, q/2)$ distribution. From (1.3.8) and from C_{xx}, C_{xy}, C_{yy} (the matrices of the corrected s.s. and s.p. of $\underline{x}$ and $\underline{y}$), it can be proved, after some algebra, that

$$t^2_{ii} = \lambda_i = \frac{|A_i|/|A_{i-1}|}{|A_i + B_i|/|A_{i-1} + B_{i-1}|} \quad (i = 1, 2, \ldots, p) \tag{6.4}$$

where A_i, B_i are, respectively, the matrices formed by the first i rows and columns of A and B, and $A = C_{xx \cdot y}$, $B = C_{xy}C^{-1}_{yy}C_{yx}$, as defined in Section 4. ($|A_o|$ and $|A_o + B_o|$ are defined as 1.) Now use (2.2), (2.3), and (2.6). We can then easily deduce [as C of (2.2) is l.t.] that

$$\lambda_i = |L_i|/|L_{i-1}| = t^2_{ii} \quad (i = 1, 2, \ldots, p), \tag{6.5}$$

where t^2_{ii}, their distribution under H_o, and their relationship with Wilks's $\Lambda = \Lambda(n,p,q)$ for testing H_o have been discussed in Section 2. In fact, from (2.14),

$$\Lambda \text{ for testing } H_o \text{ is } \prod_1^p t^2_{ii} = \prod_1^p \lambda_i, \tag{6.6}$$

and, from (6.1), $H_o = \bigcap_{i=1}^{p} H_i$, with λ_i as the test criterion for H_i. This brings out the relationship of the factors t_{ii}^2 ($i = 1, 2, \ldots, p$) of Wilks's Λ for testing H_o, with H_i the components of H_o. If H_o is true, every H_i is true and every $\lambda_i = t_{ii}^2$ has the $B(\lambda_i \mid (n-q+1-i)/2,\ q/2)$ distribution and conversely. If H_o is not true, one or more of the H_i's are not valid, which of the H_i's are not true can be found by testing each of the H_i separately by means of the factors t_{ii}^2 of Λ, the actual test being (from Theorem 5)

$$\frac{n-q+1-i}{i-1}\,\frac{1-\lambda_i}{\lambda_i} = F_{i-1,\ n-q-i+1} \quad (i = 1, 2, \ldots, p). \tag{6.7}$$

Using these F tests sequentially is thus an alternative stepwise procedure for H_o and throws more light on the relationship between $\underline{x}$ and $\underline{y}$ than Wilks's Λ, which is only an overall test for H_o. We get more information from the factors of Wilks's Λ, when H_o is not true, i.e., when there exists some real association between $\underline{x}$ and $\underline{y}$. Those variables, out of $x_1, \ldots, x_p$, that indicate some association with $\underline{y}$ are useful for prediction of $\underline{y}$; the others can be regarded as of no further use and can be deleted.

We introduced the variables $x_1, x_2, \ldots, x_p$ in that order in forming our hypotheses $H_1, H_2, \ldots, H_p$, but any other permutation of $x_1, x_2, \ldots, x_p$ is also possible, and this can be done by rearranging the rows and columns of A and B before finding the λ_i's. Which particular order of the variables is more meaningful depends on the actual problem on hand and the physical meaning of $\underline{x}$ and $\underline{y}$. It is worth noting that the d.f. associated with the F tests in (6.7) goes on decreasing from n-q to n-q-p+1, as we proceed from H_1 to H_p or as more and more variables are eliminated. (In H_2, only one variable x_1 is eliminated; in H_3 both x_1 and x_2 are eliminated, and so on.) Every eliminated variable reduces the sample size N (= n+1) effectively by one. This is a fact that we have been observing again and again in regression theory; it was noticed in the Bartlett decomposition of a Wishart matrix also.

Wilks's Λ is invariant for a transformation

$$\underline{x}^* = M\underline{x}, \tag{6.8}$$

where M is any fixed p x p nonsingular matrix. This can be seen as follows. If we consider the relationship between $\underline{x}^*$ and $\underline{y}$, the matrices A and B will have to be replaced by

$$A^* = MAM', \quad B^* = MBM', \tag{6.9}$$

respectively, as $C_{x^*x^*} = M C_{xx} M'$, $C_{x^*y} = M C_{xy}$. Wilks's Λ for $\underline{x}^*$ and $\underline{y}$ is then

$$\frac{|A^*|}{|A^* + B^*|} = \frac{|M|\ |A|\ |M'|}{|M|\ |A + B|\ |M'|} = \frac{|A|}{|A + B|},$$

which is the original Wilks's Λ for $\underline{x}$ and $\underline{y}$. One can, therefore, use $|A^*|/|A^* + B^*|$ and its factors

$$\lambda_i^* = \frac{|A_i^*|/|A_{i-1}^*|}{|A_i^* + B_i^*|/|A_{i-1}^* + B_{i-1}^*|} \tag{6.10}$$

in exactly the same way as $\lambda_i = t_{ii}^2$, to test the relationship of the elements of $\underline{x}^*$ with $\underline{y}$, if $\underline{x}^*$ is more suitable or meaningful. Thus if $\underline{m}'$ is the first row of M, x_1^* is $\underline{m}'\underline{x}$, and then

$$\lambda_1^* = \underline{m}'A\underline{m}/\underline{m}'(A+B)\underline{m} \tag{6.11}$$

can be used to test the hypothesis that $\underline{m}'\underline{x}$ has no relationship with $\underline{y}$. Note that

$$\lambda_1^* = 1 - R^2_{\underline{m}'\underline{x}(\underline{y})}, \tag{6.12}$$

where R is the multiple correlation coefficient in a sample.

Sometimes one wishes to test the hypotheses H_{k+1}, H_{k+2}, ..., H_p $(k < p)$ simultaneously. The combined hypothesis can be symbolically expressed as $\bigcap_{i=k+1}^{p} H_i$ and states that the vector $[x_{k+1}, x_{k+2}, \ldots, x_p]$ has no relationship with $\underline{y}$ when any association due to $x_1, x_2, \ldots, x_k$ is eliminated. From the logic of $H_o = \bigcap_{i=1}^{p} H_i$ and $\Lambda = \prod_1^p t_{ii}^2$, it is natural to use the statistic

$$\Lambda_{k+1,\, k+2,\, \ldots,\, p \cdot 12 \ldots k} = \prod_{i=k+1}^{p} t_{ii}^2 . \tag{6.13}$$

It will provide the necessary test for $\bigcap_{i=k+1}^{p} H_i$. This test can be called the test of the "partial" association of $x_{k+1}, \ldots, x_p$ with $\underline{y}$, when $x_1, \ldots, x_k$ are eliminated. The statistic (6.13) is constructed from the "total" Wilks's $\Lambda(n,p,q) = \prod_1^p t_{ii}^2$, by removing the factor $\prod_{i=1}^{k} t_{ii}^2$ corresponding to the eliminated variables $x_1, \ldots, x_k$ and can be called partial Λ. When H_{k+1}, H_{k+2}, ..., H_p are all true, we have already seen that t_{ii}^2 $(i = k+1, \ldots, p)$ are independently distributed as $B(t_{ii}^2 | (n-q-i+1)/2, q/2)$. This is the same as saying that

$t^2_{k+j,\ k+j}$ $(j = 1, 2, \ldots, p^*)$ are independently distributed as

$$B\left(t^2_{k+j,\ k+j} \,\middle|\, \frac{n^*-q-j+1}{2},\ \frac{q}{2}\right), \tag{6.14}$$

where $p^* = p-k$ and $n^* = n-k$. Hence (6.13), which is $\prod_{j=1}^{p-k} t^2_{k+j,k+j}$, has, by Theorem 2, the Wilks's $\Lambda(n^*,p^*,q)$ distribution when $\bigcap_{k+1}^{p} H_i$ is true. The actual test procedure is to calculate

$$-\{n^* - \tfrac{1}{2}(p^*+q+1)\} \log_e \Lambda_{k+1,\ \ldots,\ p \cdot 12 \ldots k} \tag{6.15}$$

and reject the hypothesis $\bigcap_{k+1}^{p} H_i$, at a level of significance α, if the value of (6.15) exceeds $W_\alpha(n^*, p^*, q)$ defined in (3.9). The following expressions for $\Lambda_{k+1,\ \ldots,\ p \cdot 12 \ldots k}$ may be useful for this. From (6.4),

$$\begin{aligned}\Lambda_{k+1,\ \ldots,\ p \cdot 12 \ldots k} &= \prod_{i=1}^{p} t^2_{ii} \Big/ \prod_{i=1}^{k} t^2_{ii} \\ &= \frac{|A|/|A+B|}{|A_k|/|A_k+B_k|} \\ &= \frac{\text{Wilks's } \Lambda(n,p,q)}{\text{Wilks's } \Lambda(n,k,q)},\end{aligned} \tag{6.16}$$

where the numerator is the Wilks's Λ for testing the association of $\underline{x}$ and $\underline{y}$ while the denominator is Wilks's Λ for testing the association of $x_1, \ldots, x_k$ only, with $\underline{y}$.

7. TEST OF SIGNIFICANCE OF THE RESIDUAL ROOTS

In Section 4, we saw that the hypothesis H_o of independence of two vectors $\underline{x}$ and $\underline{y}$ is the same as the hypothesis that all the true canonical correlations $\rho_1, \rho_2, \ldots, \rho_k$ (k = smaller of p, q) between $\underline{x}$ and $\underline{y}$ are zero. The test statistic Wilks's

$\Lambda(n,p,q)$ was seen from (4.19) to be also equal to $\prod_{1}^{k} (1 - r_i^2)$, where $r_1^2 > r_2^2 > \dots > r_k^2$ are the squares of the sample canonical correlation coefficients. The actual test, using Bartlett's approximation, is

$$- \{n - \tfrac{1}{2}(p+q+1)\}\log_e \prod_{1}^{k} (1 - r_i^2) \text{ is a } \chi^2 \text{ with } pq \text{ d.f.} \tag{7.1}$$

if $\rho_1 = \rho_2 = \dots = \rho_k = 0$. However, if this hypothesis is rejected, one will naturally be interested in a hypothesis of the form

$$H(s)\colon \rho_1 \neq 0,\ \rho_2 \neq 0,\ \dots,\ \rho_s \neq 0,\ \rho_{s+1} = \rho_{s+2} = \dots = \rho_k = 0 \ . \tag{7.2}$$

It is understood that ρ_1 is the largest canonical correlation, ρ_2 is the next largest, and so on. Bartlett [6], using geometrical concepts, suggested that the hypothesis $H(s)$ can be tested by the following test:

$$- \{n - \tfrac{1}{2}(p+q+1)\}\log_e \prod_{i=s+1}^{k} (1 - r_i^2) \text{ is a } \chi^2 \text{ with } (p-s)(q-s) \text{ d.f., if } H(s) \text{ is true } (s = 1,2,\dots,k-1). \tag{7.3}$$

In other words, when we remove the first s roots from the "total" criterion $\prod_{1}^{k} (1 - r_i^2)$, we are left with $\prod_{i=s+1}^{k} (1 - r_i^2)$, and this "residual" criterion provides a test of significance of the population residual roots $\rho_{s+1}, \dots, \rho_k$. The distribution of $\prod_{i=s+1}^{k} (1 - r_i^2)$, is the same as that of $\prod_{1}^{k} (1 - r_i^2)$, provided n,p,q are all replaced by $n-s$, $p-s$, and $q-s$, respectively. The effect of removing the first s roots is to reduce n,p,q by s each. The new "correction factor" is now

$$(n-s) - \frac{1}{2}\{(p-s) + (q-s) + 1\} ,$$

but it simplifies to

$$m = n - \frac{1}{2}(p+q+1) ,$$

which is the same as before, and we get the test (7.3). Lawley [38] investigated this test in considerable detail by elaborate expansion of matrices involved in the calculation of the r_i's and come to the conclusion that, for large n, not (7.3) but

$$- \left\{n-s-\frac{1}{2}(p+q+1) + \sum_{1}^{s} \frac{1}{\rho_i^2}\right\} \log_e \prod_{i=s+1}^{k} (1-r_i^2) \qquad (7.4)$$

is a χ^2 with $(p-s)(q-s)$ d.f., under H(s). This result is correct to terms of order of n^{-2}. Bartlett's test (7.3) agrees with (7.4) only when the first s canonical correlations $\rho_1, \ldots, \rho_s$ are all unity or at least very near to it. In practice, $\rho_1, \rho_2, \ldots, \rho_s$ are unknown, and one has to substitute their estimates $r_1, \ldots, r_s$ in (7.4) in order to use the test. [The result (7.4) holds when $\underline{y}$ has also a normal distribution; it requires a slight modification when $\underline{y}$ is a "fixed" vector. But when r_i^2 is substituted for ρ_i^2 in (7.4), the test so obtained comes out the same, whether $\underline{y}$ is fixed or random.] Table 4, similar to an analysis of variance table, can be prepared for this test. [The refinement of introducing $\sum_1^s 1/\rho_i^2$ is ignored and only (7.3) is considered.]

TABLE 4

Source	d.f.	χ^2
First s roots	$pq - (p-s)(q-s)$	$-m \log_e \prod_1^s (1 - r_i^2)$
Remaining roots	$(p-s)(q-s)$	$-m \log_e \prod_{i=s+1}^k (1 - r_i^2)$
Total	pq	$-m \log_e \Lambda(n,p,q)$

Here a note of warning is necessary. The first χ^2 in the above table <u>should not</u> be considered to mean that $-m \log_e \prod_1^s (1 - r_i^2)$ is a χ^2 with $pq - (p-s)(q-s)$ d.f. if H(s) is true; <u>it is not</u>. The test for H(s) is provided by the second χ^2, with $(p-s)(q-s)$ d.f. In practice, one should set up the hypotheses H(s), s = 1, 2, 3, ..., sequentially and test them. As soon as a nonsignificant result is reached, the procedure is terminated. Thus, for example, if H(1), H(2), H(3) are rejected and H(4) is not rejected, it means that r_1, r_2, r_3 are statistically significant at the chosen level of significance and the rest, viz., r_4, r_5, ..., r_k are insignificant, and one concludes that only ρ_1, ρ_2, ρ_3 are different from zero, while the rest of the true canonical correlations are all null. The dimensionality of the relationship between $\underline{x}$ and $\underline{y}$ is thus estimated to be 3, and the first three canonical variables of the $\underline{x}$-space and the $\underline{y}$-space account for the entire association between $\underline{x}$ and $\underline{y}$. When only the largest root ρ_1 is non-null, the association between $\underline{x}$ and $\underline{y}$ is said to be "linear"; when both ρ_1 and ρ_2 are non-null and the rest are null, the association is said to be "planar", and so on. The number of significant sample canonical correlations thus provides an estimate of the rank of the covariance matrix Σ_{12} between $\underline{x}$ and $\underline{y}$, or in other words, as the rank of the regression matrix $\boldsymbol{\beta} = \Sigma_{12}\Sigma_{22}^{-1}$ of $\underline{x}$ and $\underline{y}$.

When $\rho_1, \rho_2, \ldots, \rho_s$ are not zero and the rest are zero, the canonical variables corresponding to $\rho_{s+1}, \ldots, \rho_k$ (which are all zero) are sometimes referred to as "null functions". These null functions play an important role in economic analysis. Bartlett [7] considered the regression of 4 economic variables

x_1, home consumption of cotton yarn;

x_2, price of cotton yarn;

x_3, cost of production of cotton yarn;

x_4, income of consumers of cotton yarn;

on time t and its powers t^2, t^3, t^4, and t^5. The method of orthogonal polynomials showed that a good fit is obtained when terms up to and including t^5 are considered. The $\underline{y}$ variables are thus $t, t^2, \ldots, t^5$ and so $p = 4$, $q = 5$. By using (7.3), Bartlett found that the first two sample canonical correlations r_1 and r_2 between $\underline{x}$ and $\underline{y}$ are significant, while r_3 and r_4 are insignificant at the 0.01 level of significance. He then found the "null functions" of the x-space, i.e., the canonical variables corresponding to r_3 and r_4. Since $\rho_3 = \rho_4 = 0$, these null functions are uncorrelated with time t and thus have no relationship with time. They are, therefore, what are known as "structural equations" in econometrics. By a little manipulation, Bartlett derived a "demand" relation of the form

$$x_1 = ax_2 + bx_4$$

and a "supply" relation of the form

$$x_1 = cx_2 + dx_3 .$$

This shows how canonical analysis of economic variables provides a method of estimation of structural equations. Anderson and Rubin [1] have considered a general theoretical formulation of this problem, and Kshirsagar [35] has shown the applicability of null functions for prediction from Wold's [67] implicit causal chain models.

Marriott [39] gives an alternative large sample test of significance of r_1. He found that

$$-m \log_e (1 - r_1^2) \text{ is a } \chi^2 \tag{7.5}$$

with

$$D = p+q-1 + \tfrac{1}{2}\{(p-1)(q-1)\}^{2/3} \text{ d.f., when } H(1) \text{ is true.} \tag{7.6}$$

(If D is not an integer, interpolation in the table of percentage points of the χ^2 distribution should be resorted to.) Marriott also gives exact tests for r_1 in the particular case $p = 2$ and approximate tests for $p = 3$, 4, and 5. Marriott's test (7.5) can be extended to H(s) by changing p to p-s, q to q-s, and replacing $(1 - r_1^2)$ in (7.5) by $\prod_{i=s+1}^{k} (1 - r_i^2)$.

8. ALTERNATIVES TO THE WILKS'S Λ CRITERION

The hypothesis H_o of independence of two vectors $\underline{x}$ and $\underline{y}$ considered in Section 4 can be tested by any one of the following criteria also, which provide alternatives to the Wilks's Λ test:

(i) Hotelling's [23] generalized T_o^2 (see also Lawley [37]) $T_o^2 = \text{tr } BA^{-1}$.

(ii) Pillai's [43] criterion V

$V = \text{tr } B(A + B)^{-1}$.

(iii) Roy's [58] maximum root criterion

$$r_1^2, \text{ the largest root of } |-r^2(A+B)+B| = 0. \tag{8.1}$$

The matrices **A** and B are the same as in Section 4, viz., $C_{xx\cdot y}$ and $C_{xy}C_{yy}^{-1}C_{yx}$ are as defined in (4.12) and (4.13). From (4.17), therefore, it can be readily seen that

$$\text{Hotelling's } T_o^2 = \sum_1^k r_i^2/(1 - r_i^2) \text{ and Pillai's } V = \sum_1^k r_i^2, \tag{8.2}$$

where k = smaller of p,q and $r_1^2 > r_2^2 > \dots > r_k^2$ are the squares of the sample canonical correlations between $\underline{x}$ and $\underline{y}$.

Exact percentage points of T_o^2 (when H_o is true) are available, for some values of p, in Davis [14], Grubbs [21] and Pillai and Young [54]. The null and the non-null distribution of T_o^2 has been considered by Davis [13, 14], Pillai and Young [54], Krishnaiah and Chang [33], Pillai and Samson [44], Ito [24, 25, 26], Yoong-Sin Lee [68], Constantine [9], and SioTani

[62]. Ito gives an asymptotic formula for the percentage points and cumulative distribution function of T_o^2, for general values of p, q and n in terms of χ^2 distribution.

The moments, moment generating function, and approximations to the null and non-null distribution of Pillai's V criterion have been investigated thoroughly by Pillai himself, Khatri, and a number of collaborators in a series of papers.

Roy derived his maximum root criterion from his union-intersection principle. Tables of percentage points of the distribution of r_1^2, under H_o, have been prepared by Foster and Rees [16, 18] and Pillai [45] and are included in Pearson's tables for multivariate analysis. The distribution of r_1^2 (as well as of r_k^2 and any r_i^2) was first considered by Roy [57] himself, and a good discussion of its usefulness and applications can be found in his book [58]. The null and non-null distributions have also been investigated by Hayakawa [22], Sugiyama [64], Waal [65], Pillai and Sugiyama [51], Fukutomi [19] and Krishnaiah and Chang [33].

As Pearson [42] has very aptly remarked, there is at present no adequate theory to guide the choice of a statistic from among Wilks's Λ, Hotelling's T_o^2, Pillai's V, and Roy's r_1^2. Investigations about the comparison of all these criteria, from the point of view of their power against alternative hypotheses, have been carried out by Mikhail [40], Pillai and Jayachandran [46, 47], Pillai and Dotson [50] and Ito [26]. All the criteria seem to be almost equivalent to each other and perform equally well. In particular cases, one or the other criterion may have an edge over the others; for example, Roy's r_1^2 is more advantageous in the "linear" case ($\rho_1 \neq 0$, $\rho_2 = \dots = \rho_k = 0$) but, in general, no definite statement can be made regarding the superiority of one over the others. This author, however, is of the opinion that Wilks's Λ criterion throws more light on the relationship between $\underline{x}$ and $\underline{y}$ than any of the other statistics. This is because of the information we can get from various factorizations of Wilks's Λ, which we have already studied, and some others, which we shall be studying later on. This does not mean that the other criteria, V, T_o^2, and r_1^2, are not capable of bringing out this information. They may be, but no work in this direction has been done so far, to the best knowledge of this author. Another reason for preferring Wilks's Λ is that it can be very easily used, because of Bartlett's correction factor (3.3) and the resulting χ^2 test, while the use of other criteria can be made only with the help of special tables of their percentage points.

Daly [11] and Narain [41] have shown the unbiasedness of Wilks's Λ. Ghosh [20] has proved the admissibility of Hotelling's T_o^2 and Roy's maximum root criterion. Schwartz [61] has proved the admissibility of Pillai's V. Kiefer and Schwartz [27] have shown that tests based on Wilks's Λ and Pillai's V are admissible Bayes, fully invariant, similar and unbiased. Monotonicity properties of these tests have been shown by Roy [58], Roy and Mikhail [59], Das Gupta, Anderson and Mudholkar [12], and Anderson and Das Gupta [3]. Kiefer [28] has written an excellent survey article presenting these optimality results in multivariate analysis. Reference should also be made to Anderson [2], who has used the principle of invariance to obtain these test statistics.

The problem of simultaneous confidence bounds of certain parametric functions, associated with $\underline{x}$ and $\underline{y}$ has been considered by Roy [58] and Khatri [29]. We shall, however, not be able to consider this aspect in this book, due to lack of space.

The matrix L defined in (2.3) is, in a certain sense, a generalization of the multiple correlation coefficient to the correlation between two vector variables. This aspect of L and the noncentral linear case distribution of L and its application are considered by Kshirsagar [34]. Later, Pillai, Khatri, and both together, and with other collaborators (see [30, 31, 32, 48, 49, 51]), have studied the distribution of L exhaustively. All the four test statistics Λ, T_o^2, V, and r_1^2 can be expressed in terms of L very easily. The distribution of L therefore plays an important role in the distributional properties of all these criteria.

REFERENCES

[1] Anderson, T. W., and Rubin, H. (1949). "Estimation of the parameters of a single equation in a complete system of stochastic equations", Ann. Math. Statist., 20, p. 46.

[2] Anderson, T. W. (1958). Introduction to Multivariate Statistical Analysis. John Wiley and Sons, New York.

[3] Anderson, T. W., and Das Gupta, S. (1964). "Monotonicity of the power functions of some tests of independence between two sets of variates", Ann. Math. Statist., 35, p. 206.

[4] Bartlett, M. S. (1934). "The vector representation of a sample", Proc. Camb. Phil. Soc., 30, p. 327.

[5] Bartlett, M. S. (1938). "Further aspects of the theory of multiple regression", Proc. Camb. Phil. Soc., 34, p. 33.

[6] Bartlett, M. S. (1939). "A note on tests of significance in multivariate analysis", Proc. Camb. Phil. Soc., 35, p. 180.

[7] Bartlett, M. S. (1948). "A note on the statistical estimation of supply and demand relations from time series", Econometrika, 16, p. 323.

[8] Box, G. E. P. (1949). "A general distribution theory for a class of likelihood criteria", Biometrika, 36, p. 317.

[9] Constantine, A. G. (1966). "The distribution of Hotelling's generalized T_0^2", Ann. Math. Statist., 37, p. 215.

[10] Consul, P. C. (1966). "Exact distributions of the likelihood ratio", Ann. Math. Statist., 37, p. 1319.

[11] Daly, J. F., (1940). "On the unbiased character of likelihood ratio tests for independence in normal systems", Ann. Math. Statist., 11, p. 1.

[12] Das Gupta, S., Anderson, T. W., and Mudholkar, G. S. (1964). "Monotonicity of power functions of some tests of the multivariate linear hypothesis", Ann. Math. Statist., 35, p. 200.

[13] Davis, A. W. (1968). "A system of linear differential equations for the distribution of Hotelling's generalized T_0^2", Ann. Math. Statist., 39, p. 815.

[14] Davis, A. W. (1970). "Exact distribution of Hotelling's generalized T_0^2", Biometrika, 57, p. 187.

[15] Deemer, W. L., and Olkin, I. (1951). "The jacobians of certain matrix transformations useful in multivariate analysis", Biometrika, 38, p. 345.

[16] Foster, F. G., and Rees, D. H. (1957). "Upper percentage points of the generalized beta distribution, I", Biometrika, 44, p. 237.

[17] Foster, F. G. (1957). "Upper percentage points of the generalized beta distribution, II", Biometrika, 44, p. 441.

[18] Foster, F. G. (1958). "Upper percentage points of the generalized beta distribution, III", Biometrika, 45, p. 492.

[19] Fukutomi, K. (1967). "On the distributions of the extreme characteristic roots of the matrices in multivariate analysis" Rep. Stat. Appl. Res., JUSE, 14, p. 8.

[20] Ghosh, M. N. (1964). "On the admissibility of some tests of Manova", Ann. Math. Statist., 35, p. 789.

[21] Grubbs, F. E. (1954). Tables of 1% and 5% Probability Levels for Hotelling's Generalized T^2 Statistic (Bivariate Case). Technical Report No. 926, Bollistic Research Laboratories, Aberdeen Proving Ground, Maryland.

[22] Hayakawa, T. (1967). "On the distribution of the maximum latent root of a positive definite symmetric random matrix", Ann. Inst. Stat. Math., 18.

[23] Hotelling, H. (1951). "A generalized T-test and measure of multivariate dispersion", Proc. Second Berkeley Symposium, p. 23.

[24] Ito, K. (1956). "Asymptotic formulae for the distribution of Hotelling's generalized T_0^2 statistic", Ann. Math. Statist., 27, p. 109.

[25] Ito, K. (1960). "Asymptotic formulae for the distribution of Hotelling's generalized T_0^2 statistic", Ann. Math. Statist., 31, p. 1148.

[26] Ito, K. (1960). "On multivariate analysis of variance tests", Bulletin of Int. Inst. of Stat., 38, p. 88.

[27] Kiefer, J., and Schwartz, R. (1965). "Admissible Bayes character of T^2, $-R^2$, and other fully invariant tests for classical multivariate normal problems", Ann. Math. Statist., 36, p. 747.

[28] Kiefer, J. (1966). "Multivariate optimality results", Multivariate Analysis I, (edited by Krishnaiah, P. R.), Academic Press, New York.

[29] Khatri, C. G. (1965). "A note on the confidence bounds for the characteristic roots of dispersion matrices of normal variates", Ann. Inst. Stat. Math., 17, p. 175.

[30] Khatri, C. G., and Pillai, K. C. S. (1966). "On the moments of the trace of a matrix and approximations to its noncentral distribution", Ann. Math. Statist., 37, p. 1312.

[31] Khatri, C. G., and Pillai, K. C. S. (1967). "On the moments of traces of two matrices in multivariate analysis", Ann. Inst. Stat. Math., 19, p. 143.

[32] Khatri, C. G., and Pillai, K. C. S. (1968). "On the moments of elementary symmetric functions of the roots of two matrices and approximations to a distribution", Ann. Math. Statist., 39, p. 1274.

[33] Krishnaiah, P. R., and Chang, T. C. (1970). "On the exact distributions of the traces of $S_1(S_1 + S_2)^{-1}$ and $S_1S_2^{-1}$", Project No. 7071, Aerospace Research Laboratories, U. S. Air Force.

[34] Kshirsagar, A. M. (1961). "The noncentral multivariate beta distribution", Ann. Math. Statist., 32, p. 104.

[35] Kshirsagar, A. M. (1962). "Prediction from the simultaneous equations system and Wold's implicit causal chain model", Econometrica, 30, p. 801.

[36] Kshirsagar, A. M. (1964). "Wilks's Λ criterion", J. Indian Stat. Assoc., 2, p. 1.

[37] Lawley, D. N. (1938). "A generalization of Fisher's Z-test", Biometrika, 30, p. 180.

[38] Lawley, D. N. (1959). "Tests of significance in canonical analysis", Biometrika, 46, p. 59.

[39] Mariott, F. H. C. (1952). "Tests of significance in canonical analysis", Biometrika, 39, p. 58.

[40] Mikhail, M. N. (1965). "A comparison of tests of the Wilks-Lawley hypothesis in multivariate analysis", Biometrika, 52, p. 149.

[41] Narain, R. D. (1950). "On the completely unbiased character of the tests of independence in multivariate normal systems", Ann. Math. Statist., 21, p. 293.

[42] Pearson, E. S. (1971). Tables for Multivariate Analysis. Biometrika, London.

[43] Pillai, K. C. S. (1955). "Some new test criteria in multivariate analysis", Ann. Math. Statist., 26, p. 117.

[44] Pillai, K. C. S., and Samson, P. (1959). "On Hotelling's generalization of T^2", Biometrika, 46, p. 160.

[45] Pillai, K. C. S. (1960). Statistical Tables for Tests of Multivariate Hypotheses. The Statistical Center, Manila, The Philippines.

[46] Pillai, K. C. S., and Jayachandran, Kanta (1967). "Power comparisons of tests of two multivariate hypotheses based on four criteria", Biometrika, 54, p. 195.

[47] Pillai, K. C. S., and Jayachandrau, Kanta (1968). "Power comparisons of tests of equality of two covariance matrices based on four criteria", Biometrika, 55, p. 335.

[48] Pillai, K. C. S., and Gupta, A. K. (1968). "On the noncentral distribution of the second elementary symmetric function of the roots of a matrix", Ann. Math. Statist., 39, p. 833.

[49] Pillai, K. C. S., and Jouris, G. M. (1969). "On the moments of elementary symmetric functions of the roots of two matrices", Ann. Inst. Stat. Math., 21, p. 309.

[50] Pillai, K. C. S., and Dotson, C. O. (1969). "Power comparisons of tests of two multivariate hypotheses based on individual characteristic roots", Ann. Inst. Stat. Math., 21, p. 49.

[51] Pillai, K. C. S., and Sugiyama, T. (1969). "Noncentral distributions of the largest latent roots of three matrices in multivariate analysis", Ann. Inst. Stat. Math., 21, p. 321.

[52] Pillai, K. C. S. (1969). "On the exact distribution of Wilks's criterion", Biometrika, 56, p. 109.

[53] Pillai, K. C. S., and Gupta, A. K. (1969). "On the exact distribution of Wilks's criterion", Biometrika, 56, p. 109.

[54] Pillai, K. C. S., and Young, D. C. (1971). "On the exact distribution of Hotelling's generalized T_o^2", Journal of Multivariate Analysis, 1, p. 90.

[55] Rao, C. R. (1951). "An asymptotic expansion of the distribution of Wilks's criterion", Bull. Inst. Inter. Stat., 33, p. 177.

[56] Rao, C. R. (1952). Advanced Statistical Methods in Biometric Research, New York, John Wiley and Sons.

[57] Roy, S. N. (1945). "The individual sampling distribution of the maximum, the minimum and any intermediate of the p-statistics on the null hypotheses", Sankhyā, 7, p. 133.

[58] Roy, S. N. (1957). Some Aspects of Multivariate Analysis. John Wiley and Sons, New York.

[59] Roy, S. N., and Mikhail, W. F. (1961). "On the monotonic character of the power functions of two multivariate tests", Ann. Math. Statist., 32, p. 1145.

[60] Schatzoff, M. (1966). "Exact distribution of Wilks's likelihood ratio criterion", Biometrika, 53, p. 347.

[61] Schwartz, R. (1964). "Properties of a test in Manova", (Abstract), Ann. Math. Statist., 35, p. 939.

[62] Siotani, M., (1956). "On the distribution of the Hotelling's T^2-statistics", Ann. Inst. Stat. Math., 8, p. 1.

[63] Sugiura, N., and Fujikoshi, Y. (1969). "Asymptotic expansions of the non-null distributions of the likelihood ratio criteria for multivariate linear hypothesis and independence", Ann. Math. Statist., 40, p. 942.

[64] Sugiyama, T. (1967). "Distribution of the largest latent root and the smallest latent root of the generalized B statistics and F statistics in multivariate analysis", Ann. Math. Statist., 38, p. 1152.

[65] Waal, D. J. de (1969). "On the noncentral distribution of the largest canonical correlation coefficient", South African Statist. J., 3, p. 91.

[66] Wilks, S. S. (1932). "Certain generalizations in the analysis of variance", Biometrika, 24, p. 471.

[67] Wold, H. (1959). "Ends and means in econometric model building", Probability and Statistics, the Harold Cramer Volume, John Wiley and Sons, New York.

[68] Yoong, Sin Lee (1971). "Distribution of the canonical correlations and asymptotic expansions for distributions of certain independence test statistics", Ann. Math. Statist., 42, p. 526.

CHAPTER 9

MULTIVARIATE ANALYSIS OF VARIANCE AND DISCRIMINATION IN THE CASE OF SEVERAL GROUPS

1. MULTIVARIATE ANALYSIS OF VARIANCE: ONE-WAY CLASSIFICATION

Consider $k = q+1$ p-variate nonsingular normal populations π_α ($\alpha = 1, 2, \ldots, k$), with the same variance-covariance matrix Σ but different mean vectors $\underline{\mu}_\alpha$ ($\alpha = 1, 2, \ldots, k$). We wish to test the null hypothesis

$$(H_1):\quad \underline{\mu}_1 = \underline{\mu}_2 = \ldots = \underline{\mu}_k \tag{1.1}$$

of the equality of the mean vectors on the basis of k independent random samples of sizes n_α ($\alpha = 1, 2, \ldots, k$) from the populations. Let $x_{ir\alpha}$ ($i = 1, 2, \ldots, p$; $r = 1, 2, \ldots, n_\alpha$; $\alpha = 1, 2, \ldots, k$) be the rth observation on the ith variable x_i of the αth population π_α. Let X_α be the $p \times n_\alpha$ matrix of the n_α sample observations from π_α, and let

$$X = [X_1 | X_2 | \ \ldots \ | X_k] \tag{1.2}$$

be the $p \times N$ matrix of all the $N = n_1 + n_2 + \ldots + n_k = n+1$ observations;

$$\underline{\bar{x}}_\alpha = \frac{1}{n_\alpha} X_\alpha E_{n_\alpha 1} \quad (\alpha = 1, 2, \ldots, k) \tag{1.3}$$

is the vector of the sample means of observations from π_α, and

$$X_\alpha (I - \frac{1}{n_\alpha} E_{n_\alpha n_\alpha}) X'_\alpha = S_\alpha \quad (\alpha = 1, 2, \ldots, k) \tag{1.4}$$

is the matrix of the corrected s.s. and s.p. of the sample observations X_α. From (3.4.6) we know that $\bar{\underline{x}}_\alpha$ and S_α are independently distributed as

$$N_p(\bar{\underline{x}}_\alpha | \underline{\mu}_\alpha | \frac{1}{\sqrt{n_\alpha}} \Sigma) \tag{1.5}$$

and

$$W_p(S_\alpha | n_\alpha - 1 | \Sigma) , \tag{1.6}$$

respectively, for each $\alpha = 1, 2, \ldots, k$. Further, all these distributions are independent as all the k samples are independent. From the additive property of Wishart matrices proved in Section 6 of Chapter 3, the pooled matrix of corrected s.s. and s.p.,

$$A = S_1 + S_2 + \ldots + S_k , \tag{1.7}$$

has the

$$W_p(A | n-q | \Sigma) \text{ distribution,} \tag{1.8}$$

the d.f. n-q being the sum $\sum_{\alpha=1}^{k} (n_\alpha - 1)$ of the individual d.f. This distribution is obviously independent of $\bar{\underline{x}}_\alpha$ ($\alpha = 1, 2, \ldots, k$) and holds irrespective of whether the hypothesis (H_1) is true or not. Let

$$U = [\sqrt{n_1}\, \bar{\underline{x}}_1 | \sqrt{n_2}\, \bar{\underline{x}}_2 | \ldots | \sqrt{n_k}\, \bar{\underline{x}}_k] \tag{1.9}$$

and let

$$G' = [\underline{g}_1 | \underline{g}_2 | \ldots | \underline{g}_q | \underline{h}] = \underset{q \quad 1}{[G'_o | \underline{h}]} p \tag{1.10}$$

be any k x k orthogonal matrix, such that the last column of G' is

$$\underline{h} = \left[\left(\frac{n_1}{N}\right)^{1/2}, \left(\frac{n_2}{N}\right)^{1/2}, \left(\frac{n_k}{N}\right)^{1/2} \right] \tag{1.11}$$

From U, transform to the $p \times k$ matrix

$$Z = [\underline{z}_1 | \underline{z}_2 | \ \cdots \ | \underline{z}_q | \underline{z}_k] \tag{1.12}$$

by the orthogonal transformation

$$Z = U\,G' \quad \text{or} \quad U = Z\,G \quad (\text{as } G \text{ is orthogonal}). \tag{1.13}$$

The Jacobian of this transformation is $|G|^k = 1$, G being orthogonal. Hence the distribution of U, obtainable from (1.5) [see also (2.2.4)], viz.,

$$\frac{1}{(2\pi)^{pk/2}\ |\Sigma|^{k/2}} \exp\left\{-\frac{1}{2}\ \mathrm{tr}\ \Sigma^{-1}(U-E(U))(U-E(U))'\right\} dU \tag{1.14}$$

transforms, on using (1.13), to

$$\frac{1}{(2\pi)^{pk/2}\ |\Sigma|^{k/2}} \exp\left\{-\frac{1}{2}\ \mathrm{tr}\ \Sigma^{-1}(Z-E(Z))(Z-E(Z))'\right\} dZ\,. \tag{1.15}$$

Now observe that, <u>when (H_1) is true</u>, and all $\underline{\mu}_\alpha$ are $= \underline{\mu}$, say

$$\begin{aligned} E(Z) &= E(U)\ G' \\ &= [\sqrt{n_1}\ \underline{\mu} | \sqrt{n_2}\ \underline{\mu} | \ \cdots \ | \sqrt{n_k}\ \underline{\mu}]G' \\ &= \underline{\mu} \cdot [\sqrt{n_1}, \sqrt{n_2}, \ldots, \sqrt{n_k}]G' \\ &= \underline{\mu} \cdot \sqrt{N}\ \underline{h}'\underline{G}' \\ &= [\underline{0} | \underline{0} | \ \cdots \ | \underline{0} | \ \sqrt{N}\ \underline{\mu}]\ , \end{aligned} \tag{1.16}$$

as $\underline{h}$ is orthogonal to the first q columns of G' and $\underline{h}'\underline{h} = 1$. Equations (1.15) and (1.16) together show that $\underline{z}_\alpha$ ($\alpha = 1, 2, \ldots, q$) are independently distributed as

$$N_p(\underline{z}_\alpha | \underline{0} | \Sigma)\ , \tag{1.17}$$

when the null hypothesis (H_1) is true. (Note that $\underline{z}_k$, the last column of Z, also has the same distribution, but its mean is not zero and so we remove it.) We have thus the matrix A, which has thw $W_p(A|n-q|\Sigma)$ distribution, and the matrix

$$B = \sum_{\alpha=1}^{q} \underline{z}_\alpha \underline{z}'_\alpha , \tag{1.18}$$

where $\underline{z}_\alpha$ have the distribution (1.17) when (H_1) is true. Since the $\underline{z}_\alpha$'s are functions of only the $\bar{\underline{x}}_\alpha$ ($\alpha = 1, 2, \ldots, k$), they are independently distributed of A. Thus, A and B are independent. From (8.1.2), therefore,

$$\Lambda \text{ or } \Lambda(n,p,q) = \frac{|A|}{|A + B|} \tag{1.19}$$

has the Wilks's $\Lambda(n,p,q)$ distribution whenever (H_1) is true. We can, therefore, test (H_1) by this Wilks's Λ criterion. (H_1) will be rejected at the $100\,\alpha\%$ level of significance whenever $-\{n - (p+q+1)/2\} \log_e \Lambda$ exceeds $W_\alpha(n,p,q)$ defined by (8.3.9). Observe that

$$\begin{aligned} B &= \sum_{\alpha=1}^{q} \underline{z}_\alpha \underline{z}'_\alpha = ZZ' - \underline{z}_k \underline{z}'_k \\ &= UG'GU' - U\underline{h}\underline{h}'U' \\ &= UU' - \left(\sum_{\alpha=1}^{k} \frac{n_\alpha}{\sqrt{N}} \bar{\underline{x}}_\alpha \right) \left(\sum_{\alpha=1}^{k} \frac{n_\alpha}{\sqrt{N}} \bar{\underline{x}}'_\alpha \right) \\ &= \sum n_\alpha \bar{\underline{x}}_\alpha \bar{\underline{x}}'_\alpha - N\bar{\underline{x}}\bar{\underline{x}}' \\ &= \sum_{\alpha=1}^{k} n_\alpha (\bar{\underline{x}}_\alpha - \bar{\underline{x}})(\bar{\underline{x}}_\alpha - \bar{\underline{x}}_\alpha)' , \end{aligned} \tag{1.20}$$

where

$$\bar{\underline{x}} = \sum_{\alpha=1}^{k} n_\alpha \bar{\underline{x}}_\alpha / N . \tag{1.21}$$

The matrix B is known as the "between groups" matrix of s.s.

and s.p., while the matrix A is known as the "within groups" matrix of s.s. and s.p. ("Groups" correspond to the k populations π_α). The test for (H_1) can, therefore, be obtained from Table 1, known as the multivariate analysis of variance table.

TABLE 1

Source	d.f.	Matrix of s.s. and s.p.
Between groups	q	B
Within groups	n-q	A
Total	n	A + B

The test based on (1.19) and Table 1 are generalizations of the F test and the analysis of variance table in the univariate case (one-way classification) to the multivariate case.

If (H_1) is not true, $E(\underline{z}_\alpha) \neq 0$ $(\alpha = 1, 2, \ldots, k)$, and then

$$\begin{aligned}\sum_{\alpha=1}^{q} E(\underline{z}_\alpha)\, E(\underline{z}'_\alpha) &= \sum_{\alpha=1}^{k} E(\underline{z}_\alpha)\, E(\underline{z}'_\alpha) - E(\underline{z}_k)\, E(\underline{z}'_k) \\ &= E(Z)\, E(Z') - E(\underline{Z}_k)\, E(\underline{Z}'_k) \\ &= E(U)\, G'G\, E(U') - E(U)\, \underline{h}\underline{h}'\, E(U') \\ &= \sum_{\alpha=1}^{k} n_\alpha \underline{\mu}_\alpha \underline{\mu}'_\alpha - \frac{1}{N}\left(\sum_{\alpha=1}^{k} n_\alpha \underline{\mu}_\alpha\right)\left(\sum_{\alpha=1}^{k} n_\alpha \underline{\mu}'_\alpha\right) \\ &= \sum_{\alpha=1}^{k} n_\alpha (\underline{\mu}_\alpha - \bar{\underline{\mu}})(\underline{\mu}_\alpha - \bar{\underline{\mu}})' , \\ &= \Delta \text{ say,}\end{aligned} \qquad (1.22)$$

where

$$\bar{\underline{\mu}} = \sum_{\alpha=1}^{k} n_\alpha \, \underline{\mu}_\alpha / N . \tag{1.23}$$

Since $V(\underline{z}_\alpha) = \Sigma$, the matrix

$$\Omega = \Sigma^{-1}\left\{ \sum_{\alpha=1}^{k} n_\alpha (\underline{\mu}_\alpha - \bar{\underline{\mu}}) \, (\underline{\mu}_\alpha - \bar{\underline{\mu}})' \right\} = \Sigma^{-1}\Delta \tag{1.24}$$

is known as the noncentrality matrix. When (H_1) is true, all the $\underline{\mu}_\alpha$ are equal, and then Ω becomes a null matrix. Ω is thus a measure of the departure from the null hypothesis. Since A has the $W_p(A|n-q|\Sigma)$ distribution, irrespective of whether (H_1) holds or not,

$$E(A) = (n-q)\Sigma \tag{1.25}$$

by Lemma 10 of Chapter 3, but, when we come to B, we have to consider $\sum_{\alpha=1}^{q} \{\underline{z}_\alpha - E(\underline{z}_\alpha)\}\{\underline{z}_\alpha - E(\underline{z}_\alpha)\}'$ first. This has a Wishart (or pseudo-Wishart if $q < p$) distribution with q d.f., and its expectation is $q\,\Sigma$. From this it can be deduced that

$$E(B) = q\,\Sigma + \Sigma\,\Omega = q\,\Sigma + \Delta \tag{1.26}$$

and reduces to $q\,\Sigma$ only when Ω is null, i.e., all the $\underline{\mu}_\alpha$ are equal. Thus $\frac{1}{n-q}$ A and $\frac{1}{q}$ B provide independent estimates of Σ, the latter doing so only when (H_1) is true. Wilks's Λ criterion compares these two estimates and provides a test of (H_1), as in univariate analysis of variance. An unbiased estimate of Ω, when (H_1) is not true, is provided by (use Lemma 10, Chapter 3, for A^{-1})

$$(n-q-p-1)A^{-1}B - q\,I_p . \tag{1.27}$$

This multivariate analysis of variance problem, which we discussed here, can also be looked upon as a regression problem. Define q "dummy" variables $y_1, y_2, \ldots, y_q$ as follows:

$$\begin{aligned} y_\alpha &= 1, \quad \text{if an observation } \underline{x} \text{ comes from } \pi_\alpha \\ &= 0, \quad \text{otherwise} \quad (\alpha = 1, 2, \ldots, q \text{ only}). \end{aligned} \tag{1.28}$$

Then

$$E(\underline{x}) = \underline{\mu}_k + (\underline{\mu}_1 - \underline{\mu}_k)\, y_1 + \ldots + (\underline{\mu}_q - \underline{\mu}_k)\, y_q, \tag{1.29}$$

as (1.29) reduces to $\underline{\mu}_\alpha$ ($\alpha = 1, 2, \ldots, q$) whenever $\underline{x}$ comes from π_α and to $\underline{\mu}_k$ when $\underline{x}$ comes from π_k, as all the y's are zero, when $\underline{x}$ comes from π_k. We can write (1.29) as

$$E(\underline{x}|\underline{y}) = \underline{\mu}_k + \beta\underline{y}, \tag{1.30}$$

where $\underline{y}$ is the vector of the q dummy variables $y_1, \ldots, y_q$, and

$$\beta = [\underline{\mu}_1 - \underline{\mu}_k | \underline{\mu}_2 - \underline{\mu}_k | \ldots | \underline{\mu}_q - \underline{\mu}_k] \tag{1.31}$$

can be looked upon as the $p \times q$ matrix of regression coefficients in the regression of the vector $\underline{x}$ on the vector $\underline{y}$. The hypothesis (H_1) is the same as the hypothesis that β is null. We can, therefore, test this hypothesis by the criterion (8.4.16), which we discussed in the last chapter, when we were dealing with the association between two vectors. Here $\underline{y}$ is not a random vector but only a vector of fixed quantities, but this does not come in the way of using (8.4.16), as there also, $\underline{y}$ was kept fixed in deriving the distribution of (8.4.16) and it turned out to be independent of $\underline{y}$, when $\beta = 0$. Corresponding to the matrix X_α of the n_α observations on $\underline{x}$, from π_α we have now the matrix

Y_α (a $q \times n_\alpha$ matrix) of observations on $\underline{y}$. The elements of Y_α, on account of (1.28), are all zero, except those in the αth row, which consists of all unit elements. This is true for $\alpha = 1, 2, \ldots, q$ only. The matrix of all the N observations on $\underline{y}$ is then

$$Y = [Y_1 | Y_2 | \ldots | Y_q] , \qquad (1.32)$$

of order $q \times N$. One can then evaluate

$$C_{xx} = XJX', \; C_{xy} = XJY', \; C_{yy} = YJY', \text{ etc.,} \qquad (1.33)$$

easily. A little algebra shows that

$$C_{xx} = A + B, \quad C_{xy} = [n_1(\bar{\underline{x}}_1 - \bar{\underline{x}}) | \ldots | n_q(\bar{\underline{x}}_q - \bar{\underline{x}})] \quad (1.34)$$

$$C_{yy} = \text{diag}(n_1, n_2, \ldots, n_q) - \frac{1}{N}[n_1, \ldots, n_q]' \; [n_1, \ldots, n_q],$$

$$C_{yy}^{-1} = \text{diag}\left(\frac{1}{n_1}, \ldots, \frac{1}{n_q}\right) + \frac{1}{n_k} E_{qq} ,$$

$$C_{xy} C_{yy}^{-1} C_{yx} = B, \quad C_{xx \cdot y} = A , \qquad (1.34)$$

where A and B are given by (1.7) and (1.20), respectively. Table 1 of multivariate analysis of variance in this section is, therefore, exactly the same as Table 2 of Section 4, Chapter 8. We thus find that the criterion (1.19) is exactly the same as the criterion (8.4.16). Regression analysis thus gives exactly the same test as we derived, showing that multivariate analysis of variance is only a particular case of regression of a random vector $\underline{x}$ on q dummy variables $\underline{y}$ defined above. As far as hypothesis (H_1) is concerned, this analogy with regression analysis did not give any new test or information, but this analogy will be useful, when (H_1) is rejected and the data are to be analyzed further to obtain more information on the means $\underline{\mu}_\alpha$ of the k groups π_α.

When (H_1) is not true, the distribution of B and hence the distribution of $\Lambda(n,p,q)$ involves the roots δ_i^2 of the determinantal equation $|\Omega - \delta^2 I| = 0$, or, in other words, the roots of

$$|\Delta - \delta^2 \Sigma| = 0 \ . \tag{1.35}$$

Let the number of nonzero roots be

$$s = \text{rank } \Delta = \text{rank } \Omega \ . \tag{1.36}$$

The elements of the mean vector $\underline{\mu}_\alpha$ can be regarded as the coordinates of a point in a p-dimensional space. The group means $\underline{\mu}_\alpha$ thus represent k points in this space. When (H_1) is true, all these vectors $\underline{\mu}_\alpha$ $(\alpha = 1, 2, \ldots, k)$ are identical, and these k points all coincide. In geometrical terminology, this means that the k group means lie in a zero dimensional space (see Section 5 of Chapter 5 in this connection). When (H_1) does not hold, the dimensionality of the space in which these group means lie is greater than zero. It can be equal to, at most, the smaller of p and q. If (H_1) is rejected, it is interesting to determine the actual dimensionality of the space in which these k group means lie. Since

$$\Delta = [\sqrt{n_1}(\underline{\mu}_1 - \bar{\underline{\mu}}) | \ldots | \sqrt{n_k}(\underline{\mu}_k - \bar{\underline{\mu}})][\sqrt{n_1}(\underline{\mu}_1 - \bar{\underline{\mu}}) | \ldots | \sqrt{n_k}(\underline{\mu}_k - \bar{\underline{\mu}})]'$$

from a well-known result about the rank of a matrix, it follows that

$$s = \text{rank } \Delta = \text{rank of } [\sqrt{n_1}(\underline{\mu}_1 - \bar{\underline{\mu}}) | \cdots | \sqrt{n_k} - \bar{\underline{\mu}})]$$

$$= \text{rank of } [\underline{\mu}_1 - \bar{\underline{\mu}} | \cdots | \underline{\mu}_k - \bar{\underline{\mu}}] \operatorname{diag}(\sqrt{n_1}, \ldots, \sqrt{n_k})$$

$$= \text{rank of } [\underline{\mu}_1 - \bar{\underline{\mu}} | \cdots | \underline{\mu}_k - \bar{\underline{\mu}}] \tag{1.37}$$

as diag $(\sqrt{n_1}, \ldots, \sqrt{n_k})$ is a nonsingular matrix. Since $\underline{\mu}_1 - \underline{\mu}_k$, $\underline{\mu}_2 - \underline{\mu}_k, \ldots, \underline{\mu}_q - \underline{\mu}_k$ are linear combinations of $\underline{\mu}_1 - \bar{\underline{\mu}}, \ldots, \underline{\mu}_k - \bar{\underline{\mu}}$, s is also the rank of the matrix β defined by (1.31). Since s is the rank of β, only s columns of β are linearly independent, and each column of β can be expressed as a linear combination of these s columns. In other words, this means that there exist s, <u>and no less</u>, linearly independent p-vectors, say $\underline{\xi}_1, \underline{\xi}_2, \ldots, \underline{\xi}_s$ such that

$$\underline{\mu}_\alpha - \underline{\mu}_k = C_{\alpha 1}\underline{\xi}_1 + C_{\alpha 2}\underline{\xi}_2 + \cdots + C_{\alpha s}\underline{\xi}_s \quad (\alpha = 1, 2, \ldots, q). \tag{1.38}$$

This will be so if and only if

$$\underline{\mu}_\alpha = \underline{\xi}_o + C_{\alpha 1}\underline{\xi}_1 + \cdots + C_{\alpha s}\underline{\xi}_s \tag{1.39}$$

for each $\alpha = 1, 2, \ldots, k$, where $\underline{\xi}_o, \underline{\xi}_1, \ldots, \underline{\xi}_s$ are linearly independent. Any (s+1) linearly independent p-vectors determine an s-dimensional flat in a p-dimensional space. The vectors $\underline{\xi}_o, \underline{\xi}_1, \ldots, \underline{\xi}_s$ thus determine an s-flat. The vector $\underline{\xi}$ of the coordinates of any point on this s-flat is [from the theory of flats in n-dimensional geometry (see Kendall [12])] given by

$$\underline{\xi} = \underline{\xi}_o + C_1\underline{\xi}_1 + \cdots + C_s\underline{\xi}_s . \tag{1.40}$$

Equation (1.40) is known as the vectorial equation of the flat determined by the (s+1) points $\underline{\xi}_o, \underline{\xi}_1, \ldots, \underline{\xi}_s$. From (1.39), we see that the points $\underline{\mu}_\alpha$ ($\alpha = 1, 2, \ldots, k$), representing the means of the k populations π_α, satisfy (1.40) and thus lie on an s-flat. The dimensionality of the space of the group means is thus given by the rank of the matrix Δ or Ω or $\boldsymbol{\beta}$. Since $\boldsymbol{\beta}$ is a $p \times q$ matrix, it is obvious that this dimensionality cannot exceed the smaller of p and q.

Since $[\underline{\xi}_1, \underline{\xi}_2 | \ldots | \underline{\xi}_s]$ is a $p \times s$ matrix of rank s, we can always find a matrix L_1 of order $(p-s) \times p$, such that its row vectors are orthogonal to each $\underline{\xi}_i$ ($i = 1, 2, \ldots, s$). Then premultiplying (1.39) by L_1, it can be seen that

$$L_1 \underline{\mu}_\alpha = \underline{\nu} \quad (\alpha = 1, 2, \ldots, k) , \tag{1.41}$$

where $\underline{\nu}$ is $L_1 \underline{\xi}_o$. Thus, if the k group means $\underline{\mu}_\alpha$ lie on some s-dimensional flat (but not any s-1 or less dimensional flat), there exists a $(p-s) \times p$ matrix L_1 of rank $(p-s)$ (and not more) and a vector $\underline{\nu}$ such that each $\underline{\mu}_\alpha$ satisfies (1.41). Retracing the steps, it can also be shown that, if (1.41) holds for some L_1 and $\underline{\nu}$, the k group means $\underline{\mu}_\alpha$ satisfy (1.39) and thus lie on an s-dimensional flat. So, (1.36), (1.39), (1.41), and rank $\boldsymbol{\beta} = s$ are all equivalent, and any one implies the remaining.

In practice, Δ or $\boldsymbol{\beta}$ are unknown and can only be estimated from sample observations. s, the dimensionality of the space spanned by the group means, can then only be inferred from the estimate of Δ, Ω, or $\boldsymbol{\beta}$. From (1.27) we know that Ω is estimated by $(n-q-p-1)\mathbf{A}^{-1}B - qI_p$. If the rank of Ω is s, it does not mean that all the eigenvalues of $(n-q-p-1)\mathbf{A}^{-1}B - qI_p$ except the first s are zero, because $(n-q-p-1)A^{-1}B - qI_p$ is only an estimate. But if our sample is sufficiently large, i.e., if n is large, it can be reasonably regarded that the first s roots of $(n-q-p-1)A^{-1}B - qI_p$ will be significant and the last (p-s) roots will be insignificant. Thus an estimate of s will be provided by the number of significant roots of $(n-q-p-1)A^{-1}B - qI_p$. Let λ_i be the roots. Then they satisfy

$$|(n-q-p-1)A^{-1}B - qI_p - \lambda I_p| = 0 . \tag{1.42}$$

This is the same as the equation in r^2

$$| - r^2(B + A) + B| = 0 , \tag{1.43}$$

with $r^2 = (\lambda+q)/(n-p-1+\lambda)$. We can therefore base our inference about the rank of Ω on the roots of (1.43) also. The advantage in using (1.43) is that the roots $r_1^2 > r_2^2 > \ldots > r_f^2$ (f = smaller of p, q) are nothing but the squares of the sample canonical correlations between our random variables $\underline{x}$ and the dummy variables $\underline{y}$ defined in (1.28). This can be seen by noticing that $A + B = C_{xx}$, $B = C_{xy}C_{yy}^{-1}C_{yx}$ [see (1.34)], and so (1.43) is

$$| - r^2 C_{xx} + C_{xy}C_{yy}^{-1}C_{yx}| = 0$$

and is the same as (7.3.3) defining canonical correlations. We can therefore immediately use the tests of significance of canonical correlations derived in Section 7 of Chapter 8 to determine the number of significant roots. The procedure is then to prepare Table 4 of Section 8 for each value of s, starting with s = 1, and use the χ^2 tests of (8.7.3) or (8.7.4) successively until a nonsignificant result is achieved. The dimensionality of the group means is then inferred from the number of significant roots of (1.43), determined by this procedure. The criterion is thus

$$- \{n - \frac{1}{2}(p+q+1)\}\log_e \prod_{i=s+1}^{f} (1 - r_i^2) \tag{1.44}$$

for testing the hypothesis that

the dimensionality of the space spanned by $\underline{\mu}_\alpha$ ($\alpha = 1, 2, \ldots, k$) is s . (1.45)

The likelihood ratio criterion for the above hypothesis (1.45) can be derived by maximizing the likelihood of the sample observations X, unconditionally first and then subject to (1.41). The reader may refer to C. Radhakrishna Rao [23] or, for a more concise and elegant derivation of this, to Wani and Kabe [29], together with Kshirsagar [18]. The likelihood ratio criterion is then given by

$$(\nu_{s+1} + \nu_{s+2} + \cdots + \nu_f) , \tag{1.46}$$

where $\nu_1 > \nu_2 > \cdots > \nu_f$ are the roots of

$$|B - \frac{\nu}{n} A| = 0 . \tag{1.47}$$

From (1.43) and (1.47), one can easily see that

$$\nu_i = n\, r_i^2/(1 - r_i^2) ,$$

and, hence,

$$- \log_e (1 - r_i^2) = \log_e (1 + \frac{\nu_i}{n})$$

$$= \text{approximately } \frac{\nu_i}{n}, \text{ for large n.}$$

Therefore,

$$- n \log_e \prod_{i=s+1}^{f} (1 - r_i^2) \doteqdot \sum_{i=s+1}^{f} \nu_i \, . \tag{1.48}$$

According to the large sample theory of a likelihood ratio criterion, $\sum_{i=s+1}^{f} \nu_i$ is a χ^2 with (p-s)(q-s) d.f., if (1.45) is true. This likelihood ratio test is thus asymptotically equivalent to the χ^2 test based on (1.44) on account of (1.48). However the test based on (1.44) should be preferred, as Bartlett's correction factor or the modified correction factor suggested by Lawley [see (8.7.4)] improves its performance. No such investigation seems to have been done for $\sum_{s+1}^{f} \nu_i$.

2. DISCRIMINATION IN THE CASE OF SEVERAL GROUPS

We are now in a position to consider the problem of discriminating among more than two groups. The case of only two groups was thoroughly dealt with in Chapter 6, and we propose to extend those methods to k groups in this section. If the reader is expecting an elegant solution, such as Fisher's discriminant function in the case of two groups, he is in for disappointment. The moment we leave the simple case of two groups, the problem becomes more complex, and the solutions turn out to be less satisfactory.

In Section 2 of Chapter 6, it was shown that Fisher's discriminant function could be obtained by maximizing the ratio of "between groups s.s." to the "within groups s.s." in the analysis of variance of an arbitrary linear function $\underline{\ell}'\underline{x}$. Let us try this method in the case of k (> 2) groups corresponding to the k p-variate normal populations π_α (α = 1, 2, ..., k) with means $\underline{\mu}_\alpha$ and the same variance-covariance matrix Σ. We shall employ the same notation as in the previous section for these populations and the N sample observations from them. The multivariate analysis of variance table is Table 1 of the last section. From that table it follows that, for $\underline{x}$, the "between" and "within" groups s.s. for some arbitrary linear combination $\underline{\ell}'\underline{x}$ are, respectively, $\underline{\ell}'B\underline{\ell}$ and $\underline{\ell}'A\underline{\ell}$. We therefore have to find that $\underline{\ell}$ which maximizes $\underline{\ell}'B\underline{\ell}/\underline{\ell}'A\underline{\ell}$. It can be readily shown that this is also the same as maximizing the ratio $\ell'B\ell/\underline{\ell}'(A+B)\underline{\ell}$, and actual differentiation with respect to $\underline{\ell}$ will show that the "optimum" $\underline{\ell}$ must satisfy the equation

$$[-r^2(A+B)+B]\underline{\ell} = \underline{0}, \tag{2.1}$$

where r^2 is a root of the determinantal equation

$$|-r^2(A+B)+B| = 0. \tag{2.2}$$

But, if we now consider the dummy variables $\underline{y}$ of (1.28) and use (1.34), we immediately find that (2.2) has <u>not one</u> but f (= smaller of p, q) roots $r_1^2 > r_2^2 > \dots > r_f^2$. These are the squares of the canonical correlations between $\underline{x}$ and $\underline{y}$. Corresponding to each root r_i^2, there will be a vector $\underline{\ell}_i$ satisfying (2.1), and we are confronted with not one but f discriminant functions $\underline{\ell}_i'\underline{x}$ (i = 1, 2, ..., f). These linear functions $\underline{\ell}_i'\underline{x}$ are the sample canonical variables of the x-space [see (7.3.8)]. Since r_1^2 is the largest root of (2.2), the corresponding function

$\ell_1'x$ provides the maximum "separation" of the group means and is useful as a discriminant function. But it is not sufficient, as the next discriminant function $\ell_2'x$ also provides separation among these groups in a different direction and is also needed for discrimination. Similarly $\ell_3'x$, and so on. There is thus a multiplicity of discriminant functions. The question that now arises is whether all these discriminant functions really matter or whether the first few of these are adequate for practical purposes and the rest can be ignored, without sacrificing any vital information. The answer to this question naturally depends on the discriminating ability of these discriminant functions, and we therefore need a "measure" of this ability. In the case of two groups, the performance of the discriminant function was measured by Mahalanobis's D^2, which from (6.4.15) was related to R^2, the square of the multiple correlation coefficient between x and ξ, the single dummy variable there. Since R^2 is replaced by $r_1^2, r_2^2, \ldots, r_f^2$, when instead of a single ξ we have q dummy variables y, it is natural to expect that r_1^2 measures the discriminating ability of $\ell_1'x$, r_2^2 of $\ell_2'x$, and so on. This immediately supplies the answer to our question. If only $r_1^2, r_2^2, \ldots, r_s^2$ are significantly large and the rest are insignificant, we should employ only $\ell_i'x$ (i = 1, 2, ..., s) for discrimination. The statistical procedure for determining this number s of significant roots r_i^2 is already described in the last section, and also in Section 7 of Chapter 8. We thus see that the adequate number of discriminant functions is the same as the dimensionality of the space spanned by the k group means. When there are only two groups, the points representing the means in the p-dimensional space will always lie on a line or a one-dimensional flat, and so only one discriminant function orthogonal to this line, separating the two points, serves the purpose of discrimination. But when we have k points corresponding to the k group means, they will not in general lie on a line, and so one discriminant function will not, in general, be adequate. If they lie in an s-dimensional flat, we shall require s discriminant functions, which separate these k points, in s different orthogonal directions. Only when $s = 1$, i.e., when only r_1^2 is significant and the rest are not, the means are collinear and lie in a one-dimensional flat. Then $\underline{\ell}_1'\underline{x}$, the discriminant function corresponding to r_1^2, will be adequate for discrimination.

We can look at this problem in a different way also. The noncentrality matrix Δ of (1.22) is a measure of the divergence of the group means $\underline{\mu}_\alpha$ among themselves. If we consider a linear function $\underline{h}'\underline{x}$ instead of all the $\underline{x}$, the noncentrality parameter in the analysis of $\underline{h}'\underline{x}$ will be $\underline{h}'\Delta\underline{h}$. If $\underline{h}'\Delta\underline{h}$ is zero, it means [see (1.22)]

$$\sum_{\alpha=1}^{k} n_\alpha(\underline{h}'\underline{\mu}_\alpha - \underline{h}'\bar{\underline{\mu}})^2 = 0 , \qquad (2.3)$$

and this can happen only when $\underline{h}'\underline{\mu}_\alpha$ ($\alpha = 1, 2, \ldots, k$) are all equal, i.e., $E(\underline{h}'\underline{x})$ is the same in all the k groups. $\underline{h}'\underline{x}$ will therefore not be able to discriminate among the means of the k groups. Consider now the eigenvalues $\underline{h}_i$ (i = 1, 2, ..., p) corresponding to the roots $\delta_1^2 \geq \delta_2^2 \geq \cdots \geq \delta_p^2$ of (1.35), i.e., those $\underline{h}_i$ which satisfy

$$[\Delta - \delta_i^2 \Sigma]\underline{h}_i = 0 \quad (i = 1, 2, \ldots, p). \tag{2.4}$$

If the rank of the matrix Δ is s, $\delta_{s+1}^2 = \cdots = \delta_p^2 = 0$ and then

$$\Delta\, \underline{h}_i = \underline{0} \quad (i = s+1, \ldots, p). \tag{2.5}$$

Hence $\underline{h}_i' \Delta \underline{h}_i = 0$, and so from (2.3), we conclude that $\underline{h}_i'\underline{x}$ (i = s+1, ..., p) are not useful for discrimination. However, for the first s functions $\underline{h}_i'\underline{x}$ (i = 1, 2, ..., s), $\Delta\underline{h}_i \neq 0$, and they can be used for discrimination as, unlike $\underline{h}_i'\underline{x}$ ($i \geq s+1$), their means are not the same in each of the k groups. We may call $\underline{h}_i'\underline{x}$ (i = 1, 2, ..., s) as discriminators, while $\underline{h}_i'\underline{x}$ (i = s+1, ..., p) are only "covariates". They, by themselves, are not of any use for discrimination. We see once again from this discussion that the adequate number of discriminant functions is s, which is the rank of Δ, Ω, or β, and is the dimensionality of the space of the $\underline{\mu}_\alpha$'s and is also the number of nonzero eigenvalues of (1.35). In this discussion we used Δ and Σ, but they are unknown and so in practice, one must use their estimates. From (1.25) and (1.26), we know that A provides an estimate of Σ and then B provides an estimate of Δ. Therefore, when we deal with samples, we can replace (1.35) by (1.47) and (2.4) by

$$(B - \frac{\nu_i}{n} A)\, \underline{\ell}_i = 0$$

to obtain estimates of $\underline{h}_i'\underline{x}$. But this is the same as considering (2.1) and (2.2). We are thus led again to the same solution that $\underline{\ell}_i'\underline{x}$ (i = 1, 2, ..., s) corresponding to the roots r_i^2 (i = 1, 2, ..., s) of (2.2) are the sample discriminant functions if only the first s canonical correlations r_i (i = 1, 2, ..., s) are significant; the rest are insignificant statistically. In the case of two populations, there is only one "distance" between them and is measured along the line joining their means. In the case of k groups, we have s "distances", $\delta_1^2, \delta_2^2, \ldots, \delta_s^2$, along s different directions. The total "distance" can then be measured by $\sum_1^s \delta_i^2 = \mathrm{tr}\, \Sigma^{-1}\Delta = \mathrm{tr}\, \Omega$, or we have alternative measures like $\sum_1^s \delta_i^2/(1-\delta_i^2)$. The corresponding sample quantities are $\sum r_i^2$ or $\sum r_i^2/(1-r_i^2)$ or $\pi(1-r_i^2)$. When $s = p$, these are, respectively, Pillai's V, Hotelling's generalized T_o^2, or Wilks's Λ (see Section 8 of Chapter 8).

We shall now consider the method of actually utilizing these s sample discriminants $\underline{\ell}_i'\underline{x}$ for assigning a new observation $\underline{x}_o$ to one of the k groups. By substituting $\underline{x}_o$ for $\underline{x}$ in the discriminant functions $\underline{\ell}_i'\underline{x}$, we find

$$u_{o1} = \underline{\ell}_1'\underline{x}_o, \quad u_{o2} = \underline{\ell}_2'\underline{x}_o, \quad \ldots, \quad u_{os} = \underline{\ell}_s'\underline{x}_o . \tag{2.6}$$

$u_{o1}, u_{o2}, \ldots, u_{os}$ are known as the "new coordinates of the point $\underline{x}_o$. In other words, instead of considering all the p elements of $\underline{x}_o$, we reduce the dimensionality of the problem by considering only the s-flat in which the means of the k groups lie. The coordinates of this point $\underline{x}_o$, referred to the new axes

$\underline{\ell}_1, \ldots, \underline{\ell}_s$ in the s-flat, are $u_{o1}, u_{o2}, \ldots, u_{os}$. Similarly the "new" coordinates of the "estimated" mean $\bar{\underline{x}}_\alpha$ of π_α are

$$u_{\alpha 1} = \underline{\ell}_1'\bar{\underline{x}}_\alpha, \quad u_{\alpha 2} = \underline{\ell}_2'\bar{\underline{x}}_\alpha, \quad \ldots, \quad u_{\alpha s} = \underline{\ell}_s'\bar{\underline{x}}_\alpha . \tag{2.7}$$

[$\bar{\underline{x}}_\alpha$ is the mean of the sample from π_α, defined in (1.3).] The distance between the new point $\underline{x}_o$ and the mean $\bar{\underline{x}}_\alpha$ of π_α based on these new coordinates is $\Delta_{\alpha o}$, where

$$\Delta_{\alpha o}^2 = \sum_{i=1}^{s} (u_{\alpha i} - u_{oi})^2 \quad (\alpha = 1, 2, \ldots, k) . \tag{2.8}$$

The observation $\underline{x}_o$ should naturally be assigned to π_a if the point $\underline{x}_o$ is nearer to the mean of π_a than the mean of any other π_α. In other words, assign $\underline{x}_o$ to π_a if

$$\Delta_{ao}^2 = \mathrm{Min}(\Delta_{1o}^2, \Delta_{2o}^2, \ldots, \Delta_{ko}^2) . \tag{2.9}$$

The "dummy" variables $y_1, y_2, \ldots, y_q$ defined in (1.28) are a kind of index variables attached to the k populations $\pi_1, \pi_2, \ldots, \pi_k$. If an individual comes from π_a, by the definition of $\underline{y}$, the values of $y_1, y_2, \ldots, y_{a-1}, y_{a+1}, \ldots, y_q$ will be zero for him, and only y_a will be 1. (In the case of π_k, all the y's will be zero.) Theoretically, therefore, we can assign an individual to his population by knowing the value of $\underline{y}$, but this is impossible unless we know the population. The best possible alternative seems to be to utilize the association between $\underline{y}$ and the vector $\underline{x}$ of variables, which can be measured. If one can predict $\underline{y}$ from $\underline{x}$, we can then use the predicted value of $\underline{y}$ to assign the individual to one of $\pi_1, \pi_2, \ldots, \pi_k$. By the theory of canonical correlations and canonical variables, the best predictors of $\underline{y}$ are the canonical variables of the $\underline{x}$-space. If only s canonical correlations are significant, the corresponding canonical variables $\underline{\ell}_1'\underline{x}, \ldots, \underline{\ell}_s'\underline{x}$ will be the best predictors, and the remaining variables $\underline{\ell}_{s+1}'\underline{x}, \ldots, \underline{\ell}_p'\underline{x}$, being uncorrelated with $\underline{y}$, will be unable to provide any information on $\underline{y}$. This is the reason why canonical variables are sometimes also referred to as discriminant functions. In Chapter 6, Section 4, we used this logic to derive Fisher's discriminant function in the case of 2 groups only, by considering the regression of one dummy variable ξ on $\underline{x}$. In the case of k groups, we have to consider the regression of the vector $\underline{y}$ of dummy variables on $\underline{x}$. As we have seen in Chapter 7, this leads to $\underline{\ell}_1'\underline{x}, \ldots, \underline{\ell}_s'\underline{x}$ as the best predictors of $\underline{y}$. We once again obtain the same solution to this problem by this method.

Corresponding to $\underline{\ell}_1'\underline{x}, \underline{\ell}_2'\underline{x}, \ldots, \underline{\ell}_s'\underline{x}$ (the canonical variables of the $\underline{x}$-space), one also has $\underline{m}_1'\underline{y}, \underline{m}_2'\underline{y}, \ldots, \underline{m}_s'\underline{y}$, the canonical variables of the $\underline{y}$-space. They are obtainable from [see (7.3.9)]

$$[-r_i^2 C_{yy} + C_{yx}C_{xx}^{-1}C_{xy}]\underline{m}_i = \underline{0} \quad (i = 1, 2, \ldots, s) \tag{2.10}$$

or from [see (7.3.6)]

$$C_{yy}^{-1}C_{yx}\underline{\ell}_i = r_i\underline{m}_i \quad (i = 1, 2, \ldots, s)\,. \tag{2.11}$$

These functions $\underline{m}_i'\underline{y}$ are sometimes referred to as the discriminant functions from the dummy variables space. Their importance and physical interpretation can be seen from the following approach of Fisher [8] to this problem of discrimination among k groups.

Fisher considered the transformed variables Z given by (1.12). The first q columns of Z are

$$\underline{z}_\alpha = U\, \underline{g}_\alpha \quad (\alpha = 1, 2, \ldots, q), \tag{2.12}$$

and these are "contrasts" or "comparisons" among the sample means $\bar{\underline{x}}_\alpha$, because the sum of the coefficients of the $\bar{\underline{x}}_\alpha$'s in $\underline{z}_\alpha$ is

$$[\sqrt{n_1}, \sqrt{n_2}, \ldots, \sqrt{n_k}]\underline{g}_\alpha = \sqrt{N}\, \underline{h}'\underline{g}_\alpha = \underline{0}, \tag{2.13}$$

as G of (1.10) is orthogonal. Fisher then extended the definition of Mahalanobis's distance and his discriminant function, viz., $(\underline{d}'S^{-1}\underline{d})^{1/2}$ and $(S^{-1}\underline{d})'\underline{x}$, respectively [see (6.2.10) and (6.2.3)], from a simple contrast $\underline{d} = \bar{\underline{x}}_1 - \bar{\underline{x}}_2$ to a more general contrast like $\underline{z}_\alpha$. The $(\text{distance})^2$ based on $\underline{z}_\alpha$ is then $(n-q)\underline{z}_\alpha' A^{-1}\underline{z}_\alpha$ [as $[1/(n-q)]A$ is an estimate of Σ), and so the total $(\text{distance})^2$ based on all the q contrasts $\underline{z}_1, \ldots, \underline{z}_q$ is

$$\begin{aligned}\phi &= \sum_{\alpha=1}^{q} (n-q)\underline{z}_\alpha' A^{-1}\underline{z}_\alpha \\ &= (n-q)\ \mathrm{tr}\ G_o U' A^{-1} U\, G_o,\end{aligned} \tag{2.14}$$

where G_o and U are defined by (1.10) and (1.9). Fisher then chose those "optimum" contrasts that maximize this ϕ. The solution of this problem (see Binet and Watson [5]) is provided by the eigenvectors and values associated with the equation

$$[(I - \underline{h}\underline{h}')U'A^{-1}U - \theta_i I]\underline{g}_i = 0. \tag{2.15}$$

By choosing these "optimum" g_i $(i = 1, 2, \ldots)$, one obtains the optimum contrasts $\underline{z}_\alpha = U\,\underline{g}_\alpha$ of (2.12), and then the optimum discriminant functions are $(A^{-1}\underline{z}_\alpha)'\underline{x}$ $(\alpha = 1, 2, \ldots)$, analogous to

$(S^{-1}\underline{d})'\underline{x}$ in the case of two groups. But it so happens that these discriminant functions are exactly the same as (apart from a constant of proportionality) $\underline{\ell}_i'\underline{x}$ obtained earlier by other methods. The roots θ_i are also related to the canonical correlations r_i in a simple manner, and hence s, the adequate number of discriminant functions, is equal to the number of significant eigenvalues θ_i in (2.15) and is the same as before. It has been shown that the coefficient vector $\underline{g}_i$ in the optimum contrast $\underline{z}_i$, so determined and the coefficient vector $\underline{m}_i$ [of (2.11)] in the discriminant function $\underline{m}_i'\underline{y}$, from the dummy variables space, are related in the following way:

The jth element of $\underline{m}_i$ is proportional to $\dfrac{g_{ij}}{\sqrt{n_j}} - \dfrac{g_{ik}}{\sqrt{n_k}}$ ($j = 1, 2, \ldots, q$), where g_{ij} is the jth element of $\underline{g}_i$ ($j = 1, 2, \ldots, k$). (2.16)

In other words, $\underline{\ell}_i'\underline{x}$ are those linear functions which separate the k group means as much as possible, and $\underline{m}_i'\underline{y}$ provide [on account of (2.16)] those contrasts $\underline{z}_\alpha = U\,\underline{g}_\alpha$ among the group means which are significantly large. The $\underline{m}_i'\underline{y}$ or $\underline{z}_\alpha$ therefore provide quantitative "measures" to be associated with the k populations, or "scores" to be assigned to the π_α's, providing some idea of their relative separation. The score to be assigned to π_α is $m_{i\alpha}$, the αth element of $\underline{m}_i$ ($\alpha = 1, 2, \ldots, q$), the score for π_k being always zero, as each of $y_1, y_2, \ldots, y_q$ is identically zero for an individual from π_k, according to (1.28). These are the scores based on the ith discriminant function $\underline{m}_i'\underline{y}$, and we have s such sets of scores, corresponding to the s significant discriminant functions $\underline{m}_1'\underline{y}$, $\underline{m}_2'\underline{y}$, ..., $\underline{m}_s'\underline{y}$ obtainable from (2.11) corresponding to $r_1^2 > r_2^2 > \ldots > r_s^2$. Since the "score" assigned to π_k is zero, the other scores measure the distances of other population means from π_k, on some scale, and along different directions. Table 2 shows the s sets of scores to be assigned to $\pi_1, \pi_2, \ldots, \pi_k$.

TABLE 2

Populations

Set of Scores	π_1	π_2	...	π_q	π_k
1st	m_{11}	m_{12}		m_{1q}	0
2nd	m_{21}	m_{22}		m_{2q}	0
⋮					
sth	m_{s1}	m_{s2}		m_{sq}	0

When r_1^2 is the only significant canonical correlation, $\underline{\ell}_1'\underline{x}$ is the only discriminant function. It is adequate for discrimination. The corresponding discriminant function $\underline{m}_1'\underline{y}$ from the dummy variables space will give a set of scores m_{11}, m_{12}, ..., m_{1q} and 0 for the k groups, and this set of scores alone is sufficient to describe the relative positions of the group means. In other words, when r_1 is the only significant canonical correlation, the populations can be represented on a "linear scale", with these scores. Fisher [9] and later Bartlett [4] have analyzed Taylor's blood serological data consisting of twelve samples of human blood tested with twelve different sera and gave reactions represented by the symbols -, ?, w, (+) and +. They found that these symbols can be represented on a linear scale, and Bartlett has given the corresponding scores. Unfortunately, for this example, the vector $\underline{x}$ which they used is also of "dummy" variables, as $\underline{y}$ corresponding to these symbols. For the application of the tests of significance of canonical correlations, it is essential (as seen from Chapters 7 and 8) that at least one of $\underline{x}$ and $\underline{y}$ have a normal distribution. When $\underline{x}$ and $\underline{y}$ both are "dummy" or pseudo variables, the assumptions underlying the distribution theory of the r_i^2's are not strictly valid. Nevertheless, as Bartlett has remarked, the tests will be asymptotically correct and still informative. This example shows how qualitative data such as frequencies of -, ?, w, (+), + can be turned into quantitative data, by assigning optimum scores using the method of canonical analysis. It may be added here that the scores $m_{i1}, m_{i2}, \ldots, m_{iq}, 0$ are only relative, and any other set of scores such as

$$\frac{m_{i1} - a}{b}, \frac{m_{i2} - a}{b}, \ldots, \frac{m_{iq} - a}{b}, -\frac{a}{b},$$

with origin and scale changed, is equivalent to the original set. We now turn to the remaining canonical variables $\underline{m}_\alpha'\underline{y}$ and the remaining contrasts $\underline{z}_\alpha$ ($\alpha = s+1, \ldots, q$) which correspond to the statistically insignificant canonical correlations r_i $(i > s)$. The contrasts

$$\underline{z}_\alpha = \sqrt{n_1}\, g_{\alpha 1}\, \bar{\underline{x}}_1 + \sqrt{n_2}\, g_{\alpha 2}\, \bar{\underline{x}}_2 + \ldots + \sqrt{n_k}\, g_{\alpha k}\, \bar{\underline{x}}_k \qquad (\alpha = s+1, \ldots, q), \quad (2.17)$$

where $g_{\alpha 1}, g_{\alpha 2}, \ldots, g_{\alpha k}$ are elements of $\underline{g}_\alpha$, are thus insignificant. One can obtain the coefficients $\sqrt{n}\, g_{\alpha 1}, \ldots, \sqrt{n_k}\, g_{\alpha k}$ in (2.17) either from (2.15) (corresponding to the last q-s roots), or alternatively, one can find the canonical variable $\underline{m}'_\alpha \underline{y}$ $(\alpha = s+1, \ldots, q)$ corresponding to the last q-s canonical correlations r_i $(i > s)$. If $m_{\alpha 1}, m_{\alpha 2}, \ldots, m_{\alpha q}$ are the elements of $\underline{m}_\alpha$, the "insignificant" contrast is then

$$C_{xy}\underline{m}_\alpha = C_{x(\underline{m}'_\alpha \underline{y})} = m_{\alpha 1}\, n_1(\bar{\underline{x}}_1 - \bar{\underline{x}}) + \ldots + m_{\alpha q}\, n_q(\bar{\underline{x}}_q - \bar{\underline{x}}), \quad (2.18)$$

where C_{xy} is given by (1.34) and $\bar{\underline{x}}$ by (1.21). Using (1.21), (2.18) can be expressed also as

$$(m_{\alpha 1} - \bar{m}_\alpha) n_1 \bar{\underline{x}}_1 + \ldots + (m_{\alpha q} - \bar{m}_\alpha) n_q \bar{\underline{x}}_q - \bar{m}_\alpha n_k \bar{\underline{x}}_k, \quad (2.19)$$

where

$$\bar{m}_\alpha = (n_1 m_{\alpha 1} + n_2 m_{\alpha 2} + \ldots + n_q m_{\alpha q})/N. \quad (2.20)$$

One can now easily verify that (2.19) and (2.17) are the same because of (2.16).

3. SUBDIVISION OF THE "BETWEEN GROUPS" s.s. AND s.p. MATRIX

If the hypothesis (H_1), of the equality of the k groups considered in Section 1, is true, it is evident that <u>any</u> contrast among the group means $\underline{\mu}_1, \underline{\mu}_2, \ldots, \underline{\mu}_k$ is null. But if Wilks's Λ

given in (1.19) is significant, all contrasts will not be null; at least one contrast will be different from zero. It therefore becomes interesting to find out whether a given contrast, or to be more general, whether a given set of contrasts, are different from zero. This can be done by splitting the "between groups" s.s. and s.p. matrix B in Table 1. B corresponds to all the contrasts among the k means, and has, therefore, $q = k-1$ d.f., as one can have at most q linearly independent contrasts among k means.

Equation (1.20) shows that the matrix B is expressible as $\sum_{\alpha=1}^{q} \underline{z}_\alpha \underline{z}_\alpha'$, where the $\underline{z}_\alpha$ are independent and have the $N_p(\underline{z}_\alpha | \underline{0} | \Sigma)$ distribution if (H_1) is true. In (2.12), we had seen that each $\underline{z}_\alpha$ is a contrast, i.e.,

$$\underline{z}_\alpha = g_{\alpha 1} \sqrt{n_1}\, \bar{\underline{x}}_1 + g_{\alpha 2} \sqrt{n_2}\, \bar{\underline{x}}(2) + \ldots + g_{\alpha k} \sqrt{n_k}\, \bar{\underline{x}}(k) , \qquad \alpha = 1, 2, \ldots, q \tag{3.1}$$

where $g_{\alpha 1}, \ldots, g_{\alpha k}$ are elements of $\underline{\underline{g}}_\alpha$. The matrix B is thus split into q independent matrices $\underline{z}_1\underline{z}_1', \underline{z}_2\underline{z}_2', \ldots, \underline{z}_q\underline{z}_q'$, each carrying 1 d.f. This is the subdivision of B corresponding to the q contrasts $\underline{z}_\alpha = U\, \underline{\underline{g}}_\alpha$ $(\alpha = 1, 2, \ldots, q)$ given by (2.12) or (1.13). The $\underline{\underline{g}}_\alpha$ are all mutually orthogonal and also orthogonal to $\underline{h}$ of (1.11). This subdivision of B is not unique; instead of the q contrasts $\underline{z}_\alpha$, we could have considered any other set of q contrasts, by choosing a different orthogonal matrix, from G' of (1.10). The only restriction is that any such orthogonal matrix must have $\underline{h}$ as its last column.

If we wish to test whether the s contrasts ($s < q$) $\underline{z}_1, \underline{z}_2, \ldots, \underline{z}_s$ are significant [i.e., whether $E(\underline{z}_1), \ldots, E(\underline{z}_s)$ are different from zero or not], we consider only their contributions, viz., $\sum_{\alpha=1}^{s} \underline{z}_\alpha \underline{z}'_\alpha$, to B. To test its significance, construct

$$\frac{|A|}{\left|A + \sum_{\alpha=1}^{s} \underline{z}_\alpha \underline{z}'_\alpha\right|}, \tag{3.2}$$

where A is the error s.s. and s.p. matrix in Table 1. This statistic has the Wilks's $\Lambda(n-q+s, p, s)$ distribution, when $E(\underline{z}_\alpha) = 0$ ($\alpha = 1, 2, \ldots, s$). We test the significance of (3.2) by using (8.3.9).

Sometimes one wishes to test a hypothesis that only a given contrast

$$\underline{u} = a_1\underline{\mu}_1 + a_2\underline{\mu}_2 + \ldots + a_k\underline{\mu}_k, \tag{3.3}$$

where

$$a_1 + a_2 + \ldots + a_k = 0 . \tag{3.4}$$

is different from zero and that all other contrasts are null. For this, we must remove the contribution of $\underline{u}$ to B and test the significance of the remainder. $\underline{u}$ is estimated by

$$\begin{aligned}\hat{\underline{u}} &= a_1\bar{\underline{x}}_1 + a_2\bar{\underline{x}}_2 + \ldots + a_k\bar{\underline{x}}_k \\ &= g^*_{11}\sqrt{n_1}\,\bar{\underline{x}}_1 + g^*_{12}\sqrt{n_2}\,\bar{\underline{x}}_2 + \ldots + g^*_{1k}\sqrt{n_k}\,\bar{\underline{x}}_k,\end{aligned} \tag{3.5}$$

where

$$\underline{g}^{*\prime}_1 = [g^*_{11}, g^*_{12}, \ldots, g^*_{1k}], \quad g^*_{1\alpha}\sqrt{n_\alpha} = a_\alpha \quad (\alpha = 1, 2, \ldots, k). \tag{3.6}$$

Let $\underline{g}_1 = (\underline{g}^{*\prime}_1\,\underline{g}^*_1)^{-1/2}\,\underline{g}^*_1$. Then $\underline{g}'_1\underline{g}_1 = 1$ and $\underline{g}'_1\underline{h} = 0$, where $\underline{h}$

is given by (1.11). Thus $\underline{g}_1$, whose elements are $g_{11}, \ldots, g_{1k}$, can be a column of an orthogonal matrix, whose last column is $\underline{h}$. Therefore, as has been shown, for the contrast,

$$\underline{z}_1 = g_{11}\sqrt{n_1}\,\underline{\bar{x}}_1 + \cdots + g_{1k}\sqrt{n_k}\,\underline{\bar{x}}_k, \tag{3.7}$$

the contribution to B is $\underline{z}_1\underline{z}_1'$ (d.f. 1). But

$$\begin{aligned}\underline{z}_1\underline{z}_1' &= \hat{\underline{u}}\,\hat{\underline{u}}'/(\underline{g}_1^{*\prime}\,\underline{g}_1^{*}) \\ &= \left(\sum_1^k a_\alpha \underline{\bar{x}}_\alpha\right)\left(\sum_1^k a_\alpha \underline{\bar{x}}_\alpha\right)'\Big/\left(\frac{a_1^2}{n_1} + \frac{a_2^2}{n_2} + \cdots + \frac{a_k^2}{n_k}\right).\end{aligned} \tag{3.8}$$

This is known as s.s. and s.p. matrix due to $\underline{u}$ and tests the hypothesis that $E(\underline{z}_1) = \underline{0}$, which is the same as $E(\hat{\underline{u}}) = \underline{0}$. If we remove $\underline{z}_1\underline{z}_1'$ from B, the remainder is $B - \underline{z}_1\underline{z}_1'$ with q-1 d.f. and is the s.s. and s.p. matrix due to all the other q-1 contrasts among the k means, uncorrelated with $\hat{u}$. The test criterion is then

$$\frac{|A|}{|A + (B - \underline{z}_1\underline{z}_1')|} \tag{3.9}$$

and has the $\Lambda(n-1, p, q-1)$ distribution when u is the only non-null contrast and all other contrasts are null.

Another method of obtaining the s.s. and s.p. matrix due to the contrast u of (3.3) is as follows. Let

$$\underline{m}' = \left[\frac{a_1}{n_1} - \frac{a_k}{n_k},\ \frac{a_2}{n_2} - \frac{a_k}{n_k},\ \ldots,\ \frac{a_q}{n_q} - \frac{a_k}{n_k}\right]. \tag{3.10}$$

Then, because of (3.4) and (1.21), it can be readily seen that

$$\hat{u} = C_{xy}\underline{m} = C_{x(\underline{m}'\underline{y})}, \tag{3.11}$$

where C_{xy} is given by (1.34), for the q dummy variables $\underline{y}$, defined in (1.28). Also, the s.s. and s.p. matrix due to $\underline{u}$, given by (3.8), is

$$\begin{aligned} \underline{z}_1\underline{z}_1' &= C_{xy}\underline{m}\,(\underline{m}'C_{yy}\underline{m})^{-1}\underline{m}'C_{yx} \\ &= C_{\underline{x}(\underline{m}'\underline{y})}\,C^{-1}_{(m'y)(m'y)}\,C_{(\underline{m}'\underline{y})\underline{x}} \\ &= \text{matrix of regression s.s. and s.p. in the regression of } \underline{x} \text{ on } \underline{m}'\underline{y}. \end{aligned} \tag{3.12}$$

We thus observe that

> Contribution to the "between groups" s.s. and s.p. matrix B, due to a contrast $a_1\underline{\mu}_1 + \dots + a_k\underline{\mu}_k$, is obtained by considering the regression of $\underline{x}$ on $\underline{m}'\underline{y}$, where the vector $\underline{m}$ is given by (3.10). (3.13)

Barnard [2], Bartlett [3], and Rao [23] have analyzed data on the measurements of $p = 4$ variables on samples of skulls drawn from $k = 4$ series or groups of Egyptian skulls. The samples were of sizes n_α (α = 1, 2, 3, 4), the total number being $N = n+1$. A multivariate analysis of variance table such as Table 1 of Section 1 was formed, and the resulting Wilks's Λ test showed that the four series had different means. It was also known that the relative times between the four series were in the ratio 2:1:2. In view of this, the data were further analyzed to find out whether time t alone (corresponding to the 4 series) can account for these differences in the means of the series. In other words, it was to be investigated whether regression of $\underline{x}$ on t could remove the significant portion of B, the matrix of s.s. and s.p. due to the differences in the means.

This is done by taking $t = t_\alpha$ ($\alpha = 1, 2, 3, 4$) for the αth series and choosing t_α in such a way that $t_2 - t_1$, $t_3 - t_2$, $t_4 - t_3$ are in the ratio 2:1:2. This is achieved by taking $t_1 = a$, $t_2 = a + 2b$, $t_3 = a + 3b$, $t_4 = a + 5b$. a and b can be any arbitrary numbers. Since all the n_α skulls in the αth series have value t_α for the variable t, the average of t is $\bar{t} = \sum_1^4 n_\alpha t_\alpha / N$. The corrected s.s. of t is $C_{tt} = \sum_{\alpha=1}^4 n_\alpha (t_\alpha - \bar{t})^2$ and the corrected s.p. of $\underline{x}$ and t is $C_{xt} = \sum_{\alpha=1}^4 n_\alpha \bar{\underline{x}}_\alpha (t_\alpha - \bar{t})$. The regression s.s. and s.p. matrix, when the regression of $\underline{x}$ on t is considered, is

$$C_{xt} C_{tt}^{-1} C_{tx}, \qquad (3.14)$$

with 1 d.f. When this is removed from B, the remainder is $B - C_{xt} C_{tt}^{-1} C_{tx}$, with $q-1 = k-2 = 2$ d.f., and if t removes the entire significant portion, this remainder will be insignificant. To test this, the criterion is

$$\frac{|A|}{|A + B - C_{xt} C_{tt}^{-1} C_{tx}|} \qquad (3.15)$$

and has $\Lambda(n-1, p, q-1)$ d.f. If this is found to be insignificant, t alone accounts for the differences in the series. Comparing (3.14) with (3.12) or (3.11) with C_{xt}, it is evident that $\underline{m}'\underline{y}$ in this case is t, and the contrast that is significant is $C_{xt} = \sum_{\alpha=1}^4 n_\alpha \bar{\underline{x}}_\alpha (t_\alpha - \bar{t})$. This is the only significant contrast, i.e., $\sum_{\alpha=1}^4 n_\alpha (t_\alpha - \bar{t}) \underline{\mu}_\alpha$, is the only contrast among the $\underline{\mu}_\alpha$'s that is non-null, if (3.15) is found to be insignificant. In that case $\underline{m}'\underline{y} = t$ is the only discriminant function in the space of the dummy variables $\underline{y}$. This assigns the scores t_1, t_2, t_3, t_4 to the four series, and by choosing $a = -5$, $b = 2$, these scores are -5, -1, 1, 5 as Rao has chosen. The means then can be regarded as collinear, and this single set of scores is adequate. The discriminant function in terms of the $\underline{x}$ variable is $\underline{\ell}'\underline{x}$, and it can be obtained by finding that $\underline{\ell}$ which satisfies (2.1) corresponding to its largest root r_1^2, or one can utilize (7.3.6) and project t or $\underline{m}'\underline{y}$ on the $\underline{x}$-space, i.e., consider the regression of t on $\underline{x}$, yielding

$$C_{tx}C_{xx}^{-1}\underline{x} = \sum_{\alpha=1}^{4} n_\alpha(t_\alpha - \bar{t})\bar{\underline{x}}_\alpha'(A+B)^{-1}\underline{x} \tag{3.16}$$

as the discriminant function. Unfortunately, however, in the case of this example of Egyptian skulls, the statistic (3.15) was found to be significant, and thus t could not remove the entire significant portion from B. The differences among the means of the 4 series are not due to t alone. In fact, it was found that both r_1^2 and r_2^2, the first two roots of (2.2), were significant, showing thereby that no single variable, t or any other, can account for the mean differences.

Several applications of canonical analysis in biological problems of discrimination are available in the literature. It should be noted that the choice of the dummy variables $\underline{y}$ in (1.28) is arbitrary. Instead one can choose any other q suitable dummy variables, which are linear combinations of these $\underline{y}$ and and that the matrix $B = C_{xy}C_{yy}^{-1}C_{yx}$ is invariant for any such choice. For example, Williams [34] considers a problem of discrimination among 6 different groups of lamb carcasses. The six groups were formed by two factors, weight and grade. Two weight classes, medium and heavy, were considered, and 3 grades, Down, Royal, and Tallarook, were included. Instead of $\underline{y}$ of (1.28), Williams chose different dummy variables $\xi_1, \xi_2, \ldots, \xi_5$. The scores assigned to the 6 groups, by each of the ξ's, is shown in Table 3.

TABLE 3

Group	Weight Class	Grade	ξ_1	ξ_2	ξ_3	ξ_4	ξ_5
1	Medium	Down	-1	-1	1	1	-1
2	Medium	Royal	-1	0	-2	0	2
3	Medium	Tallarook	-1	1	1	-1	-1
4	Heavy	Down	1	-1	1	-1	1
5	Heavy	Royal	1	0	-2	0	-2
6	Heavy	Tallarook	1	1	1	1	1

From the scores, it is evident that Williams' dummy variables $\underline{\xi}$ are related to $\underline{y}$ of (1.28) by the relations

$$\xi_1 = -y_1 - y_2 - y_3 + y_4 + y_5 + y_6 ,$$

and similarly for $\xi_2, \ldots, \xi_5$. Those who are familiar with factorial experiments will immediately recognize that ξ_1 is the main effect of the factor weight w, ξ_2 is the linear effect of the factor grade G, ξ_3 is its quadratic effect, ξ_4 is the interaction W x linear G, and ξ_5 is the interaction W x quadratic G. This physical interpretation of $\xi_1, \ldots, \xi_5$ is more meaningful, but such a situation does not always arise. If ξ_1 alone is found to be an adequate discriminator, it means that discrimination among the different groups of carcasses is possible by the factor weight alone.

4. GOODNESS OF FIT OF A SINGLE HYPOTHETICAL DISCRIMINANT FUNCTION IN THE CASE OF SEVERAL GROUPS

Williams [30, 32, 34] and Bartlett [4] have derived an exact goodness of fit test for a single function $\underline{h}'\underline{x}$ to discriminate among k p-variate normal populations $N_p(\underline{x}|\underline{\mu}_\alpha|\Sigma)$, $(\alpha = 1, 2, \ldots, k)$. From Section 2, we see that a single discriminant function can be adequate only if the k means are collinear, i.e., the equation

$$|\Delta - \delta^2\Sigma| = 0, \tag{4.1}$$

where Δ is defined by (1.22), has only one nonzero root, say δ_1^2. The appropriate discriminant function is then $\underline{h}'\underline{x}$ if $\underline{h}$ is the eigenvector corresponding to δ_1^2, i.e., $\underline{h}$ satisfies

$$(\Delta - \delta_1^2\, \Sigma)\underline{h} = \underline{0}. \tag{4.2}$$

The hypothesis H of goodness of fit of a given discriminant function $\underline{h}'\underline{x}$, therefore, comprises two aspects: (a) collinearity of the means and (b) the direction $\underline{h}$ of the given function; whether it agrees with the direction of the true discriminant function, [i.e., $\underline{h}$ satisfies (4.2)] or not. If the hypothesis H is true, the entire relationship between $\underline{x}$ and the vector $\underline{y}$ of dummy variables in (1.28) is due to $\underline{h}'\underline{x}$, and if it is removed, there is no true residual relationship. Bartlett, therefore, considered Wilks's $\Lambda = |A|/|A+B|$ (where A and B are given in Table 1), for the relationship between $\underline{x}$ and $\underline{y}$ and removed [see (8.6.11)] $\lambda = \underline{h}'A\underline{h}/\underline{h}'(A+B)\underline{h}$, corresponding to the relationship of $\underline{h}'\underline{x}$ with $\underline{y}$. The "residual" Wilks's Λ is then

$$\Lambda_R = \frac{\Lambda}{\lambda} = \frac{|A|}{|A+B|} \frac{\underline{h}'(A+B)\underline{h}}{\underline{h}'A\underline{h}}. \tag{4.3}$$

This is the statistic to test the hypothesis H. From Section 6 of Chapter 8, Λ_R is distributed as $\prod_{i=2}^{p} t_{ii}^2$ (where t_{ii}^2 are as in Theorem 2 of Chapter 8), i.e., Λ_R has the Wilks's $\Lambda(n-1, p-1, q)$ distribution, if H is true; H can thus be tested by using (8.3.9) with appropriate changes. Bartlett then factorized Λ_R further, corresponding to the direction and collinearity aspects (a) and (b) of the given function $\underline{h}'\underline{x}$. It is given by

$$\Lambda_R = \Lambda_D \Lambda_{(C|D)} \tag{4.4}$$

where

$$\Lambda_D = \frac{1 - \underline{h}'B(A+B)^{-1}B\underline{h}/\underline{h}'B\underline{h}}{\underline{h}'A\underline{h}/\underline{h}'(A+B)\underline{h}} \tag{4.5}$$

is the "direction" factor and

$$\Lambda_{(C|D)} = \Lambda_R/\Lambda_D \tag{4.6}$$

is the "partial" "collinearity factor". Λ_D has the Wilks's $\Lambda(n-1, 1, p-1)$ distribution and $\Lambda_{(C|D)}$ has an independent Wilks's Λ (n-2, q-1, p-1) distribution when H is true. Λ_D tests the "direction" aspect of $\underline{h}'\underline{x}$, and $\Lambda_{(C|D)}$ tests the collinearity aspect, eliminating the direction aspect. The actual tests will be carried out using (8.2.24) to (8.2.27) and (8.3.9).

n, p, q are defined in the same way as in Section 1 of this chapter. Bartlett has given an alternative factorization also:

$$\Lambda_R = \Lambda_C \, \Lambda_{(D|C)} , \tag{4.7}$$

where

$$\Lambda_C = \frac{|A|}{|A+B|} \left\{ 1 + \frac{\underline{h}'B\,A^{-1}B\,\underline{h}}{\underline{h}'\,B\underline{h}} \right\} \tag{4.8}$$

is the "collinearity factor" and

$$\Lambda_{(D|C)} = \Lambda_R / \Lambda_C \tag{4.9}$$

is the "partial" "direction factor". Λ_C has the Λ(n-1, q-1, p-1) distribution and $\Lambda_{(D|C)}$ has an independent Λ(n-q, 1, p-1) distribution, when H is true. Λ_C tests the collinearity aspect and $\Lambda_{(D|C)}$ tests the direction aspect of $\underline{h}'\underline{x}$, eliminating the collinearity part. Bartlett derived these factors and their distribution by a geometrical method, but an alternative analytical method is given by Kshirsagar [13, 16]. The test and the factorization of Wilks's Λ was extended by Williams [33, 34] to test the goodness of fit of s (s > 1) hypothetical discriminant functions. Reference should also be made to Kshirsagar [14, 16] and Radcliffe [21] for additional information on this topic.

Sometimes the hypothetical discriminant in the hypothesis H is not given as $\underline{h}'\underline{x}$ but as $\underline{m}'\underline{y}$ in terms of the dummy variables $\underline{y}$. In other words, the hypothetical function comes from the space of the dummy variables. An example of this type is that of Egyptian skulls, discussed in Section 3, where time t was proposed as a discriminant function. The overall criterion in that case was seen to be [see (3.15)]

$$\Lambda_R' = \frac{|A|}{|A + B - C_{xt}C_{tt}^{-1}C_{tx}|} \tag{4.10}$$

and had the Wilks's $\Lambda(n-1, p, q-1)$ distribution under H. This can again be factorized as (see Williams [34], Kshirsagar [17])

$$\Lambda_R' = \Lambda_D' \, \Lambda_{(C|D)}' \tag{4.11}$$

or, alternatively, as

$$\Lambda_R' = \Lambda_C' \, \Lambda_{(D|C)}' , \tag{4.12}$$

where

$$\Lambda_D' = \frac{C_{tx}(A+B)^{-1} \, A(A+B)^{-1} \, C_{xt}}{C_{tx}(A+B)^{-1} \, C_{xt}} \; \frac{C_{tt}}{C_{tt} - C_{tx}(A+B)^{-1} \, C_{xt}} \tag{4.13}$$

and

$$\Lambda_C' = \frac{|A|}{|A+B|} \; \frac{C_{tx}A^{-1}C_{xt}}{C_{tx}(A+B)^{-1}C_{xt}} . \tag{4.14}$$

$\Lambda_{(C|D)}'$ and $\Lambda_{(D|C)}'$ can be obtained from (4.11) and (4.12). Λ_D', Λ_C' are the "direction" and "collinearity" factors, respectively, while $\Lambda_{(D|C)}'$ and $\Lambda_{(C|D)}'$ are the "partial direction" and "partial collinearity" factors respectively. The distributions of these factors, when the null hypothesis of goodness of fit of t is true, are given by

Factor	Distribution
Λ'_D	$\Lambda(n-1, 1, q-1)$
$\Lambda'_{(C\|D)}$	$\Lambda(n-2, p-1, q-1)$
Λ'_C	$\Lambda(n-1, p-1, q-1)$
$\Lambda'_{(D\|C)}$	$\Lambda(n-p, 1, q-1)$

(4.15)

Λ'_D and $\Lambda'_{(C|D)}$ are independently distributed and similarly Λ'_C and $\Lambda'_{(D|C)}$ are independently distributed when the null hypothesis is true. For derivation of the distributions of these factors, the reader should refer to Kshirsagar [17].

In practice, it may only rarely happen that a single discriminant function is good enough. But sometimes the experimenter is ready to ignore the effect of noncollinearity of the means and wishes to use only a single discriminant function as a good approximation. In that case, he will be more interested in the "direction" aspect of a proposed discriminant function $\underline{h}'\underline{x}$ or t, and this can be tested by using the appropriate direction factors given in this section.

5. AN ALTERNATIVE METHOD OF DISCRIMINATION

From Chapter 6, Section 1, we know that the discriminating procedure in the case of only two populations π_α: $N_p(\underline{x}|\underline{\mu}_\alpha|\Sigma)$ and π_β: $N_p(\underline{x}|\underline{\mu}_\beta|\Sigma)$ is to calculate

$$U_{\alpha\beta} = (\underline{\mu}_\alpha - \underline{\mu}_\beta)'\ \Sigma^{-1}\{\underline{x} - \tfrac{1}{2}(\underline{\mu}_\alpha + \underline{\mu}_\beta)\} \tag{5.1}$$

for a new observation $\underline{x}$ and to assign it to π_α if $U_{\alpha\beta} \geq 0$ and to π_β if $U_{\alpha\beta} < 0$. When we have k populations π_α ($\alpha = 1, 2, \ldots, k$), this method can be generalized by observing that

$$U_{\alpha\beta} = S_\alpha - S_\beta , \tag{5.2}$$

where

$$S_\alpha = \underline{\mu}_\alpha' \Sigma^{-1} \underline{x} - \frac{1}{2} \underline{\mu}_\alpha' \Sigma^{-1} \underline{\mu}_\alpha \quad (\alpha = 1, 2, \ldots, k). \tag{5.3}$$

S_α can be called the "score" of the population π_α, for discrimination. The individual with measurements $\underline{x}$ is then assigned to the population that has the largest discriminant score. In other words, assign $\underline{x}$ to π_a if

$$S_a = \text{Maximum } \{S_1, S_2, \ldots, S_k\} . \tag{5.4}$$

If prior probabilities ϕ_α ($\alpha = 1, 2, \ldots, k$) of the k populations are known, Rao [23] has shown that the discriminant score of π_α should be taken as $S_\alpha + \log_e \phi_\alpha$, and decision should be based on these scores, rather than S_α. When $\underline{\mu}_\alpha$, Σ are unknown, the scores are estimated by using $A/(n-q)$ to estimate Σ and $\bar{\underline{x}}_\alpha$ to estimate $\underline{\mu}_\alpha$.

For a discussion of this procedure, in a more general setting and for properties such as admissibility and completeness, the reader should refer to Anderson [1], Wald [28], Blackwell and Girshick [6], Mises [19], and Rao [22, 23]

6. USE OF CANONICAL ANALYSIS IN CONTINGENCY TABLES

In Chapter 7, we considered the canonical correlations between any two vectors $\underline{x}$ and $\underline{y}$. In Chapter 8, both $\underline{x}$ and $\underline{y}$ were regarded as random normal vectors, though sometimes we held

$\underline{y}$ and observations on $\underline{y}$ as fixed and considered the conditional distribution of $\underline{x}$. In this chapter, so far, $\underline{x}$ has been the vector of variables having a joint normal distribution, while $\underline{y}$ has been a vector of dummy variables, assigned to the k groups or populations under consideration and so it was a vector of fixed elements. Now, however, we shall be considering an application of canonical analysis where both $\underline{x}$ and $\underline{y}$ are not random. They are both vectors of "dummy" variables. Such a situation arises, when we have a contingency table with $p+1$ rows and $q+1$ columns, the rows corresponding to the categories $A_1, \ldots, A_{p+1}$ of an attribute A and the columns corresponding to the categories $B_1, B_2, \ldots, B_{q+1}$ of another attribute B. If a random sample of $n_{..}$ individuals is classified according to these categories, the frequencies in the different cells of the contingency table and the marginal totals will be as given in Table 4.

TABLE 4

A \ B	B_1	...	B_j		B_{q+1}	Total
A_1	n_{11}	...	n_{ij}	...	n_{1q+1}	$n_{1.}$
⋮						
A_2	n_{i1}	...	n_{ij}	...	n_{iq+1}	$n_{i.}$
⋮						
A_{p+1}	$n_{p+1,\,1}$	...	$n_{p+1,\,j}$	...	$n_{p+1,\,q+1}$	$n_{p+1.}$
Total	$n_{.1}$	...	$n_{.j}$	...	$n_{.q+1}$	$n_{..}$

Let π_{ij} be the probability that an individual belongs to the category A_i of A and the category B_j of B (i = 1, ..., p+1; j = 1, 2, ..., q+1). Then $\pi_{i.} = \sum_{j=1}^{q+1} \pi_{ij}$ is the probability of belonging to A_i and $\pi_{.j} = \sum_{i=1}^{p+1} \pi_{ij}$ is the probability of belonging to B_j, while $\sum_i \sum_j \pi_{ij} = 1$. If the two attributes are completely independent,

$$\pi_{ij} = \pi_{i.}\pi_{.j} \quad (i = 1, 2, \ldots, p+1;\ j = 1, \ldots, q+1), \tag{6.1}$$

and conversely. In order to study the relationship between these attributes, on the basis of these qualitative data, we define two sets of dummy variables x_i (i = 1, 2, ..., p), y_j (j = 1, 2, ..., q) as follows:

$$\begin{aligned} x_i &= 1, \quad \text{if an individual belongs to } A_i, \\ &= 0, \quad \text{otherwise} \quad (i = 1, 2, \ldots, p). \end{aligned} \tag{6.2}$$

$$\begin{aligned} y_j &= 1, \quad \text{if an individual belongs to } B_j, \\ &= 0, \quad \text{ootherwise} \quad (j = 1, 2, \ldots, q). \end{aligned} \tag{6.3}$$

Whenever an individual belongs to A_{p+1}, each x_i is zero, and similarly for an individual in B_{q+1}, each y_j is zero. Let $\underline{x}$ denote the $p \times 1$ column vector of the x_i's and $\underline{y}$ denote the $q \times 1$ column vector of the y_j's. The relationship between A and B is then the relationship between $\underline{x}$ and $\underline{y}$, and we can, therefore, use the theory of canonical vectors and canonical correlations between $\underline{x}$ and $\underline{y}$, discussed in Chapter 7 and subsequent chapters. By writing down the values of x_i and y_j for all the $n_{..}$ individuals it is easy to see that [see (1.34) also] the matrices of the corrected s.s. and s.p. of $\underline{x}$ and $\underline{y}$ are

$$C_{xx} = \text{diag}(n_{1.}, n_{2.}, \ldots, n_{p.}) - \frac{1}{n_{..}} [n_{1.}, \ldots, n_{p}]' [n_{1.}, \ldots, n_{p.}],$$

$$C_{yy} = \text{diag}(n_{.1}, n_{.2}, \ldots, n_{.q}) - \frac{1}{n_{..}} [n_{.1}, \ldots, n_{.q}]' [n_{.1}, \ldots, n_{.q}],$$

$$C_{xy} = N - \frac{1}{n_{..}} [n_{1.}, \ldots, n_{p.}]' [n_{.1}, n_{.2}, \ldots, n_{.q}], \tag{6.4}$$

where

$$N = [n_{ij}] \quad (i = 1, 2, \ldots, p); \ (j = 1, \ldots, q). \tag{6.5}$$

The canonical correlations between $\underline{x}$ and $\underline{y}$ are $r_1 > r_2 > \ldots > r_f$ where $f = \text{Min}(p, q)$ and r_i^2 are the roots of the determinantal equation [see (7.3.7)]

$$\left| \begin{array}{c|c} -r \ C_{xx} & C_{xy} \\ \hline C_{yx} & -r \ C_{yy} \end{array} \right| = 0 . \tag{6.6}$$

when $\underline{x}$ or $\underline{y}$ has a normal distribution, the independence of $\underline{x}$ and $\underline{y}$ is tested by using Wilks's $\Lambda = \prod_1^f (1 - r_i^2)$, or Hotelling's $T_o^2 = \sum_1^f r_i^2/(1 - r_i^2)$, or Pillai's $V = \sum_1^f r_i^2$. For large $n_{..}$, each of these yields a χ^2 test with pq d.f. However, in the present case, neither $\underline{x}$ nor $\underline{y}$ is random. Still one can use these tests as an asymptotic result, when $n_{..}$ is very large. The classical test of independence of A and B uses the statistic

$$u = \sum_{i=1}^{p+1} \sum_{j=1}^{q+1} \left\{n_{ij} - \frac{n_{i.} n_{.j}}{n}\right\}^2 \Big/ \left\{\frac{n_{i.} n_{.j}}{n}\right\}$$

$$= \sum_{i=1}^{p+1} \sum_{j=1}^{q+1} \frac{n_{ij}^2 n_{..}}{n_{i.} n_{.j}} - n_{..} . \tag{6.7}$$

When A and B are independent, i.e., (6.1) holds, u has a χ^2 distribution with pq d.f., provided $n_{..}$ and n_{ij} are not small. Now, from (7.3.8), the r_i^2 are roots of the equation,

$$|-r^2 C_{xx} + C_{xy}C_{yy}^{-1}C_{yx}| = 0 ,$$

and so

$$\text{Pillai's } V = \sum_1^f r_i^2 = \text{tr } C_{xx}^{-1}C_{xy}C_{yy}^{-1}C_{yx} . \qquad (6.8)$$

From (1.34),

$$C_{xx}^{-1} = \text{diag}\left(\frac{1}{n_{1.}}, \ldots, \frac{1}{n_{p.}}\right) + \frac{1}{n_{p+1.}} E_{pp} ,$$

$$C_{yy}^{-1} = \text{diag}\left(\frac{1}{n_{.1}}, \ldots, \frac{1}{n_{.q}}\right) + \frac{1}{n_{.q+1}} E_{qq} . \qquad (6.9)$$

From this, it can be seen that the matrix $C_{xx}^{-1}C_{xy}$ has the elements $n_{ij}/n_{i.} - n_{p+1,\ j}/n_{p+1.}$ and $C_{yy}^{-1}C_{yx}$ has the elements $n_{ij}/n_{.j} - n_{iq+1}/n_{.q+1}$ $(i = 1, 2, \ldots, p;\ j = 1, 2, \ldots, q)$ and so we find that

$$\text{Pillai's } V = \text{tr } C_{xx}^{-1}C_{xy}C_{yy}^{-1}C_{yx} = \frac{u}{n_{..}} . \qquad (6.10)$$

Thus the classical χ^2 test is the same as the test based on Pillai's V, even though the assumption of normality does not hold. If the hypothesis of independence is rejected, one can assign scores to the categories $A_1, \ldots, A_{p+1}$ and $B_1, \ldots, B_{q+1}$ by finding the canonical vectors $\underline{g}$, $\underline{h}$ corresponding to a root r of the equation (6.6), satisfying (7.3.7), viz.,

$$\left[\begin{array}{c|c} -r\,C_{xx} & C_{xy} \\ \hline C_{xy} & -r\,C_{yy} \end{array}\right] \left[\begin{array}{c} \underline{g} \\ \hline \underline{h} \end{array}\right] = 0 . \qquad (6.11)$$

The elements of $\underline{g}$ are the scores of $A_1, \ldots, A_p$ and the elements of $\underline{h}$ are the scores of $B_1, \ldots, B_q$; the scores of A_{p+1} and B_{q+1} are zero, because of our definitions (6.2) and (6.3) of $\underline{x}$ and $\underline{y}$. Since (6.6) has f roots, we need f such sets of scores to bring out the relationship between these two attributes. But all the roots may not be significant. We can determine the significant roots by using the tests given in Section 7 of Chapter 8, ignoring the fact that both $\underline{x}$, $\underline{y}$ are dummy variables and none has a normal distribution. If only the largest root r_1 of (6.6) is significant, only one set of scores is sufficient. The categories $A_1, \ldots, A_{p+1}$ can then be represented quantitatively on a linear scale, and so also $B_1, B_2, \ldots, B_{q+1}$. The association between A and B is then linear or, in other words, the departure of the probabilities π_{ij} from the "independence" probabilities $\pi_{i.}\pi_{.j}$ forms a matrix of rank one only. The relation between A and B becomes more complex when more than one root of (6.6) is significant, and one needs more sets of scores to describe it quantitatively. For the derivation and description of a test of goodness of fit of a single hypothetical set of scores for A or for B, reference may be made to Williams [31] and Kshirsagar [15]. Shrikantan [27] has proposed several measures of association between two such attributes A and B and discussed their relative merits. He has extended it to more attributes also. This type of analysis and quantification of qualitative data is important in social sciences. We had discussed the representation of the symbols -, ?, w, (+), +, representing blood reactions, on a linear scale in Section 2. This is also an illustration of the use of canonical analysis to quantify such data.

For more complicated models, involving variables and attributes both, the reader should refer to Roy and Mitra [25], Roy [26], and Kastenbaum [11].

7. THE GENERAL THEORY OF MULTIVARIATE ANALYSIS OF VARIANCE AND COVARIANCE

The general linear model (Gauss-Markoff) in the univariate case is

$$E(\underline{x}) = A\theta, \quad V(\underline{x}) = \sigma^2 I_N, \tag{7.1}$$

where $\underline{x}$ is the $N \times 1$ vector of N independent observations on a single normal variable x, A is a $N \times m$ known matrix, and $\underline{\theta}$ is an $m \times 1$ vector of unknown parameters. One then finds

$$\text{Error s.s., SSE} = \underset{\underline{\theta}}{\text{Min}}\ (\underline{x} - A\underline{\theta})'\ (\underline{x} - A\underline{\theta}) \tag{7.2}$$

(see Section 5 of Chapter 2, in this connection). This minimum turns out to be (see Rao [23]) a quadratic form $\underline{x}'R_o\underline{x}$, where the matrix R_o is $N \times N$, idempotent, and of rank N-r, and r is the rank of A. Further $E(x')R_oE(x)$ is always null, and so SSE is always a $\chi^2\sigma^2$ with N-r d.f., whatever $\underline{\theta}$ may be. A hypothesis such as $B\underline{\theta} = \underline{0}$, where B is a given $k \times m$ matrix of rank k, can be tested, if $B\underline{\theta}$ is estimable (see Rao [23]), by finding

$$\underset{\theta}{\text{Min}}\ (\underline{x} - A\underline{\theta})'\ (\underline{x} - A\underline{\theta}), \quad \text{subject to } B\underline{\theta} = \underline{0}. \tag{7.3}$$

The difference between (7.3) and (7.2) can be expressed as $\underline{x}'R_1\underline{x}$, where R_1 is an $N \times N$ idempotent matrix of rank k, satisfying

$R_oR_1 = 0$. Further $E(\underline{x}')R_1E(\underline{x})$ is zero only if $B\underline{\theta} = \underline{0}$. The matrices R_o, R_1 satisfy James's conditions [10], and so $\underline{x}'R_1\underline{x}$ is independently distributed of SSE, as a $\chi^2\sigma^2$ with k d.f. The hypothesis $B\underline{\theta} = \underline{0}$ can therefore be tested by the F test:

$$F_{k,\ N-r} = \frac{N-r}{k}\,\frac{\underline{x}'R_1\underline{x}}{\underline{x}'R_o\underline{x}}\ . \tag{7.4}$$

We thus find that the total s.s. $\underline{x}'\underline{x}$ of the N observations is split up as

$$\begin{aligned}\underline{x}'\underline{x} &= \underline{x}'R_o\underline{x} + \underline{x}'(I - R_o)\underline{x} \\ &= \underline{x}'R_o\underline{x} + \underline{x}'R_1\underline{x} + \underline{x}'(I - R_o - R_1)\underline{x}\ ,\end{aligned} \tag{7.5}$$

where R_o, R_1, $I - R_o - R_1$ satisfy Jame's requirements. Sometimes $\underline{x}'(I - R_o - R_1)\underline{x}$ can be split up further into quadratic forms with idempotent matrices, satisfying Jame's conditions, and therefore, these component quadratic forms provide appropriate s.s. to test other hypotheses about estimable linear parametric functions of $\underline{\theta}$. Usually the vector $\underline{\theta}$ of parameters consists of different groups of parameters, such as block effects and treatment effects, and the different constituents of $\underline{x}'\underline{x}$ provide tests of significance of these groups of parameters, when tested against SSE. The different models in analysis of variance, analysis of covariance, regression, and experimental designs can all be shown to be particular cases of the general Gauss-Markoff model (7.1).

This general model in univariate analysis can be easily extended to the multivariate situation involving p variables $x_1, x_2, \ldots, x_p$. The corresponding multivariate linear model is

$$E(X') = A\theta, \quad V(X') = I_N \otimes \Sigma, \tag{7.6}$$

where X is the $p \times N$ matrix of N independent observations on each of $x_1, \ldots, x_p$. A is the same matrix as before, while θ is an $m \times p$ matrix of mp unknown parameters. The N columns of X are independently distributed, each having a p-variate normal distribution, with the same but unknown variance-covariance matrix Σ. We can then borrow the matrix R_o from the error s.s. $\underline{x}'R_o\underline{x}$, in the univariate model, for any one the x's, and replace $\underline{x}'$ by X to get the "s.s. and s.p. due to error" matrix $X\,R_oX'$, for the multivariate analysis of variance. Since R_o is already known to be idempotent, of rank N-r, and satisfies $E(\underline{x}')R_oE(\underline{x}) = 0$ for any $\underline{\theta}$, it is obvious, from the results in Section 7 of Chapter 3, that $X\,R_oX'$ has the $W_p(X\,R_oX'|N-r|\Sigma)$ distribution whatever θ may be. (N-r is assumed to be > p). To test the hypothesis $B\theta = 0$, where B is as before, we now borrow the matrix R_1 from the s.s. $\underline{x}'R_1\underline{x}$ in the univariate analysis of variance for any x, replace $\underline{x}'$ by X, and obtain $X\,R_1X'$, which will now have a Wishart distribution $W_p(X\,R_1X'|k|\Sigma)$ only if $B\theta = 0$ and will be independent of $X\,R_oX'$. We then have the multivariate analysis of variance table (Table 5).

TABLE 5

Source	d.f.	s.s. and s.p. matrix
Hypothesis B = 0	k	$X\,R_1X'$
Error	N-r	$X\,R_oX'$
Total	N-r+k	XX'

The distribution of $X R_1 X'$ will be pseudo-Wishart if $k < p$, but this does not matter, and the hypothesis $B\theta = 0$ can be tested by using Wilks's Λ criterion:

$$\Lambda(N-r+k,\ p,\ k) = \frac{|X R_o X'|}{|X R_o X' + X R_1 X'|}. \tag{7.7}$$

The hypothesis will be rejected if this Λ is significant. One has to use (8.3.9) for the percentage points of this Λ. Further breakdown of $X'(I - R_o - R_1)X'$ is also possible corresponding to the breakdown of $\underline{x}'(I - R_o - R_1)\underline{x}$, and each constituent Wishart matrix can be used to test a hypothesis about some parametric functions of θ. Thus, if $X R_2 X'$ is a part of $X(I - R_o - R_1)X'$, where $R_2 R_o = 0$ and R_2 is of rank d, the hypothesis $E(X)\ R_2\ E(X') = 0$, i.e., $A\ \theta\ R_2\ \theta'\ A' = 0$ or any hypothesis equivalent to it can be tested by using Wilks's

$$\Lambda(N-r+d,\ p,\ d) = \frac{|X R_o X'|}{|X R_o X' + X R_2 X'|}. \tag{7.8}$$

In many practical situations, θ includes a general mean for each x_i, and in that case one of the R matrices is $(1/N)E_{NN}$ of rank 1. When the corresponding Wishart matrix $(1/N)\,X E_{NN} X'$ is removed from XX', we obtain $X(I - (1/N)E_{NN})X'$ as the matrix of the corrected s.s. and s.p. of the observations on $x_1, x_2, \ldots, x_p$. This then is split up into error s.s. and s.p. matrix and other matrices. The rest of the procedure is the same, except that the new corrected total s.s. and s.p. matrix $X(I - (1/N)E_{nn})X' = C_{xx}$ has $N-1 = n$ d.f.

For multivariate analysis of covariance, consider the conditional distribution [see (2.4.10)] of $x_{p_1+1}, x_{p_1+2}, \ldots, x_p$ when $x_1, x_2, \ldots, x_{p_1}$ are fixed, and partition X, θ, and Σ as

$$X' = \underset{p_1 \quad p-p_1}{[X_u' \mid X_v']} N, \qquad \underset{m\times p}{\theta} = \underset{p_1 \quad p-p_1}{[\theta_u \mid \theta_v]} m,$$

$$\Sigma = \begin{bmatrix} \Sigma_{11} & \Sigma_{12} \\ \Sigma_{21} & \Sigma_{22} \end{bmatrix} \begin{matrix} p_1 \\ p-p_1 \end{matrix} \quad . \tag{7.9}$$
$$\qquad p_1 \qquad p-p_1$$

From (7.6) and the conditional distribution, we have

$$E(X_v' \mid X_u') = A\Phi + X_u'\beta', \quad V(X_v' \mid X_u') = I_N \otimes \Sigma_{22.1}, \tag{7.10}$$

where

$$\Phi = \theta_v - \theta_u\beta', \quad \beta = \Sigma_{21}\Sigma_{11}^{-1}, \quad \Sigma_{22.1} = \Sigma_{22} - \Sigma_{21}\Sigma_{11}^{-1}\Sigma_{12} \, . \tag{7.11}$$

Then (7.10) is the analysis of covariance model for X_v, with X_u as the matrix of observations on the concomitant variables and β as the matrix of regression coefficients. The general theory outlined above for testing hypotheses about estimable parametric functions works in exactly the same way for this model.

Consider now the Wilks's Λ criterion (7.7) to test the hypothesis $B\theta = 0$, in the model (7.6). It is $|L|/|L+M|$, where $L = X R_o X'$, $M = X R_1 X'$. Partition L and M as

$$L = \begin{bmatrix} L_{11} & L_{12} \\ L_{21} & L_{22} \end{bmatrix} \begin{matrix} p_1 \\ p-p_1 \end{matrix}, \quad M = \begin{bmatrix} M_{11} & M_{12} \\ M_{21} & M_{22} \end{bmatrix} \begin{matrix} p_1 \\ p-p_1 \end{matrix} . \tag{7.12}$$
$$\qquad p_1 \qquad p-p_1 \qquad\qquad\qquad p_1 \qquad p-p_1$$

It is then interesting to note that Wilks's Λ criterion to test the hypothesis

$$H: \quad B\,\bar{\Phi} = 0 \tag{7.13}$$

in the model (7.10) is

$$\Lambda(N-r+k-p_1,\ p-p_1,\ k) = \frac{|L|/|L+M|}{|L_{11}|/|L_{11}+M_{11}|}\ . \tag{7.14}$$

This can also be expressed as

$$\frac{|L_{22} - L_{21}L_{11}^{-1}L_{12}|}{|L_{22} + M_{22} - (L_{21}+M_{21})(L_{11}+M_{11})^{-1}(L_{12}+M_{12})|}\ . \tag{7.15}$$

This result follows from (8.6.18) also, where elimination of variables and resulting "partial" Wilks's Λ were considered.

It may also be noted that the hypothesis H of (7.13) is equivalent to the conditional hypothesis that $B\theta_v = 0$, given $B\,\theta_u = 0$. In this connection, references should also be made to Rao [24], who considers some related problems, such as growth curves and a Potthoff-Roy model [20].

In the next section, we shall consider the analysis of a general incomplete block design, when p variables are measured on each plot. The model used is only a particular case of (7.6), but the analysis is worked out in order to illustrate how the R_o and R_1 matrices can be picked up from the corresponding univariate analysis to build up the Wishart matrices in the multivariate analysis of variance.

8. A GENERAL INCOMPLETE BLOCK DESIGN

Consider an experimental design with b blocks of k_1, k_2, ..., k_b plots and v treatments. Let $N = [n_{ij}]$ $(i = 1, 2, \ldots, v;$

$j = 1, \ldots, b$) be the $v \times b$ incidence matrix of the design, so that the ith treatment is applied n_{ij} times in the jth block. Let each $n_{ij} = 0$ or 1 and let N be such that the design is connected (see Chakrabarti [7]). On each plot, let each of p correlated variables $x_1, x_2, \ldots, x_p$ be measured. We shall denote by $x_{ij\alpha}$, the observation on x_α ($\alpha = 1, 2, \ldots, p$) on that plot in the jth block ($j = 1, 2, \ldots, b$) which receives the ith treatment ($i = 1, 2, \ldots, v$) when $n_{ij} = 1$. The model with fixed block and treatment effects is

$$x_{ij\alpha} = \mu_\alpha + t_{i\alpha} + \beta_{j\alpha} + \epsilon_{ij\alpha}, \quad \text{when } n_{ij} = 1, \tag{8.1}$$

$\epsilon_{ij\alpha}$ are normally distributed with zero means, variances $\sigma_{\alpha\alpha}$, and covariance given by

$$\begin{aligned} \operatorname{cov}(\epsilon_{ij\alpha}, \epsilon_{i'j'\alpha'}) &= \sigma_{\alpha\alpha'}, \quad \text{if } i = i', j = j' \\ &= 0 \quad , \quad \text{otherwise} \end{aligned}$$

$$(i = 1, 2, \ldots, v), \quad (j = 1, 2, \ldots, b), \quad (\alpha = 1, 2, \ldots, p).$$

The total number of plots in the jth block is $k_j = \sum_i n_{ij}$, and the ith treatment is replicated $r_i = \sum_j n_{ij}$ times. The total number of plots in all the blocks is $n_o = \sum_i r_i = \sum_j k_j = \sum_i \sum_j n_{ij}$. $t_{i\alpha}$ are the treatment effects, and $\beta_{j\alpha}$ are the block effects. The treatment totals and block totals are

$$T_{i\alpha} = \sum_j n_{ij} x_{ij\alpha}, \quad B_{j\alpha} = \sum_i n_{ij} x_{ij\alpha}, \tag{8.2}$$

and the grand totals are

$$g_\alpha = \sum_i T_{i\alpha} = \sum_j B_{j\alpha} . \tag{8.3}$$

We need the following vectors and matrices:

$$\underset{v\times p}{T} = [T_{i\alpha}] = [\underline{T}_1|\underline{T}_2|\cdots|\underline{T}_p],$$

$$\underset{b\times p}{B} = [B_{j\alpha}] = [\underline{B}_1|\underline{B}_2|\cdots|\underline{B}_p], \quad \underline{g}' = [g_1, g_2, \ldots, g_p],$$

$$\underset{v\times p}{t} = [t_{i\alpha}] = [\underline{t}_1|\underline{t}_2|\cdots|\underline{t}_p],$$

$$C = \text{diag}\,(r_1, \ldots, r_p) - N\,\text{diag}\left(\frac{1}{k_1}, \ldots, \frac{1}{k_p}\right)N',$$

$$\underline{Q}_\alpha = \underline{T}_\alpha - N\,\text{diag}\left(\frac{1}{k_1}, \ldots, \frac{1}{k_p}\right)\underline{B}_\alpha,$$

$$\underset{v\times b}{Q} = [\underline{Q}_1|\underline{Q}_2|\cdots|\underline{Q}_p].$$

If we consider only the variable x_α, the hypothesis of equality of treatment effects, viz.,

$$t_{1\alpha} = t_{2\alpha} = \cdots = t_{v\alpha}, \tag{8.4}$$

is tested by using (see Chakrabarti [7])

$$\text{Adjusted treatment s.s.} = \underline{Q}'_\alpha\hat{\underline{t}}_\alpha \text{ (d.f. v-1)}, \tag{8.5}$$

where $\hat{\underline{t}}_\alpha$ is any solution of the reduced normal equations

$$\underline{Q}_\alpha = C\,\hat{\underline{t}}_\alpha. \tag{8.6}$$

[One usually uses $E_{1v}\hat{\underline{t}}_\alpha = 0$ as an additional equation to solve (8.6).] The error s.s. is

$$\text{total s.s.} - \text{(adj.) treat. s.s.} - \text{(unadjusted) block s.s.}, \tag{8.7}$$

where

$$\text{total s.s.} = \sum_i \sum_j n_{ij}\, x^2_{ij\alpha} - \frac{1}{n_o} g^2_\alpha, \tag{8.8}$$

$$\text{unadjusted block s.s.} = \underline{B}'_\alpha\,\text{diag}\left(\frac{1}{k_1}, \ldots, \frac{1}{k_b}\right)\underline{B}_\alpha - \frac{1}{n_o} g^2_\alpha. \tag{8.9}$$

To express the error s.s. and treatment s.s. as quadratic forms in idempotent matrices R_o and R_1, we define

$$\begin{aligned} \xi_i &= 1, \quad \text{if the ith treatment is applied to a plot} \\ &= 0, \quad \text{otherwise} \qquad (i = 1, 2, \ldots, v), \qquad (8.10) \\ \eta_{ij} &= 1, \quad \text{if a plot is in the jth block} \\ &= 0, \quad \text{otherwise} \qquad (j = 1, 2, \ldots, b). \qquad (8.11) \end{aligned}$$

These variables are defined corresponding to each of the n_o plots. We also define $\underline{\xi}_i$ as the $n_o \times 1$ vector of the values of ξ_i corresponding to the n_o plots, and similarly $\underline{\eta}_j$. Let

$$\begin{aligned} \xi &= [\underline{\xi}_1 | \underline{\xi}_2 | \cdots | \underline{\xi}_v], \quad \text{an } n_o \times v \text{ matrix}, \\ \eta &= [\underline{\eta}_1 | \underline{\eta}_2 | \cdots | \underline{\eta}_b], \quad \text{an } n_o \times b \text{ matrix}. \qquad (8.12) \end{aligned}$$

Then it is easy to see that

$$T_{i\alpha} = \underline{\xi}_i' \underline{x}_\alpha, \quad B_{j\alpha} = \underline{\eta}_j' \underline{x}_\alpha, \qquad (8.13)$$

where $\underline{x}_\alpha$ is the $n_o \times 1$ vector of the yields $x_{ij\alpha}$ (for the variable x_α), on the n_o plots in some order (but same as $\underline{\xi}_i$ or $\underline{\eta}_j$). Then, from the definition of $\underline{Q}_\alpha$, we can write it as

$$\underline{Q}_\alpha = \left\{\xi' - N \operatorname{diag}\left(\frac{1}{k_1}, \ldots, \frac{1}{k_b}\right)\eta'\right\}\underline{x}_\alpha. \qquad (8.14)$$

From (8.6), a solution $\hat{\underline{t}}_\alpha$ is given by

$$\begin{aligned} \hat{\underline{t}}_\alpha &= C^- \underline{Q}_\alpha \\ &= C^-\left\{\xi' - N \operatorname{diag}\left(\frac{1}{k_1}, \ldots, \frac{1}{k_b}\right)\eta'\right\} \underline{x}_\alpha, \qquad (8.15) \end{aligned}$$

where C^- is a generalized inverse (see Rao [23]) of C. Since C is of rank v-1 (for a connected design, the rank is v-1; see Chakrabarti [7]), it has no inverse and we have to use C^-. The adjusted treatment s.s. is then, from (8.5), the quadratic form

$$\underline{x}'_\alpha \left\{\xi - \eta \operatorname{diag}\left(\frac{1}{k_1}, \ldots, \frac{1}{k_b}\right)N'\right\} C^-$$

$$X \left\{\xi' - N \operatorname{diag}\left(\frac{1}{k_1}, \ldots, \frac{1}{k_b}\right)\eta'\right\} \underline{x}_\alpha$$

$$= \underline{x}'_\alpha R_1 \underline{x}_\alpha . \tag{8.16}$$

This gives the R_1 matrix as

$$\left\{\xi - \eta \operatorname{diag}\left(\frac{1}{k_1}, \ldots, \frac{1}{k_b}\right)N'\right\} C^- \left\{\xi' - N \operatorname{diag}\left(\frac{1}{k_1}, \ldots, \frac{1}{k_b}\right)\eta'\right\}. \tag{8.17}$$

The unadjusted block s.s. from (8.9) is

$$\underline{x}'_\alpha\left\{\eta \operatorname{diag}\left(\frac{1}{k_1}, \ldots, \frac{1}{k_b}\right)\eta' - \frac{1}{n_o} E_{n_o n_o}\right\} \underline{x}_\alpha$$

$$= \underline{x}'_\alpha R_2 \underline{x}_\alpha, \text{ say.} \tag{8.18}$$

The total s.s. is [see (8.8)]

$$\underline{x}'_\alpha\left(I - \frac{1}{n_o} E_{n_o n_o}\right)\underline{x}_\alpha . \tag{8.19}$$

The error s.s. is, therefore [from (8.7)], $\underline{x}'_\alpha R_o \underline{x}_\alpha$, where

$$R_o = \left(I - \frac{1}{n_o} E_{n_o n_o}\right) - R_1 - R_2 . \tag{8.20}$$

Once these matrices R_o, R_1, R_2 are found, multivariate analysis of variance table for testing the hypothesis

$$t_{1\alpha} = t_{2\alpha} = \cdots = t_{v\alpha} \quad \text{for all } \alpha = 1, 2, \ldots, p,$$

i.e.,

$$\left(I - \frac{1}{v} E_{vv}\right)t = 0 , \tag{8.21}$$

can be easily constructed. As explained in Section 7, the trick is to replace $\underline{x}'_\alpha$ by the matrix, X, where

$$X' = [\underline{x}_1 | \underline{x}_2 | \cdots | \underline{x}_p] \tag{8.22}$$

in each s.s. Thus the adjusted treatment s.s. and s.p. matrix is

$$X\,R_1X' = Q'\hat{t}\ , \tag{8.23}$$

where $\hat{t}$ is any solution of $Q = C\,\hat{t}$, Q and $\hat{t}$ being $v \times p$ matrices defined earlier. Similarly, the unadjusted block s.s. and s.p. matrix is $X\,R_2\,X'$, and it simplifies to

$$B'\ \mathrm{diag}\left(\frac{1}{k_1}, \ldots, \frac{1}{k_b}\right)B - \frac{1}{n_o}\ \underline{\underline{g}}\underline{\underline{g}}'\ , \tag{8.24}$$

while the total s.s. and s.p. matrix is

$$X\left(I - \frac{1}{n_o}\,E_{n_o n_o}\right)X' = XX' - \frac{1}{n_o}\ \underline{\underline{g}}\underline{\underline{g}}'\ . \tag{8.25}$$

The multivariate analysis of variance table for testing the hypothesis (8.21) is given by Table 6.

TABLE 6

Source	d.f.	s.s. and s.p. matrix
Blocks (unadjusted)	$b-1$	$B'\ \mathrm{diag}\left(\frac{1}{k_1}, \ldots, \frac{1}{k_b}\right)B - \frac{1}{n_o}\ \underline{\underline{g}}\underline{\underline{g}}'$
Treatments (adjusted)	$v-1$	$Q'\hat{t}$
Error	$n_o-b-v+1$	Obtained by subtraction as H say
Total	n_o-1	$XX' - \frac{1}{n_o}\ \underline{\underline{g}}\underline{\underline{g}}'$

Wilks's Λ criterion for testing (8.21) is then

$$\Lambda((n_o-b-v+1) + (v-1),\ p,\ v-1) = \frac{|H|}{|H + Q'\hat{t}|}\ . \tag{8.26}$$

Wilks's Λ test then can be used for testing the hypothesis.

In practice, the total s.s. and s.p. matrix can be more easily calculated from

$$d_{\alpha\alpha'} = \sum_i \sum_j n_{ij} x_{ij\alpha} x_{ij\alpha'} - \frac{1}{n_o} g_\alpha g'_\alpha \quad (\alpha, \alpha' = 1, 2, \ldots, p),$$

$$XX' - \frac{1}{n_o} \underline{\underline{gg'}} = [d_{\alpha\alpha'}], \text{ the } p \times p \text{ matrix of the elements } d_{\alpha\alpha'}. \tag{8.27}$$

To test the hypothesis about block effects, we first compute

unadjusted treatment s.s. and s.p.

$$= T' \operatorname{diag}\left(\frac{1}{r_1}, \ldots, \frac{1}{r_v}\right) T - \frac{1}{n_o} \underline{\underline{gg'}} \tag{8.28}$$

and obtain

adjusted block s.s. and s.p. = total s.s. and s.p. matrix - unadjusted treatments s.s. and s.p. matrix - error s.s. and s.p. matrix = XX'

$$- T' \operatorname{diag}\left(\frac{1}{r_1}, \ldots, \frac{1}{r_v}\right) T - H = K, \text{ say.} \tag{8.29}$$

Wilks's Λ criterion for the null hypothesis about block effects is then

$$\Lambda((n_o-b-v+1) + (b-1), p, b-1) = \frac{|H|}{|H+K|}. \tag{8.30}$$

All incomplete block designs used in practice are particular cases of the general design considered here and correspond to different incidence matrices N. The appropriate s.s. and s.p. matrices can thus be easily obtained, once the s.s. in the univariate analysis are known, by the above trick of replacing x'_α by X.

REFERENCES

[1] Anderson, T. W. (1958). *An Introduction to Multivariate Statiatical Analysis*. John Wiley and Sons, New York.

[2] Barnard, M. M. (1935). "The secular variations of skull characters in four series of Egyptian skulls", Ann. Eugen., 6, p. 352.

[3] Bartlett, M. S. (1947). "Multivariate analysis", Suppl. J. Roy. Statist. Soc., 9, p. 176.

[4] Bartlett, M. S. (1951). "The goodness-of-fit of a single hypothetical discriminant function in the case of several groups", Ann. Eugen., 16, p. 199.

[5] Binet, F. E., and Watson, G. S. (1956). "Algebraic theory of the computing routine for tests of significance on the dimensionality of normal multivariate systems", J. Roy. Stat. Soc. B, 18, p. 70.

[6] Blackwell, D., and Girshick, M. A. (1954). *Theory of Games and Statistical Decisions*. John Wiley and Sons, New York.

[7] Chakrabarti, M. C. (1962). *Mathematics of Design and Analysis of Experiments*. Asia Publishing House, New York.

[8] Fisher, R. A. (1938). "The statistical utilization of multiple measurements", Ann. Eugen., 8, p. 376.

[9] Fisher, R. A. (1946). *Statistical Methods for Research Workers*. Oliver and Boyd, Edinburgh.

[10] James, G. S. (1952). "Notes on a theorem of Cochran", Proc. Camb. Phil. Soc., 48, p. 443.

[11] Kastenbaum, M. A. (1970). "Review of contingency tables", *Essays in Probability and Statistics*, p. 407, University of North Carolina Press, Chapel Hill.

[12] Kendall, M. G. (1958). *A Course in the Geometry of n Dimensions*. Charles Griffin and Company, London.

[13] Kshirsagar, A. M. (1964). "Distributions of the direction and collinearity factors in discriminant analysis", Proc. Camb. Phil. Soc., 60, p. 217.

[14] Kshirsagar, A. M. (1969). "Distributions associated with Wilks's Λ", J. Austral. Math. Soc., 10, p. 269.

[15] Kshirsagar, A. M. (1970). "Goodness-of-fit of an assigned set of scores for the analysis of association in a contingency table", Ann. Inst. Stat. Math., 22, p. 295.

[16] Kshirsagar, A. M. (1970). "An alternative derivation of the direction and collinearity statistics in discriminant analysis" Cal. Stat. Assoc. Bull., 19, p. 123.

[17] Kshirsagar, A. M. (1971). "Goodness-of-fit of a discriminant function from the vector space of dummy variables", J. Roy. Stat. Soc., B, 33, p.

[18] Kshirsagar, A. M. (1971). "Degrees of freedom of the χ^2 test of dimensionality", Letter to the Editor, American Statistician, 25, p. 59.

[19] Mises, R. Von (1945). "On the classification of observation data into distinct groups", Ann. Math. Statist., 16, p. 68.

[20] Potthoff, R. F., and Roy, S. N. (1964). "A generalized multivariate analysis of variance model useful especially for growth curves problems", Biometrika, 51, p. 313.

[21] Radcliffe, J. (1966). "Factorizations of the residual likelihood criterion in discriminant analysis", Proc. Camb. Phil. Soc., 62, p. 743.

[22] Rao, C. Radhakrishna (1962). "Use of discriminant and allied functions in multivariate analysis", Sankhyā, 24A, p. 149.

[23] Rao, C. Radhakrishna (1965). Linear Statistical Inference and Its Applications. John Wiley and Sons, New York.

[24] Rao, C. Radhakrishna (1966). "Covariance adjustments and related problems in multivariate analysis", p. 87, Multivariate Analysis I, (edited by Krishnaiah, P. R.), Academic Press.

[25] Roy, S. N., and Mitra, S. K. (1956). "An introduction to some nonparametric generalizations of analysis of variance and multivariate analysis", Biometrika, 43, p. 361.

[26] Roy, S. N. (1957). Some Aspects of Multivariate Analysis. John Wiley and Sons, New York.

[27] Srikantan, K. S. (1970). "Canonical association between nominal measurements", J. Am. Stat. Assoc., 65, p. 284.

[28] Wald, A. (1944). "On a statistical problem arising in the classification of an individual into one of two groups", Ann. Math. Statist., 15, p. 145.

[29] Wani, J. K., and Kabe, D. G. (1970). "On a certain minimization problem", American Statistician, 24, p. 29.

[30] Williams, E. J. (1952). "Some exact tests in multivariate analysis", Biometrika, 38, p. 17.

[31] Williams, E. J. (1952). "Use of scores for the analysis of association in contingency tables", Biometrika, 39, p. 274.

[32] Williams, E. J. (1955). "Significance tests for discriminant functions and linear functional relationships", Biometrika, 42, p. 360.

[33] Williams, E. J. (1961). "Tests for discriminant functions", J. Austral. Math. Soc., 2, p. 243.

[34] Williams, E. J. (1967). "The analysis of association among many variates", J. Roy. Stat. Soc., B, 20, p. 199.

CHAPTER 10

LIKELIHOOD RATIO TESTS

1. LIKELIHOOD RATIO CRITERIA FOR CERTAIN HYPOTHESES

We shall derive likelihood ratio criteria for certain hypotheses about the parameters of k independent p-variate normal populations $N_p(\underline{x}|\underline{\mu}_\alpha|\Sigma_\alpha)$, $\alpha = 1, 2, \ldots, k$. The hypotheses are

$$H_a:\ \Sigma_1 = \Sigma_2 = \ldots = \Sigma_k\ ,$$

$$H_b:\ \underline{\mu}_1 = \underline{\mu}_2 = \ldots = \underline{\mu}_k,\ \text{given that}\ \Sigma_1 = \Sigma_2 = \ldots = \Sigma_k\ ,$$

$$H_c:\ \underline{\mu}_1 = \underline{\mu}_2 = \ldots = \underline{\mu}_k;\ \ \Sigma_1 = \Sigma_2 = \ldots = \Sigma_k\ .$$

We shall assume that samples of sizes $n_1, n_2, \ldots, n_k$ are available from the k normal populations $N_p(\underline{x}|\underline{\mu}_\alpha|\Sigma_\alpha)$ $(\alpha = 1, 2, \ldots, k)$. Let X_α $(\alpha = 1, 2, \ldots, k)$ denote the $p \times n_\alpha$ matrix of the n_α sample observations on the p variables $\underline{x}$, from the αth population. The matrix of the corrected s.s. and s.p. of these observations is then $S_\alpha = X_\alpha\left(I - \frac{1}{n_\alpha} E_{n_\alpha n_\alpha}\right)X'_\alpha$, $\bar{\underline{x}}_\alpha = \frac{1}{n_\alpha} X_\alpha E_{n_\alpha 1}$ is the vector of the sample means and

$$A = S_1 + S_2 + \ldots + S_k \tag{1.1}$$

is the pooled matrix of the corrected s.s. and s.p. of all the sample observations. We shall denote by $\bar{\underline{x}} = \sum_{\alpha=1}^{k} n_\alpha \bar{\underline{x}}_\alpha \Big/ \sum_{\alpha=1}^{k} n_\alpha$ the vector of the means of all the sample observations. The matrix

$$\sum_{\alpha=1}^{k} n_\alpha(\bar{\underline{x}}_\alpha - \bar{\underline{x}})(\bar{\underline{x}}_\alpha - \bar{\underline{x}})' = B \tag{1.2}$$

will also be required subsequently. We shall also use

$$f_\alpha = n_\alpha - 1 \quad (\alpha = 1, 2, \ldots, k) \quad \text{and} \quad N = \sum_{\alpha=1}^{k} n_\alpha, \quad f = \sum_{\alpha=1}^{k} f_\alpha. \tag{1.3}$$

The joint distribution of all the sample observations is

$$\prod_{\alpha=1}^{k} \left[\frac{1}{(2\pi)^{n_\alpha p/2} |\Sigma_\alpha|^{n_\alpha/2}} \exp\left\{ -\frac{1}{2} \operatorname{tr} \Sigma_\alpha^{-1} \left(X_\alpha - \underline{\mu}_\alpha E_{1n_\alpha} \right) \text{ into } \left(X_\alpha - \underline{\mu}_\alpha E_{1n_\alpha} \right)' \right\} dX_\alpha \right]. \tag{1.4}$$

The parameters $\underline{\mu}_\alpha$, $\underline{\Sigma}_\alpha$ ($\alpha = 1, 2, \ldots, k$) belong to the space Ω defined by

$$\Omega: \quad -\infty < \underline{\mu}_\alpha < \infty; \quad \Sigma_\alpha > 0 \quad (\alpha = 1, 2, \ldots, k). \tag{1.5}$$

The number of parameters is

$$m = pk + \frac{p(p+1)}{2} k, \tag{1.6}$$

as each $\underline{\mu}_\alpha$ has p elements and each Σ_α has only $p(p+1)/2$ distinct elements, Σ_α being symmetric. From (1.4), the logarithm of the likelihood is

$$\log_e L(\Omega) = -\frac{Np}{2} \log_e(2\pi) - \frac{1}{2} \sum_{\alpha=1}^{k} n_\alpha \log_e |\Sigma_\alpha| - \frac{1}{2} \operatorname{tr} \sum_{\alpha=1}^{k} \Sigma_\alpha^{-1} S_\alpha - \frac{1}{2} \operatorname{tr} \sum_{\alpha=1}^{k} n_\alpha \Sigma_\alpha^{-1} (\bar{\underline{x}}_\alpha - \underline{\mu}_\alpha)(\bar{\underline{x}}_\alpha - \underline{\mu}_\alpha)'. \tag{1.7}$$

In Chapter 3 (Section 1), the maximum likelihood estimates (m.l.e.) of $\underline{\mu}_\alpha$ and $\underline{\Sigma}_\alpha$ were found to be, respectively, $\bar{\underline{x}}_\alpha$ and $\dfrac{S_\alpha}{n_\alpha}$. Since the k populations are independent, the likelihood of all the sample observations is simply the product of the separate likelihoods, and so maximizing (1.7) is the same as maximizing the likelihoods of the individual samples separately, and this will therefore yield the m.l.e.'s as

$$\hat{\underline{\mu}}_\alpha = \bar{\underline{x}}_\alpha , \quad \hat{\Sigma}_\alpha = \frac{1}{n_\alpha} S_\alpha \quad (\alpha = 1, 2, \ldots, k) . \tag{1.8}$$

Substituting these back in (1.7), we find

$$\text{Max} \log_e L(\Omega) = - \frac{Np}{2} \log_e (2\pi) - \frac{1}{2} \sum_{\alpha=1}^{k} n_\alpha \log_e \left| \frac{1}{n_\alpha} S_\alpha \right| - \frac{Np}{2} . \tag{1.9}$$

For testing H_a, we set the common value of the Σ_α's to be Σ in (1.4) and from that obtain

$$\log_e L(w_a) = - \frac{Np}{2} \log_e (2\pi) - \frac{N}{2} \log_e |\Sigma| - \frac{1}{2} \operatorname{tr} \Sigma^{-1} A$$
$$- \frac{1}{2} \operatorname{tr} \Sigma^{-1} \sum_{\alpha=1}^{k} n_\alpha (\bar{\underline{x}}_\alpha - \underline{\mu}_\alpha)(\bar{\underline{x}}_\alpha - \underline{\mu}_\alpha)' , \tag{1.10}$$

where w_a denotes the space of the parameters $\underline{\mu}_\alpha$ $(\alpha = 1, 2, \ldots, k)$ and Σ, defined by

$$w_a: \quad -\infty < \underline{\mu}_\alpha < \infty , \quad (\alpha = 1, 2, \ldots, k); \quad \Sigma > 0 . \tag{1.11}$$

The number of parameters in (1.10) is

$$m_a = pk + \frac{p(p+1)}{2} . \tag{1.12}$$

Using results in Chapter 3 (Section 1) to differentiate (1.10) with respect to the parameters, we find their m.l.e.'s to be

$$\hat{\underline{\mu}}_\alpha = \bar{\underline{x}}_\alpha \quad (\alpha = 1, 2, \ldots, k), \ \hat{\Sigma} = \frac{1}{N} A . \tag{1.13}$$

Substituting these back in (1.10),

$$\max \cdot \log_e L(w_a) = -\frac{Np}{2}\log_e(2\pi) - \frac{N}{2}\log_e\left|\frac{1}{N}A\right| - \frac{Np}{2}. \tag{1.14}$$

To test the hypothesis H_c, we set all the $\underline{\mu}_\alpha$'s equal to $\underline{\mu}$ and all the Σ_α's equal to Σ in (1.4) and obtain

$$\begin{aligned}\log_e L(w_c) &= -\frac{Np}{2}\log_e(2\pi) - \frac{N}{2}\log_e|\Sigma| - \frac{1}{2}\operatorname{tr}\Sigma^{-1}A \\ &\quad - \frac{1}{2}\operatorname{tr}\Sigma^{-1}\sum_{\alpha=1}^{k} n_\alpha(\bar{\underline{x}}_\alpha - \underline{\mu})(\bar{\underline{x}}_\alpha - \underline{\mu})' \\ &= -\frac{Np}{2}\log_e(2\pi) - \frac{N}{2}\log_e|\Sigma| - \frac{1}{2}\operatorname{tr}\Sigma^{-1}(A+B) \\ &\quad - \frac{N}{2}\operatorname{tr}\Sigma^{-1}(\bar{\underline{x}} - \underline{\mu})(\bar{\underline{x}} - \underline{\mu})',\end{aligned} \tag{1.15}$$

where w_c is the space of the parameters $\underline{\mu}$ and Σ only, and is

$$w_c: \quad -\infty < \underline{\mu} < \infty, \quad \Sigma > 0. \tag{1.16}$$

The number of parameters in (1.15) is only

$$m_c = p + \frac{1}{2}p(p+1). \tag{1.17}$$

Differentiating (1.15) with respect to $\underline{\mu}$ and Σ, we find that the m.l.e.'s of $\underline{\mu}$ and Σ are

$$\hat{\underline{\mu}} = \bar{\underline{x}}, \quad \hat{\Sigma} = \frac{1}{N}(A+B). \tag{1.18}$$

Substituting these back in (1.15),

$$\max \cdot \log_e L(w_c) = -\frac{Np}{2}\log_e(2\pi) - \frac{N}{2}\log_e\left|\frac{1}{N}(A+B)\right| - \frac{Np}{2}. \tag{1.19}$$

The likelihood ratio criterion for testing H_a is

$$\lambda_a^* = \frac{\operatorname{Max} L(w_a)}{\operatorname{Max} L(\Omega)} \tag{1.20}$$

From (1.9) and (1.14),

$$\lambda_a^* = \frac{\prod_{\alpha=1}^{k} \left| \frac{1}{n_\alpha} S_\alpha \right|^{n_\alpha/2}}{\left| \frac{1}{N} A \right|^{N/2}} . \tag{1.21}$$

From a general result about likelihood ratio criteria [see 11],

$-2 \log_e \lambda_a^*$ is asymptotically distributed as a χ^2 with $m - m_a = p(p+1)(k-1)/2$ d.f. if H_a is true and N is large. (1.22)

The likelihood ratio criterion for testing H_c is

$$\lambda_c^* = \frac{\text{Max. } L(w_c)}{\text{Max. } L(\Omega)} . \tag{1.23}$$

From (1.9) and (1.19),

$$\lambda_c^* = \frac{\prod_{\alpha=1}^{k} \left| \frac{1}{n_\alpha} S_\alpha \right|^{n_\alpha/2}}{\left| \frac{1}{N} (A+B) \right|^{N/2}} . \tag{1.24}$$

Again,

$-2 \log_e \lambda_c^*$ is asymptotically distributed as a χ^2 with $m - m_c = p(k-1) + [p(p+1)/2](k-1)$ d.f. if H_c is true and N is large. (1.25)

To test H_b, we observe that H_b is a "conditional" hypothesis. When the condition $\Sigma_1 = \ldots = \Sigma_k$ is imposed, the likelihood is $L(w_a)$ and, in this, if we further use $\underline{\mu}_1 = \ldots = \underline{\mu}_k$ for testing H_b, we get $L(w_c)$. Hence the likelihood ratio criterion for H_b is

$$\lambda_b^* = \frac{\text{Max. } L(w_c)}{\text{Max. } L(w_a)} . \tag{1.26}$$

From (1.14) and (1.19),

$$\lambda_b^* = \frac{|A|^{N/2}}{|A+B|^{N/2}} = \frac{\lambda_c^*}{\lambda_a^*} \quad . \tag{1.27}$$

When H_b is true and N is large,

$$-2\log_e\lambda_b^* \text{ is a } \chi^2 \text{ with } m_a - m_c = p(k-1) \text{ d.f.} \tag{1.28}$$

Instead of these large sample approximate χ^2 tests of (1.22), (1.25), and (1.28), better approximations are obtained by modifying the likelihood ratio criteria. The modified criteria are obtained by using f_α, the d.f. of S_α, rather than n_α, the sample size. We thus employ

$$\lambda_a = \frac{\prod_{\alpha=1}^{k} |S_\alpha|^{f_\alpha/2}}{|A|^{f/2}} \ , \quad \lambda_b = \frac{|A|^{f/2}}{|A+B|^{f/2}} \ , \quad \lambda_c = \frac{\prod_{\alpha=1}^{k} |S_\alpha|^{f_\alpha/2}}{|A+B|^{f/2}} \tag{1.29}$$

instead of λ_a^*, λ_b^*, λ_c^*, respectively for H_a, H_b, and H_c. Observe that

$$\lambda_a\lambda_b = \lambda_c \ . \tag{1.30}$$

We shall now obtain the moments of these criteria under the respective null hypotheses and use them to approximate their distributions by χ^2 distributions.

2. MOMENTS OF λ_a, λ_b, λ_c

From Section 4 of Chapter 3, we know that S_α has the $W_p(S_\alpha|f_\alpha|\Sigma_\alpha)$ distribution and is independent of $\bar{\underline{x}}_\alpha$, which has the $N_p(\bar{\underline{x}}_\alpha|\underline{\mu}_\alpha|(1/n_\alpha)\Sigma_\alpha)$ distribution. This is true for each $\alpha = 1, 2, \ldots, k$, and all these distributions are independent.

We shall now assume $\Sigma_1 = \Sigma_2 = \cdots = \Sigma_k = \Sigma$ and $\underline{\mu}_1 = \underline{\mu}_2 = \cdots = \underline{\mu}_k = \underline{\mu}$. In (9.1.18), we have proved that

$$B = \sum_{\alpha=1}^{k} n_\alpha(\bar{\underline{x}}_\alpha - \bar{\underline{x}})(\bar{\underline{x}}_\alpha - \bar{\underline{x}})'$$

is distributed as $\sum_{\alpha=1}^{k} \underline{z}_\alpha \underline{z}_\alpha'$, where $\underline{z}_\alpha$ ($\alpha = 1, 2, \ldots, k$) are independent and have the $N_p(\underline{z}_\alpha|0|\Sigma)$ distribution. Since B is a function of $\bar{\underline{x}}_\alpha$'s only, and since S_α's and $\bar{\underline{x}}_\alpha$'s are independent, $S_1, S_2, \ldots, S_k$ and B are independent. The joint distribution of $S_1, S_2, \ldots, S_k$ is

$$\prod_{\alpha=1}^{k} \left\{\frac{k(p, f_\alpha)}{|\Sigma|^{n_\alpha/2}} |S_\alpha|^{(f_\alpha-p-1)/2} e^{-\operatorname{tr}\Sigma^{-1}S_\alpha/2}\, dS_\alpha\right\}, \quad (2.1)$$

where $k(p, f_\alpha)$ is defined by (3.2.20). Transform from $S_1, \ldots, S_k$ to $A, L_1, L_2, \ldots, L_{k-1}$ by the transformation

$$A = \sum_{\alpha=1}^{k} S_\alpha, \quad S_\alpha = A^{1/2} L_\alpha A^{1/2} \quad (\alpha = 1, 2, \ldots, k-1). \quad (2.2)$$

The Jacobian of this transformation is

$$J(S_k \to A) \prod_{\alpha=1}^{k-1} J(S_\alpha \to L_\alpha) = |A|^{[(p+1)/2](k-1)}, \quad (2.3)$$

as $J(S_k \to A) = 1$ and $J(S_\alpha \to L_\alpha) = |A|^{(p+1)/2}$ (Deemer and Olkin [4]). Also, from (2.2), for $\alpha = 1, 2, \ldots, k-1$,

$$|S_\alpha| = |A|\,|L_\alpha| \quad \text{but} \quad |S_k| = |A^{1/2}(I - \sum_{\alpha=1}^{k-1} L_\alpha)A^{1/2}|$$
$$= |A|\,|I - \sum_{1}^{k-1} L_\alpha|. \quad (2.4)$$

Using (2.2), (2.3), and (2.4) in (2.1), the joint distribution of $L_1, \ldots, L_{k-1}, A$ is

$$\phi(L_1, L_2, \ldots, L_{k-1})\, W_p(A|f|\Sigma)\, dL_1 \ldots dL_{k-1}\, dA, \tag{2.5}$$

where

$$\phi(L_1, \ldots, L_{k-1}) = \frac{\prod_{\alpha=1}^{k} K(p, f_\alpha)}{K(p, f)} \left|I - \sum_{\alpha=1}^{k-1} L_\alpha\right|^{(f_k-p-1)/2}$$

$$\prod_{\alpha=1}^{k-1} |L_\alpha|^{(f_\alpha-p-1)/2}, \quad L_\alpha > 0, \quad I - \sum_{1}^{k-1} L_\alpha > 0$$

$$(\alpha = 1, 2, \ldots, k-1) . \tag{2.6}$$

Expression (2.5) shows that $(L_1, \ldots, L_{k-1})$ and A are independent, the distribution of $(L_1, \ldots, L_{k-1})$ being $\phi(L_1, \ldots, L_{k-1}) \prod_{1}^{k-1} dL_\alpha$, that of A being $W_p(A|f|\Sigma)$. This distribution of A is obvious from the additive property of Wishart distributions of S_α's also. Since $L_1, \ldots, L_{k-1}$ and A are functions of S_α's only, it is obvious that they are independent of B also. Thus we have three independent distributions, viz., (i) of $L_1, \ldots, L_{k-1}$, (ii) A, and (iii) B. Observe that [using (2.4) and (1.29)]

$$\lambda_a = \frac{\prod_{\alpha=1}^{k} |S_\alpha|^{f_\alpha/2}}{|A|^{f/2}} = \prod_{\alpha=1}^{k} |A^{-1/2} S_\alpha A^{-1/2}|^{f_\alpha/2}$$

$$= \prod_{\alpha=1}^{k-1} |L_\alpha|^{f_\alpha/2} \left|I - \sum_{1}^{k-1} L_\alpha\right|^{f_k/2} . \tag{2.7}$$

λ_a is thus a function of the $L_1, \ldots, L_{k-1}$ only. λ_b, however, is a function of A and B. Therefore λ_a is independently distributed of λ_b, as $(L_1, \ldots, L_{k-1})$, A, and B are independent. This result holds when H_c is true, i.e., when all the k normal populations are identical.

Note that H_b is the multivariate analysis of variance hypothesis considered in Chapter 9 and λ_b is the (f/2)th power of Wilks's Λ criterion $|A|/|A+B|$ of (9.1.19). The hth moment of λ_b is thus the (fh/2)th moment of Wilks's Λ and can be easily written, from (8.2.22), as

$$E(\lambda_b^h) = \prod_{i=1}^{p} \left\{ \frac{\Gamma\left(\frac{N-i}{2}\right)}{\Gamma\left(\frac{N-i+fh}{2}\right)} \cdot \frac{\Gamma\left(\frac{N-k+1-i+fh}{2}\right)}{\Gamma\left(\frac{N-k+1-i}{2}\right)} \right\}$$

$$= \prod_{i=1}^{p} \left\{ \frac{\Gamma\left(\frac{f+k-i}{2}\right)}{\Gamma\left(\frac{f+fh+k-i}{2}\right)} \cdot \frac{\Gamma\left(\frac{f+fh+1-i}{2}\right)}{\Gamma\left(\frac{f+1-i}{2}\right)} \right\}. \qquad (2.8)$$

The distribution of $\Lambda = \lambda_b^{2/f}$ has been considered in detail in Chapter 8, and the test based on Λ for H_b was described there and again in Chapter 9. It is, therefore, not necessary to discuss it here.

We shall therefore now consider the moments of λ_a. Since $\phi(L_1, \ldots, L_{k-1})$ is the p.d.f. of the L_α's,

$$\int \phi(L_1, \ldots, L_{k-1}) \prod_{\alpha=1}^{k-1} dL_\alpha = 1, \qquad (2.9)$$

where the integral is a multiple integral and the range of integration is such that $L_\alpha > 0$, $I - \sum_{1}^{k-1} L_\alpha > 0$ is satisfied. Substituting for $\phi(L_1, \ldots, L_{k-1})$, from (2.6), we obtain

$$\int \left|I - \sum_{1}^{k-1} L_\alpha\right|^{(f_k-p-1)/2} \prod_{\alpha=1}^{k-1} \left\{ |L_\alpha|^{(f_\alpha-p-1)/2} dL_\alpha \right\} = \frac{K(p,f)}{\prod_{\alpha=1}^{k} K(p,f_\alpha)}. \qquad (2.10)$$

This is an identify in f_α's. We can therefore change f_α to $f_\alpha + hf_\alpha$ ($\alpha = 1, 2, \ldots, k$) on both sides of (2.10). Then multiply both sides by $\prod_{\alpha=1}^{k} K(p, f_\alpha)/K(p, f)$ to restore the constant in the p.d.f. of $L_1, \ldots, L_{k-1}$. Performing this and noting (2.7), we obtain

$$\int \lambda_a^h \phi(L_1, \ldots, L_{k-1}) \prod_{\alpha=1}^{k-1} dL_\alpha = \frac{K(p, f+fh)}{\prod_{\alpha=1}^{k} K(p, f_\alpha + f_\alpha h)} \; \frac{\prod_{\alpha=1}^{k} K(p, f_\alpha)}{K(p, f)} . \tag{2.11}$$

Equation (2.11) thus gives the hth moment of λ_a. Substituting for the $K(p, f_\alpha)$'s, from (3.2.20), we obtain

$$E(\lambda_a^h) = \prod_{i=1}^{p} \left\{ \prod_{\alpha=1}^{k} \frac{\Gamma\left(\frac{f_\alpha + hf_\alpha + 1 - i}{2}\right)}{\Gamma\left(\frac{f_\alpha + 1 - i}{2}\right)} \right\} \frac{\Gamma\left(\frac{f+1-i}{2}\right)}{\Gamma\left(\frac{f+fh+1-i}{2}\right)} . \tag{2.12}$$

Since $\lambda_c = \lambda_a \lambda_b$ and λ_a, λ_b are independently distributed,

$$E(\lambda_c^h) = E(\lambda_a^h)\, E(\lambda_b^h)$$

$$= \prod_{i=1}^{p} \left\{ \prod_{\alpha=1}^{k} \frac{\Gamma\left(\frac{f_\alpha + hf_\alpha + 1 - i}{2}\right)}{\Gamma\left(\frac{f_\alpha + 1 - i}{2}\right)} \right\} \frac{\Gamma\left(\frac{f+k-i}{2}\right)}{\Gamma\left(\frac{f+fh+k-i}{2}\right)} . \tag{2.13}$$

Since $0 \le \lambda_a \le 1$, $0 \le \lambda_b \le 1$, and $0 \le \lambda_c \le 1$, their moments determine their distributions uniquely. By expanding the gamma functions in (2.12), using (8.3.2) and some laborious algebra, Box [2] has shown that

$$\text{Prob}\left\{ -2\rho_a \log_e \left[\frac{f^{pf/2}}{\prod_{\alpha=1}^{k} f_\alpha^{pf_\alpha/2}} \lambda_a \right] \le z \right\}$$

$$= C_g(z) + w_a[C_{g+4}(z) - C_g(z)] + O(f^3) \tag{2.14}$$

where $C_g(z)$ is defined in (8.3.5) and is the c.d.f. of a χ^2 with g d.f. Also,

$$\rho_a = 1 - \left(\sum_{\alpha=1}^{k} \frac{1}{f_\alpha} - \frac{1}{f}\right) \frac{2p^2 + 3p - 1}{6(p+1)(k-1)} ,$$

$$48\rho^2 w_a = p(p+1)[(p-1)(p+2)\left(\sum_\alpha \frac{1}{f_\alpha^2} - \frac{1}{f^2}\right) - 6(k-1)(1-\rho)^2] ,$$

$$g = \frac{1}{2}(k-1)p(p+1) . \qquad (2.15)$$

The first term on the right-hand side of (2.14) shows that the distribution of $-2\rho \log_e \left[f^{pf/2} \lambda_a / \prod_{\alpha=1}^{k} f_\alpha^{pf_\alpha/2}\right]$ can be approximated fairly well by a χ^2 distribution with g d.f. The hypothesis H_a can thus be tested by using λ_a. If more accuracy is needed, the second term in the right-hand side of (2.14) must also be included.

Similarly, from the moments of λ_c, it has been shown that

$$\text{Prob}\left\{-2\rho_c \log_e \left[\frac{f^{pf/2}}{\prod_{\alpha=1}^{k} f_\alpha^{pf_\alpha/2}} \lambda_c\right] \le z\right\}$$

$$= C_h(z) + w_c[C_{h+4}(z) - C_h(z)] + O(f^{-3}) , \qquad (2.16)$$

where

$$1 - \rho_c = \left(\sum_\alpha \frac{1}{f_\alpha} - \frac{1}{f}\right) \frac{2p^2 + 3p - 1}{6(k-1)(p+3)} + \frac{1}{f} \frac{p-k+2}{p+3} ,$$

$$w_c = \frac{p}{288 \rho_c^2} \left[6\left(\sum_\alpha \frac{1}{f_\alpha^2} - \frac{1}{f^2}\right)(p+1)(p-1)(p+2)\right.$$

$$- \left(\sum \frac{1}{f_\alpha} - \frac{1}{f}\right)^2 \frac{(2p^2 + 3p - 1)^2}{(k-1)(p+3)}$$

$$- 12\left(\sum \frac{1}{f_\alpha} - \frac{1}{f}\right)\frac{(2p^2 + 3p - 1)(p-k+2)}{f(p+3)}$$

$$- 36 \frac{(k-1)(p-k+2)^2}{f^2(p+3)} - \frac{12(k-1)}{f^2} (-2k^2 + 7k + 3pk$$

$$\left. -2p^2 - 6p - 4)\right] , \qquad (2.17)$$

$$h = \frac{1}{2}(k-1)\,p(p+3)\ .$$

The hypothesis H_c can be tested by using λ_c and the above distributional result associated with it, in the same way as λ_a.

3. INDEPENDENCE OF k SETS OF VARIATES

Let $\underline{x}$ be a column vector of p random variables having the $N_p(\underline{x}|\underline{\mu}|\Sigma)$ distribution. Let $\underline{x}$, $\underline{\mu}$, and Σ be partitioned as

$$\underline{x}' = [\underset{p_1}{\underline{x}_1'}|\underset{p_2}{\underline{x}_2'}|\ \cdots\ |\underset{p_k}{\underline{x}_k'}]' , \qquad (3.1)$$

$$\underline{\mu}' = [\underline{\mu}_1'|\underline{\mu}_2'|\ \cdots\ |\underline{\mu}_k']' , \qquad (3.2)$$

$$\Sigma = \begin{bmatrix} \Sigma_{11} & \Sigma_{12} & \cdots & \Sigma_{1k} \\ \Sigma_{21} & \Sigma_{22} & \cdots & \Sigma_{2k} \\ & & & \\ \Sigma_{k1} & \Sigma_{k2} & \cdots & \Sigma_{kk} \end{bmatrix}, \quad \Sigma_{ij} \text{ is of order } p_i \times p_j \quad (i, j = 1, 2, \ldots, k). \qquad (3.3)$$

We shall derive the likelihood ratio criterion for testing H, the hypothesis of independence of the k vectors $\underline{x}_1, \underline{x}_2, \ldots, \underline{x}_k$ of $p_1, p_2, \ldots, p_k$ components, respectively, $p_1 + \ldots + p_k$ being p. Since the p.d.f. $N_p(\underline{x}|\underline{\mu}|\Sigma)$ of $\underline{x}$ factorizes as the product of the p.d.f.'s of the $\underline{x}_1, \underline{x}_2, \ldots, \underline{x}_k$, i.e.,

$$N_p(\underline{x}|\underline{\mu}|\Sigma) = \prod_{i=1}^{k} N_{p_i}(\underline{x}_i|\underline{\mu}_i|\Sigma_{ii}) \qquad (3.4)$$

if and only if $\Sigma_{ij} = 0$ for all i, j = 1, 2, ..., k such that $i \neq j$, it is evident that the hypothesis H can be mathematically expressed as

$$H:\ \Sigma_{ij} = 0,\quad i \neq j;\quad i, j = 1, 2, \ldots, k\ . \qquad (3.5)$$

If H is true, Σ can be expressed as

$$\mathrm{diag}(\Sigma_{11}, \Sigma_{22}, \ldots, \Sigma_{kk}),$$

Σ_{ii} being the $p_i \times p_i$ submatrix along the diagonal.

Let X be the $p \times N$ matrix of N independent sample observations on $\underline{x}$. The likelihood of the sample observations is denoted by $L(\Omega)$, where Ω is the space of the parameters $\underline{\mu}$ and Σ [of (3.3)] and is defined by

$$\Omega: \quad -\infty < \underline{\mu} < \infty, \quad \Sigma > 0. \tag{3.6}$$

The m.l.e., of $\underline{\mu}$ and Σ, and Max $L(\Omega)$ are given by (see Section 1 of Chapter 3) $\hat{\underline{\mu}} = \bar{\underline{x}}$, $\hat{\Sigma} = (1/N)S$, and

$$\mathrm{Max}\ \log_e L(\Omega) = -\frac{Np}{2}\log_e(2\pi) - \frac{N}{2}\log_e\left|\frac{1}{n}S\right| - \frac{Np}{2}, \tag{3.7}$$

respectively, where

$$\bar{\underline{x}} = \frac{1}{N} X E_{N1}, \quad S = X(I - \frac{1}{N} E_{NN})X'. \tag{3.8}$$

Partition X, $\bar{\underline{x}}$, and S as

$$X' = \underset{\quad\ p_1\quad p_2 \qquad\quad p_k}{[X_1' | X_2' | \cdots | X_k']N}, \tag{3.9}$$

$$\bar{\underline{x}}' = \underset{\quad\ p_1\quad p_2 \qquad\quad p_k}{[\bar{\underline{x}}_1' | \bar{\underline{x}}_2' | \cdots | \bar{\underline{x}}_k']1}, \tag{3.10}$$

$$S = \begin{bmatrix} S_{11} & S_{12} & \cdots & S_{1k} \\ S_{21} & S_{22} & \cdots & S_{2k} \\ & & & \\ S_{k1} & S_{k2} & \cdots & S_{kk} \end{bmatrix}, \quad S_{ij} \text{ is of order } p_i \times p_j \ (i, j = 1, 2, \ldots, k). \tag{3.11}$$

Since $N_p(\underline{x}|\underline{\mu}|\Sigma)$ factorizes as $\prod_{i=1}^{k} N_{p_i}(\underline{x}_i|\underline{\mu}_i|\Sigma_{ii})$ when H is true, the likelihood of the sample observations X can be expressed as

$$L(\omega) = \prod_{i=1}^{k}\left[\frac{1}{(2\pi)^{Np_i/2}|\Sigma_{ii}|^{N/2}} \exp\left\{-\frac{1}{2}\Sigma_{ii}^{-1}(X_i - \underline{\mu}_i E_{1n}) \times (X_i - \underline{\mu}_i E_{1N})'\right\}\right]. \quad (3.12)$$

The parameters involved are $\underline{\mu}_i$ and Σ_{ii} $(i = 1, 2, \ldots, k)$ only, and the parameter space is now

$$\omega: \quad -\infty < \underline{\mu}_i < \infty \; ; \quad \Sigma_{ii} > 0 \quad (i = 1, 2, \ldots, k) . \quad (3.13)$$

Maximizing the likelihood $L(\omega)$ with respect to the parameters, one can show (using the same method as in Section 1 of Chapter 3), that the m.l.e. are

$$\hat{\underline{\mu}}_i = \bar{\underline{x}}_i , \quad \hat{\Sigma}_{ii} = \frac{1}{N} S_{ii} \quad (i = 1, 2, \ldots, k) . \quad (3.14)$$

Substituting these in $L(\omega)$, we find

$$\text{Max} \log_e L(\omega) = -\frac{Np}{2}\log_e(2\pi) - \frac{N}{2}\sum_{i=1}^{k}\log_e\left|\frac{1}{N}S_{ii}\right| - \frac{Np}{2} . \quad (3.15)$$

The likelihood ratio criterion for testing H is thus

$$\lambda^* = \frac{\text{Max } L(\omega)}{\text{Max } L(\Omega)} = \frac{|S|^{N/2}}{\prod_{i=1}^{k}|S_{ii}|^{N/2}} , \quad (3.16)$$

The number of parameters estimated by maximizing $L(\Omega)$ is $p + (p+1)/2$ as $\underline{\mu}$ has p elements and Σ has $p(p+1)/2$ distinct elements. Each Σ_{ii} has $p_i(p_i+1)/2$ distinct elements and $\underline{\mu}_i$ has p_i elements. Therefore, the number of parameters estimated by maximizing $L(\omega)$ is

$$\sum_{i=1}^{k} p_i + \sum_{i=1}^{k} p_i(p_i+1)/2 = p + \sum_{i=1}^{k} p_i(p_i+1)/2 .$$

Hence, by the large sample general result about likelihood ratio criteria, $-2 \log_e \lambda^*$ is distributed asymptotically as a χ^2 with

$$f = \frac{p(p+1)}{2} - \sum_1^k \frac{p_i(p_i+1)}{2} = \frac{1}{2}\left(p^2 - \sum_1^k p_i^2\right) \qquad (3.17)$$

d.f., this being the difference in the number of parameters in Ω and ω. To improve this asymptotic result, we need the moments of λ^* or of $\lambda = \lambda^{*2/N}$, when H is true. This, in turn, needs the following results.

When H is true, S has the $W_p(S|n|\Sigma)$ distribution, where $n = N-1$ and $\Sigma = \text{diag}(\Sigma_{11}, \Sigma_{22}, \ldots, \Sigma_{kk})$. Also, as S_{ii} is the matrix of the corrected s.s. and s.p. of observations on $\underline{x}_i$, S_{ii} has the $W_{p_i}(S_{ii}|n|\Sigma_{ii})$ distribution. When H is true, $\underline{x}_i$ $(i = 1, 2, \ldots, k)$ are independent, and therefore, the S_{ii} $(i = 1, 2, \ldots, k)$ are also independent. Since S_{ii} are submatrices of S, their distribution can also be derived (at least theoretically) from that of S, by integrating out the other elements of S, viz., of S_{ij} $(i \neq j, i, j = 1, \ldots, k)$. This gives

$$\int \frac{K(p, n)}{|\Sigma|^{n/2}} |S|^{(n-p-1)/2} \exp(-\tfrac{1}{2}\text{tr}\,\Sigma^{-1} S) \prod_{\substack{i,j=1\\ i\neq j}}^{k} dS_{ij}$$
$$= \prod_{i=1}^{k} W_{p_i}(S_{ii}|n|\Sigma_{ii}), \qquad (3.18)$$

where the integration is over the entire range of values of the elements of S_{ij} $(i \neq j)$, such that $S > 0$. Since (3.18) is an identity in n, we can change n to $n+2h$ on both sides. This gives

$$\int \frac{K(p,\ n+2h)}{|\Sigma|^{n/2+h}} |S|^{(n-p-1)/2+h} \exp\left(-\frac{1}{2}\operatorname{tr}\Sigma^{-1}S\right) \prod_{\substack{i,j=1\\ i\neq j}}^{k} dS_{ij}$$

$$= \prod_{i=1}^{k} W_{p_i}(S_{ii}|n+2h|\Sigma_{ii}) . \qquad (3.19)$$

Moreover, it is easy to verify that

$$|S|^h W_p(S|n|\Sigma) = \frac{K(p,\ n)}{K(p,\ n+2h)} |\Sigma|^h W_p(S|n+2h|\Sigma) . \qquad (3.20)$$

We now use (3.20) and (3.19) to find the hth moment of λ. It is

$$E(\lambda^h) = \int_{S>0} \lambda^h W_p(S|n|\Sigma)dS$$

$$= \int_{S>0} \prod_{i=1}^{k} |S_{ii}|^{-h} \cdot |S|^h W_p(S|n|\Sigma)dS$$

$$= \frac{K(p,\ n)|\Sigma|^h}{K(p,\ n+2h)} \int_{S>0} \prod_{i=1}^{k} |S_{ii}|^{-h} W_p(S|n+2h|\Sigma)dS$$

$$= \frac{K(p,\ n)|\Sigma|^h}{K(p,\ n+2h)} \int \prod_{i=1}^{k} |S_{ii}|^{-h}$$

$$\left\{\int W_p(S|n+2h|\Sigma) \prod_{\substack{i,j=1\\ i\neq j}}^{k} dS_{ij}\right\} \prod_{i=1}^{k} dS_{ii} , \qquad (3.21)$$

where dS is split as $\prod_{i=1}^{k} dS_{ii} \cdot \prod_{\substack{i,j=1\\ i\neq j}}^{k} dS_{ij}$ and the range of integration for the integral in the large bracket is over values of S_{ij} such that $S > 0$, and then the first integral with respect to S_{ii}'s, have $S_{ii} > 0$ $(i = 1, 2, \ldots, k)$ as the range of integration. Using (3.19) we get

$$E(\lambda^h) = \frac{K(p, n)|\Sigma|^h}{K(p, n+2h)} \int_{\text{each } S_{ii}>0} \prod_{i=1}^{k} \{|S_{ii}|^{-h}$$

$$\times W_{p_i}(S_{ii}|n+2h|\Sigma_{ii})dS_{ii}\}$$

$$= \frac{K(p, n)|\Sigma|^h}{K(p, n+2h)} \prod_{i=1}^{k} \left\{ \int_{S_{ii}>0} \frac{K(p_i, n+2h)}{K(p_i, n)} |\Sigma_{ii}|^{-h} \right.$$

$$\left. \times W_{p_i}(S_{ii}|n|\Sigma_{ii})dS_{ii} \right\}$$

$$= \frac{K(p, n)}{K(p, n+2h)} \prod_{i=1}^{k} \left\{ \frac{K(p_i, n+2h)}{K(p_i, n)} \right\}, \tag{3.22}$$

as $|\Sigma| = \prod_{i=1}^{k} |\Sigma_{ii}|$, when H is true, and $\int_{S_{ii}>0} W_{p_i}(S_{ii}|n|\Sigma_{ii})dS_{ii} = 1$. Finally, therefore, from (3.2.20),

$$E(\lambda^h) = \frac{\prod_{i=1}^{p} \Gamma\left(\frac{n+1-i}{2} + h\right) \prod_{i=1}^{k} \prod_{j=1}^{p_i} \Gamma\left(\frac{n+1-j}{2}\right)}{\prod_{i=1}^{p} \Gamma\left(\frac{n+1-i}{2}\right) \prod_{i=1}^{k} \prod_{j=1}^{p_i} \Gamma\left(\frac{n+1-j}{2} + h\right)}. \tag{3.23}$$

By expanding the gamma functions in (3.23), Box [2] has proved that

$$\text{Prob}\{-m \log_e \lambda \leq z\} = C_f(z) + \frac{\gamma_2}{m^2} \{C_{f+4}(z) - C_f(z)\} + O(m^{-3}). \tag{3.24}$$

Where f is given by (3.17) and

$$\gamma_2 = \frac{p^4 - \sum_i p_i^4}{48} - \frac{5\left(p^2 - \sum_i p_i^2\right)}{96} - \frac{\left(p^3 - \sum_i p_i^3\right)^2}{72\left(p^2 - \sum_i p_i^2\right)}. \tag{3.25}$$

$-m \log_e \lambda$ is thus approximately a χ^2 with f d.f., and this provides a test of the hypothesis of independence of $\underset{-}{x}_1, \ldots, \underset{-}{x}_k$. For more accuracy, the term $\frac{\gamma_2}{m^2}\{C_{f+4}(z) - C_f(z)\}$ should also be included in obtaining the critical value λ_α such that

$$\text{Prob}\{-m \log_e \lambda \geq -m \log_e \lambda_\alpha\} = \alpha ,$$

when H is true. H is then rejected, whenever $-m \log \lambda$ exceeds $-m \log \lambda_\alpha$, the level of significance being α.

From (4.3.2), R, the correlation matrix (of sample correlations) of the p variables $\underline{x}$, is given by

$$R = \text{diag}\left(s_{11}^{-1/2}, s_{22}^{-1/2}, \ldots, s_{pp}^{-1/2}\right) S \,\text{diag}\left(s_{11}^{-1/2}, s_{22}^{-1/2}, \ldots, s_{pp}^{-1/2}\right), \tag{3.26}$$

where s_{ii} $(i = 1, 2, \ldots, p)$ are the diagonal elements of S. If this matrix R is partitioned as

$$R = \left[\begin{array}{c|c|c|c} R_{11} & R_{12} & \cdots & R_{1k} \\ \hline R_{21} & R_{22} & \cdots & R_{2k} \\ \hline & & & \\ \hline R_{k1} & R_{k2} & \cdots & R_{kk} \end{array}\right],$$

where R_{ij} is of order $p_i \times p_j$ $(i, j = 1, 2, \ldots, k)$, one can readily verify that the criterion λ can be expressed as

$$\lambda = |R| / \prod_{i=1}^{k} |R_{ii}| . \tag{3.27}$$

Also, observe that, when $k = 2$, λ reduces to

$$\frac{|S|}{|S_{11}|\;|S_{22}|}, \tag{3.28}$$

and by (7.3.14), this is Wilks's $\Lambda = \prod_{i=1}^{t} (1 - r_i^2)$, where $t = \min(p_1, p_2)$ and tests the independence of $\underline{x}_1$ and $\underline{x}_2$, as seen in Chapter 8, Section 4.

4. THE SPHERICITY TEST

Given a random sample of size N from a $N_p(\underline{x}|\underline{\mu}|\Sigma)$ population, the hypothesis

$$H: \quad \Sigma = \sigma^2 I_p \quad (\sigma^2 \text{ unspecified})$$

is known as the hypothesis of sphericity, because the ellipsoid

$$(\underline{x} - \underline{\mu})' \Sigma^{-1} (\underline{x} - \underline{\mu}) = \text{constant}, \tag{4.1}$$

(where the left-hand side is the exponent, apart from -(1/2), in the p.d.f. of $\underline{x}$) reduces to a sphere $(\underline{x} - \underline{\mu})'(\underline{x} - \underline{\mu})$ = constant. The probability density at any point on the ellipsoid (4.1) is constant. If X is the $p \times N$ matrix of sample observations and $S = X(I - (1/N)E_{NN})X'$ is the matrix of the corrected s.s. and s.p. of the sample observations. Max $\log_e L(\Omega)$, where Ω is the space of the parameters $\underline{\mu}$ and Σ [and is defined by (3.6)], is given by (3.7). When H is true, the likelihood is given by

$$\log_e L(\omega) = -\frac{Np}{2}\log_e(2\pi) - \frac{pN}{2}\log_e \sigma^2$$
$$- \frac{1}{2\sigma^2}\,\text{tr}(X - \underline{\mu}\,E_{1N})(X - \underline{\mu}\,E_{1N})' \tag{4.2}$$

where ω is the space of the parameters $\underline{\mu}$ and σ^2 only and is

$$\omega: \quad -\infty < \underline{\mu} < \infty, \quad \sigma^2 > 0 . \tag{4.3}$$

By maximizing (4.2) with respect to $\underline{\mu}$ and σ^2, it is easy to find that their m.l.e.'s are

$$\hat{\underline{\mu}} = \bar{\underline{x}} = \frac{1}{N} X\,E_{N1}, \quad \hat{\sigma}^2 = \frac{1}{N}\,\text{tr}(X - \bar{\underline{x}}\,E_{1N})(X - \bar{\underline{x}}\,E_{1N})'$$
$$= \text{tr}\, S/Np . \tag{4.4}$$

Substituting these back in (4.2),

$$\text{Max } \log_e L(\omega) = -\frac{Np}{2} \log_e (2\pi) - \frac{pN}{2} \log_e \hat{\sigma}^2 - \frac{Np}{2}. \quad (4.5)$$

The likelihood ratio criterion for testing H is thus

$$\lambda^* = \frac{\text{Max. } L(\omega)}{\text{Max. } L(\Omega)} = \frac{|S|^{N/2}}{(\text{tr } S/p)^{Np/2}} \quad (4.6)$$

when H is true, $\Sigma = \sigma^2 I$ and all the eigenvalues of Σ are the same, viz., σ^2. Σ is estimated by $(1/N)S$ and so we expect all the eigenvalues of S to be not much different from each other. In other words, their arithmetic mean and geometric mean should be approximately the same. Since tr S is the sum of the eigenvalues of S and $|S|$ is their product, the arithmetic mean is tr S/p and the geometric mean is $|S|^{1/p}$. Their ratio is $|S|^{1/p}/(\text{tr } S/p)$ and is expected to be near 1, when H is true. The criterion (4.6) is the [(pN)/2]th power of the ratio.

We shall now find the moments of this criterion, when H is true. When H is true, $x_1, \ldots, x_p$ (the elements of $\underline{x}$) are all normal independent with the same variance σ^2. The criterion λ^* can be expressed as [using (3.26)]

$$\lambda = \lambda^{*2/N} = \frac{|R| \prod_{i=1}^{p} s_{ii}}{\left\{\frac{1}{p} \sum_{i=1}^{p} s_{ii}\right\}^p}, \quad (4.7)$$

where R is the correlation matrix, and s_{ii} are the diagonal elements of S. In Chapter 4, Section 3, we had proved that R and s_{ii} $(i = 1, \ldots, p)$ are independently distributed when x_1, $x_2, \ldots, x_p$ are independent. Hence

$$E(\lambda^h) = E|R|^h, \quad E\left\{\frac{\prod_{i=1}^{p} s_{ii}}{\left(\sum_{i=1}^{p} s_{ii}/p\right)^p}\right\}^h . \tag{4.8}$$

$E|R|^h$ is given by (4.3.17). It only remains to find

$$U = E\left\{\frac{\prod_{i=1}^{p} s_{ii}^h}{\left(\frac{1}{p}\sum_{i=1}^{p} s_{ii}\right)^{hp}}\right\} . \tag{4.9}$$

Since s_{ii} is the corrected s.s. of observations of x_i, it has a $\chi^2\sigma^2$ distribution with n = N-1 d.f. Further, as the x_i are independent (when H is true), the s_{ii} are all independent. Hence

$$U = \int_0^\infty \cdots \int_0^\infty \left(\frac{1}{p}\sum_i s_{ii}\right)^{-hp} \prod_{i=1}^{p}\left\{s_{ii}^h \cdot \chi_n^2\left(\frac{s_{ii}}{\sigma^2}\right) d\left(\frac{s_{ii}}{\sigma^2}\right)\right\} , \tag{4.10}$$

where $\chi_f^2(v)$ is defined in (2.5.9). Letting $s_{ii}/\sigma^2 = t_i$ (i = 1, ..., p), the above becomes

$$U = \int_0^\infty \cdots \int_0^\infty p^{hp} \left(\sum_i t_i\right)^{-hp} \prod_{i=1}^{p}\left\{t_i^h \chi_n^2(t_i) d\, t_i\right\}$$

$$= \int_0^\infty \cdots \int_0^\infty p^{hp} \left(\sum_i t_i\right)^{-hp} \prod_{i=1}^{p}\left\{\frac{2^h\, \Gamma\left(\frac{n}{2}+h\right)}{\Gamma\left(\frac{n}{2}\right)} \chi_{n+2h}^2(t_i) dt_i\right\}$$

$$= 2^{hp}\, p^{hp}\left\{\frac{\Gamma\left(\frac{n}{2}+h\right)}{\Gamma\left(\frac{n}{2}\right)}\right\}^p \cdot v \tag{4.11}$$

where v is the expected value of the (hp)th power of $1/(\Sigma t_i)$, with t_i's having independent χ^2 distributions with n + 2h d.f. But then $\sum_1^p t_i$ has the χ^2 distribution with p(n + 2h) d.f. (due to the additive property of χ^2's). Hence, setting $w = \sum_1^p t_i$, we obtain

$$v = \int_0^\infty w^{-hp} \chi^2_{p(n+2h)}(w)\,dw$$

$$= 2^{-hp} \frac{\Gamma\left(\frac{np+2hp}{2} - hp\right)}{\Gamma\left(\frac{np+2hp}{2}\right)}$$

$$= 2^{-hp}\, \Gamma\left(\frac{np}{2}\right) \Big/ \Gamma\left(\frac{np+2hp}{2}\right) . \tag{4.12}$$

Putting all these results together, we finally obtain

$$E(\lambda^h) = \frac{p^{hp}\, \Gamma\left(\frac{np}{2}\right)}{\Gamma\left(\frac{np+2hp}{2}\right)} \cdot \prod_{i=1}^{p} \left\{ \frac{\Gamma\left(\frac{n+1-i}{2} + h\right)}{\Gamma\left(\frac{n+1-i}{2}\right)} \right\} . \tag{4.13}$$

By expanding the gamma functions in (4.13), it has been shown that

$$\text{Prob}\{-np \log_e \lambda \le z\} = C_f(z) + w_2(C_{f+4}(z) - C_f(z)) + O(n^{-3}), \tag{4.14}$$

where

$$\rho = 1 - \frac{2p^2 + p + 2}{6pn} ,$$

$$w_2 = \frac{(p+2)(p-1)(p-2)(2p^3 + 6p^3 + 6p^2 + 3p + 2)}{288\, p^2 n^2 \rho^2} ,$$

$$f = \frac{p(p+1)}{2} - 1 . \tag{4.15}$$

In Ω, we had $p + [p(p+1)]/2$ parameters, while in w, we had only $\underline{\mu}$ and σ^2, i.e., $p+1$ parameters. The difference in their number is f, given above. Equation (4.14) enables us to test H by using λ, in the same way as other likelihood ratio criteria considered previously.

A hypothesis of the type

$$H^*: \quad \Sigma = \sigma^2 B \quad (\sigma^2 \text{ unspecified, } B \text{ specified}) \tag{4.16}$$

can always be reduced to H by transforming from $\underline{x}$ to $\underline{y} = B^{-1/2}\underline{x}$. (We have assumed B to be nonsingular.) The variance-covariance matrix of $\underline{y}$ is

$$\begin{aligned} \Sigma_y &= B^{-1/2} \, \Sigma \, B^{-1/2} \\ &= \sigma^2 I, \quad \text{if } H^* \text{ is true.} \end{aligned} \tag{4.17}$$

Thus H^* for $\underline{x}$ reduces to H, considered earlier. Since S_y, the matrix of the corrected s.s. and s.p. for observations on $\underline{y}$ are related to S by the relation

$$S_y = B^{-1/2} \, S \, B^{-1/2} \; , \tag{4.18}$$

the criterion (4.6) for the hypothesis $\Sigma_y = \sigma^2 I$ is $|S_y|^{N/2}/(\text{tr } S_y/P)^{Np/2}$, and it reduces to

$$\lambda^* = \frac{|S|^{N/2}}{|B|^{N/2}(\text{tr } B^{-1} \; S/P)^{Np/2}} \, . \tag{4.19}$$

Its moments and distribution are the same as before.

Likelihood criteria for some other hypotheses such as $\Sigma = I$ or $\underline{\mu} = \underline{\mu}_0$ and $\Sigma = I$ have been considered in the literature. Exact distributions in some particular cases have also been worked out for some of the criteria we considered earlier. Reference may be made to Anderson's book [1] for these and other details, and to other references listed at the end of this chapter.

REFERENCES

[1] Anderson, T. W. (1958). _An Introduction to Multivariate Statistical Analysis_. John Wiley and Sons, New York.

[2] Box, G. E. P. (1949). "A general distribution theory for a class of likelihood criteria", Biometrika, 36, p. 317.

[3] Davis, A. W. (1971). "Percentile approximations for a class of likelihood ratio criteria", Biometrika, 58, p. 349.

[4] Deemer, W. L., and Olkin, I. (1951). "The jacobians of certain matrix transformations useful in multivariate analysis", Biometrika, 38, p. 345.

[5] Korin, B. P. (1968). "On the distribution of a statistic used for testing a covariance matrix", Biometrika, 55, p. 171.

[6] Korin, B. P. (1969). "On testing the equality of k covariance matrices", Biometrika, 56, p. 216.

[7] Lawley, D. N. (1956). "A general method of approximating to the distribution of likelihood ratio criteria", Biometrika, 43, p. 295.

[8] Mauchly, J. W. (1940). "Significance test for sphericity of normal n-variate distribution", Ann. Math. Statist., 11, p. 204.

[9] Wilks, S. S. (1935). "On the independence of k sets of normally distributed statistical variables", Econometrica, 3, p. 309.

[10] Wilks, S. S. (1946). "Sample criteria for testing equality of means, equality of variances, and equality of covariances in a normal multivariate distribution", Ann. Math. Statist., 17, p. 257.

[11] Wilks, S. S. (1938). "The large-sample distribution of the likelihood ratio for testing composite hypotheses", Ann. Math. Statist., 9, p. 60.

CHAPTER 11

PRINCIPAL COMPONENTS

1. PRINCIPAL COMPONENTS OF A RANDOM VECTOR

Let $x_1, x_2, \ldots, x_p$ be the elements of a p-component random vector $\underline{x}$, with mean $\underline{0}$ and variance-covariance matrix $\Sigma = (\sigma_{ij})$. Σ is a real positive semidefinite matrix. Let its eigenvalues be $\delta_1 \geq \delta_2 \geq \ldots \geq \delta_p$ (≥ 0). From the theory of matrices, it is well known that there exists a p x p orthogonal matrix

$$\Gamma = [\underline{\gamma}_1 | \underline{\gamma}_2 | \cdots | \underline{\gamma}_p] \qquad (1.1)$$

such that

$$\Sigma\,\Gamma = \Gamma\,\Delta \quad \text{or} \quad \Sigma = \Gamma\,\Delta\,\Gamma' , \qquad (1.2)$$

where $\Delta = \text{diag}(\delta_1, \delta_2, \ldots, \delta_p)$. Consider the orthogonal transmation

$$\underline{v} = \Gamma'\underline{x} . \qquad (1.3)$$

Then $v_1, v_2, \ldots, v_p$, the elements of $\underline{v}$, are called the principal components of $\underline{x}$. v_1, which corresponds to the maximum eigenvalue δ_1, is called the first principal component, v_2 is the second, and so on. If the eigenvalues of Σ are repeated, and only t are distinct, the matrix Γ is not uniquely determined. It could be multiplied by a post-factor

$$A = \text{diag}\,(A_1, A_2, \ldots, A_t) \qquad (1.4)$$

where A_j is any orthogonal matrix of order m_j, m_j's being the multiplicities of the eigenvalues. The principal components of $\underline{x}$ are therefore unique except for a pre-factor A' of the type (1.4) applied to $\underline{v}$. From (1.3) and (1.2), it is easy to see that

$$\begin{aligned} V(\underline{v}) &= \Gamma'\Sigma\Gamma \\ &= \Delta\ . \end{aligned} \tag{1.5}$$

The principal components are thus uncorrelated and their variances are $\delta_1, \delta_2, \ldots, \delta_p$. An overall measure of the variability of $\underline{x}$ can be taken as tr Σ or the generalized variance $|\Sigma|$ if $\Sigma > 0$. On account of (1.2)

$$\begin{aligned} \operatorname{tr}\Sigma &= \operatorname{tr}\Gamma\Delta\Gamma' \\ &= \operatorname{tr}\Gamma'\Gamma\Delta \\ &= \operatorname{tr}\Delta = \sum_1^p \delta_i \quad (\text{as } \Gamma'\Gamma = I,\ \Gamma \text{ being orthogonal}) \end{aligned} \tag{1.6}$$

and

$$\begin{aligned} |\Sigma| &= |\Gamma\Delta\Gamma'| \\ &= |\Delta| = \prod_1^p \delta_i, \quad \text{if all } \delta\text{'s are} > 0\ . \end{aligned} \tag{1.7}$$

But tr Δ, $|\Delta|$ are the corresponding measures of variability for the principal components $\underline{v}$. This shows that the total variation remains the same, even after transforming from $\underline{x}$ to the principal components. Then what do we achieve by considering the principal components? The answer to this lies in certain optimality properties of the principal components. We discuss these in the next section.

2. OPTIMALITY PROPERTIES OF PRINCIPAL COMPONENTS

(a) Let us suppose we wish to replace $x_1, \ldots, x_p$ by as few a number of linear combinations of them as possible, without sacrificing much information about their total variation. Let us first see whether it could be done by only one suitably chosen linear combination $\underline{h}_1'\underline{x}$. Obviously $\underline{h}_1'\underline{x}$ must be normalized, i.e., $\underline{h}_1$ must satisfy $\underline{h}_1'\underline{h}_1 = 1$. Subject to this condition, we maximize its variance viz. $\underline{h}_1'\Sigma\underline{h}_1$. The standard maximization procedure using a Lagrangian multiplier yields the solution $\underline{h}_1 = \underline{\gamma}_1$ or that the desired linear combination is $v_1 = \underline{\gamma}_1'\underline{x}$, the first principal component. Its variance is, by (1.5), δ_1. The total variation, as seen earlier in (1.6), is $\sum_1^p \delta_i$. If δ_1 accounts for a sufficiently large portion of $\delta_1 + \delta_2 + \ldots + \delta_p$, or in other words, if $\delta_2 + \ldots + \delta_p$ is negligible and can be sacrificed, the total variation of $\underline{x}$ can be explained by only one variable viz. $\underline{\gamma}_1'\underline{x}$. Hence the name first principal component. However if $\delta_2 + \ldots + \delta_p$ is not negligible, we must try to find another linear combination, say $\underline{h}_2'\underline{x}$, which can account for the residual variation $\sum_2^p \delta_i$, as much as possible. Obviously $\underline{h}_2'\underline{x}$ must be uncorrelated to $\underline{\gamma}_1'\underline{x}$ and must be normalized by $\underline{h}_2'\underline{h}_2 = 1$. We, therefore, maximize $V(\underline{h}_2'\underline{x}) = \underline{h}_2'\Sigma\underline{h}_2$, subject to the conditions $\underline{h}_2'\underline{h}_2 = 1$, $\mathrm{cov}(\underline{\gamma}_1'\underline{x}, \underline{h}_2'\underline{x}) = \underline{\gamma}_1'\Sigma\underline{h}_2 = 0$. This will yield $\underline{h}_2 = \underline{\gamma}_2$ and the desired linear combination is $v_2 = \underline{\gamma}_2'\underline{x}$, the second principal component of $\underline{x}$. These two principal components thus account for a portion $\delta_1 + \delta_2$ of the total variation $\sum_1^p \delta_i$; the unaccounted portion of the variation is only $\sum_3^p \delta_i$. If it is negligible, we have succeeded in replacing the p-variables $\underline{x}$ by only two viz. v_1 and v_2, the first two principal components. However if $\sum_3^p \delta_i$ is not negligible, the process must be continued further and it will yield, in succession, the remaining principal components $v_3, v_4, \ldots, v_p$ in that order. This justifies the name principal components. In practice, oftentimes, only two or three principal components are adequate and a great reduction in the dimensionality is achieved.

Equation (1.2) can alternatively be written as

$$\Sigma = \delta_1 \underline{\gamma}_1 \underline{\gamma}_1' + \delta_2 \underline{\gamma}_2 \underline{\gamma}_2' + \ldots + \delta_p \underline{\gamma}_p \underline{\gamma}_p' \, . \tag{2.1}$$

This is known as the spectral decomposition of the matrix Σ (see Bartlett [6] or Good [14]). This clearly shows that the contribution of the first principal component to Σ is $\delta_1 \underline{\gamma}_1 \underline{\gamma}_1'$; when that is removed, the "residual" matrix of variances and covariances is

$$\Sigma - \delta_1 \underline{\gamma}_1 \underline{\gamma}_1' = \sum_2^p \delta_i \underline{\gamma}_i \underline{\gamma}_i' \, . \tag{2.2}$$

The second principal component contributes $\delta_2 \underline{\gamma}_2 \underline{\gamma}_2'$, and so on.

(b) Suppose we wish to replace $x_1, x_2, \ldots, x_p$ by some q $(q < p)$ suitably chosen linear combinations, the objective being to account for the total variation of $\underline{x}$ viz. tr Σ, as much as possible. Obviously the linear combinations, say, $\underline{h}_i'\underline{x}$ $(i = 1, 2, \ldots, q)$ can be chosen, without loss of generality, to satisfy

$$\underline{h}_i'\underline{h}_i = 1; \quad \underline{h}_i'\underline{h}_j = 0 \ (i \neq j), \quad i, j = 1, 2, \ldots, q \, . \tag{2.3}$$

The variance of $\underline{h}_i'\underline{x}$ is $\underline{h}_i' \Sigma \underline{h}_i$ and hence the q linear combinations account for a portion, $\sum_{i=1}^{q} \underline{h}_i' \Sigma \underline{h}_i$, of the total variation of $\underline{x}$. We now prove the following result:

$$\max_{\underline{h}_1, \ldots, \underline{h}_q} \sum_{1}^{q} \underline{h}_i' \Sigma \underline{h}_i = \delta_1 + \delta_2 + \ldots + \delta_q \tag{2.4}$$

and the maximum is attained when $\underline{h}_i = \underline{\gamma}_i$ $(i = 1, 2, \ldots, q)$. To prove this, we observe that there cannot be more than p linearly independent p-component orthogonal vectors, and so $\underline{h}_1, \ldots, \underline{h}_q$ must be linear combinations of $\underline{\gamma}_1, \ldots, \underline{\gamma}_p$. Hence we can write

$$\underline{h}_i = C_{i1}\underline{\gamma}_1 + \ldots + C_{ip}\underline{\gamma}_p \quad (i = 1, 2, \ldots, q). \tag{2.5}$$

It therefore follows, from the orthogonality of the $\underline{\gamma}$'s and (2.1), that

$$\sum_{1}^{q} \underline{h}_i' \Sigma \underline{h}_i = \sum_{j=1}^{q} \delta_j (C_{1j}^2 + C_{2j}^2 + \ldots + C_{qj}^2). \tag{2.6}$$

Equation (2.5) can be expressed as

$$[\underline{h}_1 | \underline{h}_2 | \ldots | \underline{h}_q] = [\underline{\gamma}_1 | \underline{\gamma}_2 | \ldots | \underline{\gamma}_p] C' \tag{2.7}$$

where C is the q x p matrix of the elements C_{ij} $(i = 1, 2, \ldots, q;$ $j = 1, 2, \ldots, p)$. From the orthogonality of the $\underline{h}$'s and $\underline{\gamma}$'s, it follows that $CC' = I_q$, i.e., C is a semiorthogonal matrix, as $q < p$. Hence $C_{1j}^2 + C_{2j}^2 + \ldots + C_{qj}^2$ occurring in (2.6) is less than or equal to one, for each j. Also observe that $\delta_1 \geq \delta_2 \geq \ldots \geq \delta_p$. The left-hand side of (2.6) is, therefore, clearly maximized if we choose $C_{11} = C_{22} = \ldots = C_{qq} = 1$ and the rest of the C_{ij}'s to be zero. This means $\underline{h}_1 = \underline{\gamma}_1$, $\underline{h}_2 = \underline{\gamma}_2$, $\ldots$, $\underline{h}_q = \underline{\gamma}_q$, and (2.4) follows.

In (a), we tried to replace $x_1, \ldots, x_p$ by one linear combination only, and if that was not satisfactory we tried to find one more linear combination and so on. Here (in the beginning itself) we tried to find q linear combinations to replace $\underline{x}$. As is intuitively obvious, the solutions turn out to be the same.

(c) Instead of explaining the total variation tr Σ, let us try to approximate Σ, the variance-covariance matrix itself, by another matrix of order p, say B, of a smaller rank, say q ($<p$). The goodness of approximation of the closeness of $B = [b_{ij}]$ to Σ can be measured by the norm of $\Sigma - B$, which is defined as

$$\text{Norm } (\Sigma - B) = \left\{ \sum_{i=1}^{p} \sum_{j=1}^{p} (\sigma_{ij} - b_{ij})^2 \right\}^{1/2} . \tag{2.8}$$

Among all possible $p \times p$ matrices of rank q, we desire to find the one for which (2.8) is minimum. To do this, we see that (as Γ is orthogonal)

$$\begin{aligned} \{\text{Norm } (\Sigma - B)\}^2 &= \text{tr}(\Sigma - B)(\Sigma - B)' \\ &= \text{tr } \Gamma\, \Gamma'(\Sigma - B)\, \Gamma\, \Gamma'(\Sigma - B)' \\ &= \text{tr}(\Gamma' \Sigma \Gamma - \Gamma' B \Gamma)(\Gamma' \Sigma \Gamma - \Gamma' B \Gamma)' \\ &= \text{tr}(\Delta - G)(\Delta - G)' , \end{aligned} \tag{2.9}$$

where

$$G = \Gamma' B \Gamma . \tag{2.10}$$

We used (1.2) and (2.2.5) in obtaining (2.9). Since Γ is an orthogonal matrix, G is of the same rank as B, viz. q. If the elements of G are g_{ij}, it follows from (2.9) that

$$\{\text{Norm } (\Sigma - B)\}^2 = \sum_1^p (\delta_i - g_{ii})^2 + \sum_{i \neq j}\sum g_{ij}^2 . \qquad (2.11)$$

To minimize this, we must obviously choose $g_{ij} = 0$ for all i, j such that $i \neq j$, i.e., G is a diagonal matrix. But G is of rank q. Therefore, p-q elements on the diagonal of G must be zero and the remaining elements must be different from zero. In order to satisfy this and minimize the quantity $\sum_1^p (\delta_i - g_{ii})^2$, it is obvious that the choice must be $g_{ii} = \delta_i$ ($i = 1, 2, \ldots, q$). From (2.10), therefore, it follows that the required matrix B is

$$\delta_1 \underline{\gamma}_1 \underline{\gamma}_1' + \delta_2 \underline{\gamma}_2 \underline{\gamma}_2' + \cdots + \delta_q \underline{\gamma}_q \underline{\gamma}_q' . \qquad (2.12)$$

This result shows that, if Σ is to be approximated by a matrix of order p and rank q, choose the first q terms in the spectral decomposition of Σ. These correspond to the first q principal components.

(d) Suppose we wish to approximate the p-component vector $\underline{x}$ by a linear form $B\underline{y} + C$, of some random $q \times 1$ vector $\underline{y}$ ($q < p$). The error of approximation may be measured by the eigenvalues of the matrix

$$M = E(\underline{x} - B\underline{y} - C)(\underline{x} - B\underline{y} - C)' . \qquad (2.13)$$

Therefore, the optimum choice of B, $\underline{y}$, and C is obtained by minimizing these eigenvalues. Okamoto and Kanazawa [34] have shown that the solution of this problem is

$$B = [\underline{\gamma}_1 | \underline{\gamma}_2 | \cdots | \underline{\gamma}_q], \quad \underline{y}' = [v_1, v_2, \ldots, v_q], \quad C = 0 . \qquad (2.14)$$

This again shows that the first q principal components provide the necessary replacement of $\underline{x}$, if $\underline{x}$ is to be replaced by q new variables.

If one uses tr M or norm M as the criterion of goodness of approximation and minimizes it, the solution is still (2.14), as proved by Darroch [9] or Rao [36].

We now state various other optimality properties of the principal components. The interested reader should refer to Okamoto [35] or Rao [36] for further details and references.

(e) $$\text{Sup} \frac{\underline{h}'\Sigma\underline{h}}{\underline{h}'\underline{h}} = \delta_{q+1}, \text{ subject to } \underline{\gamma}_i'\underline{h} = 0 \quad (i = 1, 2, \ldots, q)$$

and is attained when $\underline{h} = \underline{\gamma}_{q+1}$. Similarly,

$$\text{Inf} \frac{\underline{h}'\Sigma\underline{h}}{\underline{h}'\underline{h}} = \delta_{p-q}, \text{ subject to } \underline{\gamma}_i'\underline{h} = 0 \quad (i = 1, 2, \ldots, q)$$

and is attained when $\underline{h} = \underline{\gamma}_{p-q}$.

(f) $$\inf_{L} \sup_{L'\underline{h}=0} \frac{\underline{h}'\Sigma\underline{h}}{\underline{h}'\underline{h}} = \delta_{q+1},$$

where the infimum is with respect to a p x q matrix L, while the supremum is with respect to a p x 1 vector $\underline{h}$, satisfying $L'\underline{h} = 0$. The infimum is attained when

$$L = [\underline{\gamma}_1 | \underline{\gamma}_2 | \cdots | \underline{\gamma}_q] .$$

This is known as the Courant-Fischer min-max theorem. It holds even if L is not p x q but any p x α matrix of rank = q, at most.

(g) If we maximize the generalized variance, viz. $|V(T'\underline{x})|$, where T is any p x q matrix ($q \leq p$), subject to the condition that all diagonal elements of T'T are unity, we shall find the solution to be

$$T = [\underline{\gamma}_1 | \underline{\gamma}_2 | \cdots | \underline{\gamma}_q] .$$

This solution is not unique and any other matrix TQ, where Q is any $q \times q$ orthogonal matrix, will also be equally good. This result shows that, even if generalized variance is taken as the overall measure of variation of a set of variables, the principal components still provide the necessary optimum replacement of $\underline{x}$.

(h) Suppose we wish to predict x_i, using some $q(<p)$ linear functions $\underline{h}_1'\underline{x}$, $\underline{h}_2'\underline{x}$, ..., $\underline{h}_q'\underline{x}$. The best predictor will be (see Section 3, Chapter 1) the regression of x_i on $\underline{h}_1'\underline{x}$, ..., $\underline{h}_q'\underline{x}$ and the performance of this predictor will be judged by the residual variance. This residual variance can be easily seen to be (there is no loss of generality in assuming that the $\underline{h}_i'\underline{x}$ are all uncorrelated and have unit variance; if not, we can always make a transformation to achieve this)

$$\sigma_{ii} - \sum_{j=1}^{q} \underline{h}_j' \underline{\sigma}_i \underline{\sigma}_i' \underline{h}_j ,$$

where σ_{ii} is $V(x_i)$ and $\underline{\sigma}_i$ is the ith column of Σ. If we now find this residual variance, for the prediction of each x_i $(i = 1, 2, ..., p)$, the overall performances of the predictors $\underline{h}_j'\underline{x}$ $(j = 1, 2, ..., q)$ can be judged by

$$\sum_{i=1}^{p} \left\{\sigma_{ii} - \sum_{j=1}^{q} \underline{h}_j'\underline{\sigma}_i\underline{\sigma}_i'\underline{h}_j\right\} = \operatorname{tr} \Sigma - \sum_{j=1}^{q} \underline{h}_j' \Sigma^2 \underline{h}_j .$$

It is therefore natural to choose the vectors $\underline{h}_1$, ..., $\underline{h}_q$ in such a way that this expression is minimum (subject, of course, to the conditions that $\underline{h}_j'\underline{x}$ are uncorrelated and have unit variance). It can be shown by a little algebra that this again leads to the solution

$$\underline{h}_1 = \underline{\gamma}_1, \quad \underline{h}_2 = \underline{\gamma}_2, \quad \ldots, \quad \underline{h}_q = \underline{\gamma}_q .$$

In other words, the first q principal components $\underline{\gamma}_j'\underline{x}$ $(j = 1, 2, \ldots, q)$ are the best predictors of $\underline{x}_i$ $(i = 1, 2, \ldots, p)$ in the least square sense.

3. SINGULAR NORMAL DISTRIBUTION

Suppose the characteristic function of the distribution of the p-component vector $\underline{x}$ is

$$E\left(e^{\sqrt{-1}\,\underline{t}'\underline{x}}\right) = \psi(\underline{t}) = \exp\left\{-\frac{1}{2}\,\underline{t}'\Sigma\underline{t}\right\} \tag{3.1}$$

where $\underline{t}$ is the column vector of the arguments $t_1, t_2, \ldots, t_p$. If Σ is positive definite, i.e., if $\delta_1, \delta_2, \ldots, \delta_p$ are all > 0, by the inversion theorem, it can be readily seen that the p.d.f. of $\underline{x}$ is $N_p(\underline{x}|\underline{0}|\Sigma)$. If, however, Σ is not of full rank, but only of rank $q < p$, the distribution of $\underline{x}$ does not admit a p.d.f. $\psi(\underline{t})$ is still its characteristic function and determines the distribution. The distribution of $\underline{x}$ in such a situation is said to be a singular, p-variate normal distribution with mean vector $\underline{0}$ and variance-covariance matrix Σ. In the case of a nonsingular distribution, the probability content of any region in the p-dimensional space of $\underline{x}$ can be found by integrating the p.d.f. of $\underline{x}$ over that region, but this is not possible when the distribution is singular, as the p.d.f. does not exist. The question then arises, how do we find the probability? The answer to this lies again in the principal components of $\underline{x}$. From $\underline{x}$, transform to the principal components $\underline{v}$ by the transformation (1.3). To obtain the characteristic function of $\underline{v}$, we replace $\underline{t}$ by $\Gamma\,\underline{\theta}$ (where $\underline{\theta}$ is the column vector of θ_1, θ_2, ..., θ_p, the arguments in the characteristic function of $\underline{v}$), on both sides of the identity (3.1). This gives

$$E\left(e^{\sqrt{-1}\ \underline{\theta}'\Gamma'\underline{x}}\right) = e^{-\frac{1}{2}\underline{\theta}'\Gamma'\Sigma\Gamma\underline{\theta}} . \tag{3.2}$$

On using (1.2), (1.3), and the fact that

$$\delta_{q+1} = \cdots = \delta_p = 0$$

when Σ is of rank $q < p$, equation (3.2) becomes

$$\begin{aligned} E\left(e^{\sqrt{-1}\ \underline{\theta}'\underline{v}}\right) &= e^{-\frac{1}{2}\underline{\theta}'\Delta\,\underline{\theta}} \\ &= e^{-\frac{1}{2}\sum_{1}^{p}\delta_i\theta_i^2} \\ &= e^{-\frac{1}{2}\delta_1\theta_1^2} \,.\, e^{-\frac{1}{2}\delta_2\theta_2^2} \cdots \\ &\quad .\, e^{-\frac{1}{2}\delta_q\theta_q^2} .\,1\,.\,1\,.\,1\ldots 1 . \end{aligned} \tag{3.3}$$

This shows that v_1, v_2, ..., v_p are independently distributed, the characteristic function of the distribution of v_j being $\exp\left(-\frac{1}{2}\delta_j\theta_j^2\right)$ if $j = 1, 2, \ldots, q$ and 1 if $j > q$. In other words, v_1, v_2, ..., v_q are normal independent variables with zero means, variances $\delta_1, \delta_2, \ldots, \delta_q$, respectively, while $v_{q+1}, v_{q+2}, \ldots, v_p$ are identically equal to zero, with probability one (as their characteristic function is unity). Whenever we wish to find the probability content of any region in the p-dimensional space of $\underline{x}$, we find the corresponding region in the space of the v's and then use the distribution of $v_1, v_2, \ldots, v_q$ and the fact $v_{q+1} = \ldots = v_p = 0$ (with probability one), to find the required probability. We shall illustrate this with a simple example. Let Σ be given by

$$\Sigma = (1 - \rho)I + \rho\, E_{pp} \tag{3.4}$$

where $\rho = -1/(p-1)$. Since the eigenvalues of Σ are $(1 - \rho)$ repeated $(p - 1)$ times and $1 + (p-1)\rho$ once, it is obvious that Σ is singular and of rank $p - 1$ when $\rho = -1/(p - 1)$. The eigenvector corresponding to $1 + (p-1)\rho = 0$ is $E_{p1}/\sqrt{p}$. Any set of $(p - 1)$ mutually orthogonal and unit vectors that are orthogonal to E_{p1} can be taken as the eigenvectors corresponding to the eigenvalue $1 - \rho$, and so the principal components of $\underline{x}$ can be taken as

$$\begin{aligned} v_1 &= (x_1 - x_2)/\sqrt{(1)(2)} \\ v_2 &= (x_1 + x_2 - 2x_3)/\sqrt{(2)(3)} \\ &\vdots \\ v_{p-1} &= (x_1 + \ldots + x_{p-1} - (p-1)x_p)/\sqrt{(p-1)(p)} \end{aligned}$$

and

$$v_p = (x_1 + \ldots + x_p)/\sqrt{p}\,. \tag{3.5}$$

v_p corresponds to the zero eigenvalue. Hence $v_1, v_2, \ldots, v_{p-1}$ are normal independent variables with zero means and variance $(1-\rho) = (p - 2)/(p - 1)$, while v_p is identically zero with probability one and independent of $v_1, \ldots, v_{p-1}$. If we therefore wish to find, say,

$$\text{Prob}\ (M = 3x_1 + x_2 - x_3 + x_4 + \ldots + x_p > 0),$$

we must first express the linear function M in terms of the principal components $v_1, \ldots, v_p$. It is easy to see that

$$M = \sqrt{2}\, v_1 + \sqrt{6}\, v_2 + \sqrt{p}\, v_p$$

and hence Prob $(M > 0) =$ Prob $(\sqrt{2}\, v_1 + \sqrt{6}\, v_2 > 0)$, as $v_p = 0$ identically. The probability of $\sqrt{2}\, v_1 + \sqrt{6}\, v_2 > 0$ can now be found from the fact that v_1, v_2 are independent normal variables with zero means and variances $(p-2)/(p-1)$.

In general, any function of $\underline{x}$ can be expressed as a function of $\underline{v}$, by the reciprocal transformation $\underline{x} = \Gamma\underline{v}$ (obtainable from (1.3), as Γ is orthogonal) and the probability then can be found by first putting $v_{q+1} = \ldots = v_p = 0$ (when rank $\Sigma = q$) and then using the distribution of $v_1, v_2, \ldots, v_q$.

Certain probability integrals associated with the singular normal distribution having (3.4) as its variance-covariance matrix have been considered by Bland and Owen [7]. For additional information about singular normal distributions, reference should also be made to Miller [33] and Rao [37].

4. SAMPLE PRINCIPAL COMPONENTS

Let us now assume that $\underline{x}$ has the $N_p(\underline{x}|\underline{\mu}|\Sigma)$ distribution (nonsingular) and that a random sample of size $N = n + 1 > p$ is available from this distribution. If X denotes the $p \times N$ matrix of sample observations, we have seen in Chapter 3 that the maximum likelihood estimate (m.l.e.) of Σ is given by the matrix $A = (1/n)S$, where $S = X(I - N^{-1} E_{NN})X'$ is the matrix of the corrected s.s. and s.p. of the sample observations. The matrix S is positive definite with probability one. Let its eigenvectors be

$$f_1 > f_2 > \dots > f_p \ . \tag{4.1}$$

Then there exists an orthogonal matrix

$$C = [\underline{c}_1 | \underline{c}_2 | \cdots | \underline{c}_p] \tag{4.2}$$

such that

$$S = C F C' \ , \tag{4.3}$$

where

$$F = \operatorname{diag}(f_1, f_2, \dots, f_p) \ . \tag{4.4}$$

The vectors $\underline{c}_1, \underline{c}_2, \dots, \underline{c}_p$ are the eigenvectors of S, corresponding to the eigenvalues $f_1, \dots, f_p$. The eigenvectors of A are the same, but its eigenvalues are $d_i = f_i/n$ $(i = 1, 2, \dots, p)$. The sample principal components are, therefore, defined as

$$u_i = \underline{c}_i' x \quad (i = 1, 2, \dots, p) \tag{4.5}$$

or, which is the same as,

$$\underline{u} = C'\underline{x}, \quad \underline{x} = C\underline{u} \ , \tag{4.6}$$

$\underline{u}$ being the column vector of $u_1, \dots, u_p$. From (1.3) and (4.6) the relation between the sample principal components and the population principal components is given by

$$\underline{v} = \Gamma' C \underline{u}$$
$$= L \underline{u} \tag{4.7}$$

where $L = [\ell_{ij}] = \Gamma'C$ is an orthogonal matrix, as both Γ and C are orthogonal. The matrix of sample observations on the true principal components $\underline{v}$ is, by (1.3),

$$V = \Gamma'X \tag{4.8}$$

and the corrected s.s. and s.p. of these observations is

$$V(I - N^{-1}E_{NN})V' = \Gamma' S \Gamma$$
$$= T, \text{ say.} \tag{4.9}$$

Observe that (from (4.3)

$$T = \Gamma' S \Gamma = \Gamma' C F C' \Gamma = L F L' \tag{4.10}$$

where $LL' = L'L = I$, L being orthogonal. The distribution of S is $W_p(S|n|\Sigma)ds$ and hence the logarithm of the likelihood, in order to obtain the m.l.e.'s of Γ and Δ, is (apart from a constant)

$$- \frac{n}{2} \log_e |\Sigma| - \frac{1}{2} \operatorname{tr} \Sigma^{-1} S$$
$$= - \frac{n}{2} \log_e |\Gamma \Delta \Gamma'| - \frac{1}{2} \operatorname{tr}(\Gamma \Delta \Gamma')^{-1}(C F C')$$
$$= - \frac{n}{2} \log |\Delta| - \frac{1}{2} \operatorname{tr} \Delta^{-1} L F L' . \tag{4.11}$$

From this, Anderson [2] has shown that $(1/n)F$ is the m.l.e. of Δ and C is the m.l.e. of Γ, provided the eigenvalues of Σ, viz., $\delta_1, \delta_2, \ldots, \delta_p$, are all distinct. In other words, the sample principal components are the m.l.e.'s of the corresponding true principal components. When, however, δ_i's are not all distinct, we assume

$$\delta_1 = \delta_2 = \cdots = \delta_{q_1} = \lambda_1 ,$$
$$\delta_{q_1+1} = \delta_{q_1+2} = \cdots = \delta_{q_1+q_2} = \lambda_2 ,$$
$$\cdots$$
$$\delta_{p-q_r+1} = \cdots = \delta_p = \lambda_r , \qquad (4.12)$$

where $\lambda_1 > \lambda_2 > \cdots > \lambda_r > 0$. $\lambda_1, \ldots, \lambda_r$ are thus distinct roots of Σ and their multiplicities are $q_1, q_2, \ldots, q_r$. Anderson [2] has proved in this case that the m.l.e. of λ_k is the average of

$$\frac{1}{n} f_{a+1}; \ \frac{1}{n} f_{a+2}, \ldots, \frac{1}{n} f_{a+q_k} ,$$

where $a = q_1 + \cdots + q_{k-1}$ $(k = 1, 2, \ldots, r)$. If Γ and C are both partitioned as

$$\Gamma = \underset{q_1 \quad q_2 \qquad\quad q_r}{[\Gamma_1 | \Gamma_2 | \cdots | \Gamma_r]} p$$

$$C = \underset{q_1 \quad q_2 \qquad\quad q_r}{[C_1 | C_2 | \cdots | C_r]} p ,$$

it is obvious that all the q_k columns of Γ_k are eigenvectors of Σ, corresponding to the repeated eigenvalue λ_k. However, Γ_k is not unique. It can be post-multiplied by any arbitrary $q_k \times q_k$ orthogonal matrix. Therefore, no unique m.l.e. for Γ_k exists. We can take C_k, post-multiplied by any arbitrary $q_k \times q_k$ orthogonal matrix (such that the diagonal elements of the product are all positive), as an m.l.e.

5. DISTRIBUTIONS ASSOCIATED WITH THE SAMPLE PRINCIPAL COMPONENTS (NULL CASE)

In this section we shall derive the distribution of the eigenvalues and the eigenvectors of the Wishart matrix S. We shall, however, consider only the null case, i.e., we assume $\Sigma = I$, so that all the δ's are unity and the matrix Γ corresponding to the true principal components is also I. From (4.10), therefore, we obtain

$$S = L F L' \,, \tag{5.1}$$

$$LL' = I \,. \tag{5.2}$$

The columns of L, therefore, are the eigenvectors of S, and the elements $f_1 > f_2 > \dots > f_p$ (>0) of the diagonal matrix F are the eigenvalues of S. We can represent the orthogonal matrix L in terms of the rotational angles θ_{ij} $(i = 1, 2, \dots, p-1;\ j \geq i)$, as in Section 4 of Chapter 7 and then transform from S to F and the θ_{ij}'s. The Jacobian of the transformation is, by Theorem 2 of Section 4, Chapter 7,

$$\prod_{i=1}^{p} \prod_{j>i} (f_i - f_j) \prod_{i=1}^{p-2} \prod_{j=i}^{p-2} \sin^{p-j-1} \theta_{ij} \,. \tag{5.3}$$

Also, from (5.1) and (5.2),

$$|S| = |F| \quad \text{and} \quad \operatorname{tr} S = \operatorname{tr} F \,. \tag{5.4}$$

Hence, the distribution of S, viz., $W_p(S|n|I)ds$, transforms into

$$K(p, n)\, |F|^{(n-p-1)/2} \exp\left(-\frac{1}{2} \operatorname{tr} F\right) \prod_{i<j} (f_i - f_j)\, dF$$

$$\prod_{i,j} \sin^{p-j-1} \theta_{ij} \prod_{i,j} d\theta_{ij} \,. \tag{5.5}$$

The constant K(p, n) is defined by (3.2.20). The range of the θ_{ij}'s is given by (7.4.4). Equation (5.5) shows that F and L are independently distributed. By splitting the constant K(p, n) suitably, with the help of (7.4.6), we find the distribution of F to be

$$\frac{\pi^{p/2}\,|F|^{(n-p-1)/2}\exp\left(-\frac{1}{2}\operatorname{tr}F\right)\prod_{i<j}(f_i - f_j)\,dF}{2^{np/2}\prod_{i=1}^{p}\left\{\Gamma\left(\frac{n+1-i}{2}\right)\Gamma\left(\frac{p+1-i}{2}\right)\right\}} \tag{5.6}$$

and that of L, represented in terms of the θ_{ij}'s, to be

$$\frac{\pi^{p(p+1)/4}}{\prod_{i=1}^{p}\Gamma\left(\frac{p+1-i}{2}\right)}\prod_{i,j}\sin^{p-j-1}\theta_{ij}\prod_{i,j}d\theta_{ij}\,. \tag{5.7}$$

The distribution of L is called the Haar invariant distribution (see Anderson [1]). Tumura [45] has shown that every vector of the orthogonal matrix L is uniformly distributed on the sphere in the p-dimensional space, but their distributions are not jointly independent. If we take an arbitrary vector, it can rotate freely in the p-dimensional space and thus can be represented by p-1 angles $(\theta_{11}, \ldots, \theta_{1,\,p-1})$. Then, the second vector can rotate freely in the (p-1)-dimensional space orthogonal to the first vector, and so it can be represented by p-2 angles $(\theta_{22}, \ldots, \theta_{2}, p-1)$, and so on. This is what is implied by the distribution (5.7).

The non-null case, i.e., the case when $\delta_1, \delta_2, \ldots, \delta_p$ are not all unity, has been dealt with by James [19]. But it is beyond the scope of this book. However, in the next section we shall briefly describe certain asymptotic results about the distributions of the eigenvalues and vectors of S, in the non-null case. These are due to Anderson [2] and Lawley [27, 28].

6. ASYMPTOTIC DISTRIBUTIONS

From the Wishart distribution $W_p(S|n|\Sigma)dS$, we know (see 3.6.6) that $E(S) = n\,\Sigma$ and

$$\mathrm{cov}(s_{ij}, s_{gh}) = n(\sigma_{ih}\sigma_{jg} + \sigma_{ig}\sigma_{jh}) , \tag{6.1}$$

where s_{ij}, σ_{ij} are, respectively, the elements of S and Σ. By the central limit theorem, therefore, $n^{-1/2}(S - n\,\Sigma)$ is asymptotically normally distributed, with zero means and covariances given by the right-hand side of (6.1), divided by n. From this, and from (4.10), we find that

$$n^{-1/2}(\Gamma' S \Gamma - n\,\Gamma' \Sigma \Gamma) = n^{1/2}\,(T - \Delta) = U, \text{ say} \tag{6.2}$$

has an asymptotic normal distribution with zero means and covariances given by (replace elements of Σ by those of Δ in (6.1))

$$\mathrm{cov}(U_{ij}, U_{gh}) = \delta_i\delta_j(\delta_{ih}\delta_{jg} + \delta_{ig}\delta_{jh}) , \tag{6.3}$$

where U_{ij} are elements of U, δ_{ij} is the Kronecker delta (which is one if i = j and zero otherwise), and δ_i are the elements of the diagonal matrix Δ. From (6.3), we find

$$V(U_{ii}) = 2\,\delta_i^2 \quad (i = 1, 2, \ldots, p) \tag{6.4}$$

$$V(U_{ij}) = \delta_i\delta_j \quad (i, j = 1, 2, \ldots, p;\ i \neq j) \tag{6.5}$$

$$\mathrm{CoV}(U_{ig}, U_{gh}) \neq 0, \quad \text{only if } i = h,\ j = g \text{ or } i = g,\ j = h . \tag{6.6}$$

From these results and a theorem of Rubin [39] on limiting distributions, Anderson [2] has proved the following result about the column vectors of the matrix C, which yields the sample principal components. Consider $\underline{C}_i$, the ith column of C, and $\underline{\gamma}_i$, the corresponding column of Γ. If δ_i is of multiplicity one, $n^{1/2}(\underline{C}_i - \underline{\gamma}_i)$ has a limiting p-variate normal distribution with zero means. The asymptotic covariance between the gth element and the hth element of $n^{1/2}(\underline{C}_i - \underline{\gamma}_i)$ is

$$\sum_{j\neq i} [\delta_j\delta_i/(\delta_j - \delta_i)^2]\gamma_{gj}\gamma_{hj} , \tag{6.7}$$

where γ_{gj} is the gth element of $\underline{\gamma}_j$ (j = 1, 2, ..., p; g = 1, 2, ..., p). Similarly, if δ_i and δ_j are of multiplicity one, the asymptotic covariance between the gth element of $\underline{C}_i$ and the hth element of $\underline{C}_j$ is

$$- [\delta_i\delta_j/(\delta_i - \delta_j)^2] \gamma_{gj}\gamma_{hi} , \quad i \neq j , \tag{6.8}$$

suppose the last q_r roots of Σ are all equal, the common value being λ_r. Then its m.l.e., as seen earlier, is

$$\frac{1}{n} \bar{f} = \sum_{i=p-q_r+1}^{p} f_i/n\, q_r . \tag{6.9}$$

$n^{1/2}(\frac{1}{n} \bar{f} - \lambda_r)$ has an asymptotic normal distribution with mean 0 and variance $2 \lambda_r^2/q_r$. (6.10)

By expanding the matrix S about Σ, Lawley [27] has obtained the following results, for large n (assuming the eigenvalues of Σ to all be distinct)

$$E\left(\frac{1}{n} f_r\right) = \delta_r + \frac{\delta_r}{n} \sum_{i \neq r} \frac{\delta_i}{(\delta_r - \delta_i)} + O\left(\frac{1}{n^2}\right), \qquad (6.11)$$

$$V\left(\frac{1}{n} f_r\right) = \frac{2\delta_r^2}{n} \left\{1 - \frac{1}{n} \sum_{i \neq r} \left(\frac{\delta_i}{\delta_r - \delta_i}\right)^2\right\} + O\left(\frac{1}{n^3}\right) . \qquad (6.12)$$

$$\mathrm{Cov}\left(\frac{1}{n} f_r, \frac{1}{n} f_s\right) = \frac{2}{n^2} \left(\frac{\delta_r \delta_s}{\delta_r - \delta_s}\right)^2 + O\left(\frac{1}{n^3}\right) . \qquad (6.13)$$

with the help of these asymptotic results, we shall now consider, in the next section, statistical tests for some hypotheses about the principal components and their variances.

7. TESTS OF CERTAIN HYPOTHESES ABOUT PRINCIPAL COMPONENTS

(1) In Chapter 10, Section 4, we considered the sphericity test. It was derived to test the hypothesis of equality of all the eigenvalues of Σ. If all the eigenvalues of Σ are the same, all the principal components have the same variance. The variation in any direction in the p-dimensional space is then the same: it is isotropic. Any orthogonal transformation of $\underline{x}$ will retain this property. The test criterion based on the likelihood ratio was seen to be the ratio of the geometric and arithmetic means of the eigenvalues of the m.l.e. of Σ. If the hypothesis is accepted, all the variables are equally important as they contribute equally to the total variation, and no reduction in the dimensionality is possible as no variables, nor any linear combination thereof, can be ignored. If, however, the hypothesis is rejected, it is still possible that at least some of the eigenvalues of Σ are equal. So we consider the hypothesis

H_k: The last k eigenvalues of Σ are equal. The likelihood ratio criterion for this hypothesis can be found out in the same manner as for the sphericity test, and it yields the criterion

$$M_k = \left\{ \prod_{j=p-k+1}^{p} f_j \Big/ \left(\frac{1}{k} \sum_{j=p-k+1}^{p} f_j \right)^k \right\}, \tag{7.1}$$

which is again the kth power of the ratio of the geometric and arithmetic means of the last k eigenvalues of the m.l.e. of Σ. By the general theorem about likelihood ratio criteria, $-N \log_e M_k$ is asymptotically a χ^2 with $-1 + k(k+1)/2$ d.f., if H_k is true. To see how these d.f. are calculated, we consider the spectral decomposition of Σ, given by (2.1). Under H_k, it reduces to

$$\Sigma = \delta_1 \underline{\gamma}_1 \underline{\gamma}_1' + \cdots + \delta_{p-k} \underline{\gamma}_{p-k} \underline{\gamma}_{p-k}' + \delta \sum_{j=p-k+1}^{p} \underline{\gamma}_j \underline{\gamma}_j' \tag{7.2}$$

where δ is the common value of $\delta_{p-k+1}, \ldots, \delta_p$. But Γ is an orthogonal matrix, and so

$$\sum_{j=p-k+1}^{p} \underline{\gamma}_j \underline{\gamma}_j' = I - \sum_{j=1}^{p-k} \underline{\gamma}_j \underline{\gamma}_j' . \tag{7.3}$$

Thus Σ is completely determined by the p-k+1 eigenvalues $\delta_1, \ldots \delta_{p-k}, \delta$ and the p-k vectors $\underline{\gamma}_1, \underline{\gamma}_2, \ldots, \underline{\gamma}_{p-k}$, which however satisfy the orthogonality conditions, $\underline{\gamma}_i'\underline{\gamma}_j = 0$ if $i \neq j$, and 1, otherwise. The number of independent elements, therefore, is (p-k+1), which is the number of eigenvalues plus p(p-k), the number of elements of the p-component vectors $\underline{\gamma}_1, \ldots, \underline{\gamma}_{p-k}$ minus (p-k)(p-k+1)/2, the number of orthogonality conditions on the γ's.

This becomes

$$(p-k+1) + (p-k)(p+k-1)/2 . \qquad (7.4)$$

If, however, none of the eigenvalues of Σ are equal, Σ is determined by all the δ's and all the $\underline{\gamma}$'s, subject of course to the orthogonality conditions. The number of parameters in this case is, therefore (obtained by putting $k = 1$),

$$p + (p-1)p/2 . \qquad (7.5)$$

The number of degrees of freedom of the χ^2 test is the difference between (7.5) and (7.4) and is $-1 + k(k+1)/2$, as stated earlier.

Bartlett [3, 4, 5] and later Lawley [27] have investigated this test in considerable detail, with the help of some elaborate matrix expansions. Lawley has shown that, under the hypothesis H_k,

$$M_k' = -\left\{n - (p-k) - \frac{1}{6}\left(2k + 1 + \frac{2}{k}\right) + \delta^2 \sum_{j=1}^{p-k} \frac{1}{(\delta_j - \delta)^2}\right\} \log_e M_k \cdots \qquad (7.6)$$

is a χ^2 with $(k^2 + k-2)/2$ d.f. This is a better approximation than the χ^2 test based on $-N \log_e M_k$. Reference should also be made to James's [20, 21] work in this respect. He uses a different inferential standpoint using different mathematical techniques, providing thus more information on the accuracy of the approximation. However he confirms Lawley's correction factor to $\log_e M_k$. James also uses an asymptotic non-null distribution of the eigenvalues and vectors but his distribution is presumably not "as asymptotic", as Anderson's. In other words, it is probably better even if n is not as large as is required for the validity of Anderson's results. However δ_j's and δ in (7.6) are unknown, and one should therefore employ their estimates. δ_j $(j = 1, 2, \ldots, p-k)$ will be estimated by f_j/n while δ will be estimated by $\bar{f}/n$, the average of the last k eigenvalues of S. In practice, it is easier to find the first p-k eigenvalues f_1, $f_2, \ldots, f_{p-k}$ of S and then M_k can be seen to be

$$\frac{|S|/(f_1 f_2 \cdots f_k)}{\{(\operatorname{tr} S - f_1 - \cdots - f_k)/k\}^k}, \tag{7.7}$$

as $|S| = \prod_1^p f_i$ and $\operatorname{tr} S = \sum_1^p f_i$.

If this hypothesis is acceptable, and if we find δ (or rather its estimate) to be negligible, we can decide to replace $x_1, x_2, \ldots, x_p$ by only the first p-k principal components and throw away the remaining k. Most of the variation of the x's is explained, in such a situation, by the first p-k principal components only. What remains is insignificantly small and can be sacrificed to reduce the dimensionality of the problem from p to p-k. In practice, therefore, we test the hypotheses,

H_p: All the eigenvalues of Σ are equal,

H_{p-1}: Only the last p-1 eigenvalues of Σ are equal,

H_{p-2}: Only the last p-2 eigenvalues of Σ are equal,

....

....

sequentially, in this order to find out the value of k. If H_p, ..., H_{k+1} are rejected and H_k is accepted, we take the corresponding value k, as the one by which the dimensionality can be reduced, provided δ is negligible. From (6.10), we observe that (if H_k is true)

$$\left(\frac{nk}{2\delta^2}\right)^{1/2} (n^{-1}\,\bar{f} - \delta) \tag{7.8}$$

is a N(0, 1) variable (asymptotically). This gives the asymptotic confidence interval

$$n^{-1}\bar{f}/[1 \pm (2/nk)^{1/2} g_\alpha] \tag{7.9}$$

for δ, with asymptotic condifence coefficient $(1-\alpha)$, g_α being the $100\,\alpha\%$ point of a standard normal variable, i.e.,

$$\int_{-\infty}^{g_\alpha} \frac{1}{\sqrt{2\pi}}\, e^{-u^2/2}\, du = 1 - \frac{1}{2}\,\alpha\ . \tag{7.10}$$

This confidence interval will help in deciding whether δ is negligible or not. One may even use a one-sided confidence interval, viz.,

$$\text{Prob}(\delta \leq n^{-1}\bar{f}/[1 - (2/nk)^{1/2} g_{2\alpha}]) = 1 - \alpha \tag{7.11}$$

for this purpose.

(2) We now consider the hypothesis

H_k^*: the last k eigenvalues of Σ are all equal and the common value is δ_o (which is specified).

The likelihood ratio criterion for this hypothesis turns out to be

$$M_k^{*N/2} = \left\{\frac{\prod_{j=p-k+1}^{p} (f_j/N)}{\delta_o^k}\right\}^{N/2} e^{-\frac{N}{2}\left[\frac{1}{N}\sum_{j=p-k+1}^{p} f_j/\delta_o - k\right]} \tag{7.12}$$

and if H_k^* is true, $-N\log_e M_k^*$ is asymptotically a χ^2 with $k(k+1)/2$

d.f. It is instructive to express M_k^* as

$$M_k^* = M_k \cdot B_k \tag{7.13}$$

where M_k is given by (7.1) and

$$B_k = \left\{\frac{\sum_{j=p-k+1}^{p} f_j}{N_k \delta_o}\right\}^k e^{-\left[\sum_{j=p-k+1}^{p} f_j/N\delta_o - k\right]} . \tag{7.14}$$

M_k tests the hypothesis H_k that the last k eigenvalues are equal, while B_k tests the hypothesis that the common value is δ_o, as specified by the hypothesis.

(3) Goodness of fit test of a hypothetical principal component. Suppose we are given a p-component vector $\underline{a}$, such that $\underline{a}'\underline{a} = 1$ and we wish to test the hypothesis H that the eigenvector of Σ corresponding to the largest root δ_1 (assumed to be of multiplicity one only) is $\underline{a}$. In other words, the hypothesis states that

$$H: \ \underline{\gamma}_1 = \underline{a} . \tag{7.15}$$

From (6.7), $\sqrt{n}(\underline{c}_1 - \underline{\gamma}_1) = \underline{u}$ has an asymptotic normal distribution with zero means and variance-covariance matrix

$$\sum_{j=2}^{p} \frac{\delta_1 \delta_j}{(\delta_1 - \delta_j)^2} \underline{\gamma}_j \underline{\gamma}_j' = P, \text{ say} . \tag{7.16}$$

Observe that P can be written as $\Gamma_2 Q^2 \Gamma_2$, where Γ_2 is the p x (p-1) matrix obtained from Γ by deleting $\underline{\gamma}_1$ and Q^2 is the diagonal matrix with diagonal elements

$$\delta_1 \delta_i/(\delta_1 - \delta_i)^2 \quad (i = 2, 3, \ldots, p) . \tag{7.17}$$

It can be readily verified, therefore, that $\underline{w} = Q^{-1}\Gamma_2'\underline{u}$ has the

$N_{p-1}(\underline{w}|\underline{0}|I)$ distribution (asymptotic), and so $\underline{w}'\underline{w}$ is a χ^2 with p-1 d.f. A little algebra will show that

$$\underline{w}'\underline{w} = n(\underline{C}_1 - \underline{\gamma}_1)'(\delta_1\Sigma^{-1} - 2I + \delta_1^{-1}\Sigma)(\underline{C}_1 - \underline{\gamma}_1) , \qquad (7.18)$$

which has a χ^2 distribution with p-1 d.f. To test the hypothesis H of goodness of fit of the hypothetical first principal component $\underline{a}$, we replace $\underline{\gamma}_1$ in (7.18) by the hypothetical value $\underline{a}$ and also replace Σ and δ_1 which are unknown by their maximum likelihood estimates $n^{-1}S$ and $n^{-1}f_1$. The test criterion, therefore, becomes

$$n(\underline{C}_1 - \underline{a})'(f_1S^{-1} - 2I + f_1^{-1}S)(\underline{C}_1 - a)$$
$$= n(f_1\underline{a}'S^{-1}\underline{a} + f_1^{-1}\underline{a}'S\,\underline{a} - 2) . \qquad (7.19)$$

This statistic will have a χ^2 distribution with p-1 d.f., provided n is large enough and the hypothesis is true. This provides the required test of the hypothesis. In obtaining the expressions (7.18) and (7.19), we have used the facts that

$$\Sigma^{-1} = \frac{1}{\delta_1}\underline{\gamma}_1\underline{\gamma}_1' + \dots + \frac{1}{\delta_p}\underline{\gamma}_p\underline{\gamma}_p' , \qquad (7.20)$$

$$S^{-1} = \frac{1}{f_1}\underline{C}_1\underline{C}_1' + \dots + \frac{1}{f_p}\underline{C}_p\underline{C}_p' , \qquad (7.21)$$

and the orthogonality of the vectors $\underline{\gamma}_i$, as well as of the vectors $\underline{C}_i$. Equations (7.20) and (7.21) are the spectral decompositions of Σ^{-1} and S^{-1}, respectively, and follow from the fact that S, S^{-1} have the same eigenvectors but the eigenvalues are reciprocals of each other.

In this test, the value of the first principal component was specified by the hypothesis, but the test is quite general and can be used for any other principal component γ_i, provided we replace f_1 by f_i and $\underline{C}_1$ by $\underline{C}_i$ in (7.19). It is, of course, necessary to assume that δ_i is of multiplicity one.

A hypothesis similar to the one considered here is the hypothesis of goodness of fit of a single nonisotropic assigned principal component. If Σ has δ_1 as the only eigenvalue different from all others, while $\delta_2, \ldots, \delta_p$ are all equal, the principal component $\underline{\gamma}_1'\underline{x} = v_1$ is the only one that is important, for variation in any direction orthogonal to $\underline{\gamma}_1$ is the same or isotropic. So $\underline{\gamma}_1$ is called the single nonisotropic principal component. Kshirsagar [24] has derived an exact test for the hypothesis $\underline{\gamma}_1 = \underline{a}$ in such a situation. The hypothesis in this case consists of two parts, one that $\delta_2 = \ldots = \delta_p$, i.e., Σ admits a single nonisotropic principal component, and the second part states that the eigenvector corresponding to δ_1 is the assigned vector $\underline{a}$. The χ^2 statistic for the overall hypothesis can be partitioned into two parts corresponding to the two parts of the hypothesis. The reader may refer to [25, 26] for further details of this, the power of the test, and extensions. This is very much similar to the direction and collinearity parts of Wilks's Λ criterion, considered in Chapter 9, for a discriminant function.

(4) Mallows [30] has considered the following hypotheses:

H_a: k given vectors $\underline{\gamma}_1^o, \underline{\gamma}_2^o, \ldots, \underline{\gamma}_k^o$ are eigenvectors of Σ. (They are unit and mutually orthogonal.)

H_b: k given vectors $\underline{\gamma}_1^o, \underline{\gamma}_2^o, \ldots, \underline{\gamma}_k^o$ are eigenvectors of Σ, corresponding eigenvalues being $\delta_1^o, \delta_2^o, \ldots, \delta_k^o$.

He has derived likelihood ratio criteria for these two hypotheses. By the usual procedure of maximizing the likelihood, the criteria turn out to be, respectively,

$$-2\log_e\lambda_a = n\log_e\left\{\prod_{i=1}^{k}\left(\frac{1}{n}\Gamma_1' S\Gamma_1\right)_{ii}\left|\frac{1}{n}\Gamma_2' S\Gamma_2\right|\left|\frac{1}{n}S\right|^{-1}\right\}, \tag{7.22}$$

and

$$-2\log_e\lambda_b = n\log\left\{\left|\frac{1}{n}S\right|^{-1}\prod_{i=1}^{k}\delta_i^o\left|\frac{1}{n}S_{22}^*\right|\right\} + \sum_{1}^{k}\frac{s_{ii}^*}{\delta_i^o} - nk$$

$$= n\log\left\{\left|\frac{1}{n}S\right|^{-1}|\Delta_1|\;\left|\frac{1}{n}\Gamma_2' S\Gamma_2\right|\right\} + \operatorname{tr}(\Delta_1^{-1}\Gamma_1' S\Gamma_1) - nk, \tag{7.23}$$

where

$$\Gamma_1 = [\underline{\gamma}_1^o|\cdots|\underline{\gamma}_k^o], \tag{7.24}$$

$$\Gamma_2 = \text{any } p\times(p-k) \text{ matrix such that } \Gamma_2'\Gamma_2 = I,\ \Gamma_2'\Gamma_1 = 0, \tag{7.25}$$

$$S^* = [s_{ij}^*] = [\Gamma_1|\Gamma_2]' S[\Gamma_1|\Gamma_2], \tag{7.26}$$

$$\Delta_1 = \operatorname{diag}(\delta_1^o, \delta_2^o, \ldots, \delta_k^o). \tag{7.27}$$

If H_a is true, $-2\log_e\lambda_a$ is asymptotically a χ^2 with

$$f_a = k(p-1) - k(k-1)/2 \tag{7.28}$$

d.f. and if H_b is true, $-2\log_e\lambda_b$ is asymptotically a χ^2 with $f_b = f_a + k$ d.f. These degrees of freedom can be calculated in exactly the same way as for the hypothesis H_k in this section. The reader may refer to [30] or to Gupta [15] for more details.

(5) Even if the hypothesis H_k of the equality of the last k eigenvalues of Σ is rejected, we may still like to consider only the first p-k principal components and ignore the rest, provided the total contribution of these last k principal components, viz., $\sum_{i=p-k+1}^{p} \delta_i$ is negligible. For this, we consider the asymptotic normal distribution of $\bar{f}$, the average of the last k eigenvalues of S, obtainable from the results in Section 6. The asymptotic variance of $\bar{f}$ can be estimated by $2\sum_{i=p-k+1}^{p} f_i^2/k^2$. The two-sided and one-sided asymptotic confidence intervals for $\sum_{i=p-k+1}^{p} \delta_i/k$ then turn out to be (in the same way as in 7.9 or 7.11)

$$\frac{1}{n}\bar{f} \pm g_\alpha\left[\left(\frac{2}{n^2}\sum_{p-k+1}^{p} f_i^2/nk^2\right)\right]^{1/2}, \tag{7.29}$$

and

$$\frac{1}{n}\bar{f} + g_{2\alpha}\left[\frac{2}{n^2}\sum_{p-k+1}^{p} f_i^2/nk^2\right]^{1/2}. \tag{7.30}$$

If (7.30) indicates that $\sum_{p-k+1}^{p} \delta_i/k$ is very small, the investigator may wish to ignore the last k principal components.

Sometimes, the relative magnitude of the contribution of the last k principal components as compared to the total variation, viz.,

$$\sum_{i=p-k+1}^{p} \delta_i / \sum_1^p \delta_i , \tag{7.31}$$

is taken as a measure of the effectiveness of the last k principal components in "explaining" the total variation of $\underline{x}$. It this measure indicates that it is not worthwhile to include the last p-k principal components, we ignore them. Thus, for example, if the quantity in (7.31) is only 5% or less, we might decide to consider only the first k principal components. For this, a hypothesis of the type

$$\sum_{p-k+1}^{p} \delta_i / \sum_1^p \delta_i = h \quad (0 < h < 1) \tag{7.32}$$

can be set up and this can be tested by the statistic

$$\frac{1}{n} \sum_{p-k+1}^{p} f_i - \frac{h}{n} \sum_1^p f_i = - \frac{h}{n} \sum_1^{p-k} f_i + \frac{(1-h)}{n} \sum_{p-k+1}^{n} f_i . \tag{7.33}$$

This has an asymptotic normal distribution, with mean

$$\sum_{p-k+1}^{p} \delta_i - h \sum_1^p \delta_i ,$$

and variance

$$\frac{2h^2}{n} \sum_1^{p-k} \delta_i^2 + \frac{2(1-h)^2}{n} \sum_{p-k+1}^{n} \delta_i^2$$

(see 6.10). One can determine a confidence interval for $\sum_{p-k+1}^{p} \delta_i - h \sum_1^p \delta_i$, from this distribution in the same way as in (7.29), and if this confidence interval does not include the value 0, one may decide to reject the hypothesis (7.32).

Before we conclude this section, we would like to refer to Sugiyama [42, 43], who has investigated the non-null distribution of the largest eigenvalue f_1 and the corresponding eigenvector $\underline{C}_1$. However, more research on lines similar to James's [20, 21] work is essential before these distributional results can be used for statistical inference.

8. CORRELATION MATRIX

So far we considered the principal components of $\underline{x}$, based on the variance-covariance matrix Σ. In psychology, however, the various tests that a psychologist uses must be standardized first. It means that we have to deal with the correlation matrix P, instead of the variance-covariance matrix Σ. The principal components will then be obtained from the eigenvectors of P, and their effectiveness will be measured by the eigenvalues of P. The matrices P and Σ are related by

$$\Sigma = \text{diag}\left(\sigma_{11}^{1/2}, \sigma_{22}^{1/2}, \ldots, \sigma_{pp}^{1/2}\right) P \,\text{diag}\left(\sigma_{11}^{1/2}, \sigma_{22}^{1/2}, \ldots, \sigma_{pp}^{1/2}\right) \quad (8.1)$$

and their determinants by

$$|P| = |\Sigma|(\sigma_{11}\sigma_{22}\cdots\sigma_{pp}), \quad (8.2)$$

where σ_{ii} are the diagonal elements of Σ. Let $\lambda_1 > \lambda_2 > \cdots > \lambda_p$ be the eigenvalues of P, the corresponding unit and orthogonal eigenvectors being $\underline{\xi}_1, \underline{\xi}_2, \ldots, \underline{\xi}_p$. The spectral decomposition of P will then be

$$P = \lambda_1\underline{\xi}_1\underline{\xi}_1' + \cdots + \lambda_p\underline{\xi}_p\underline{\xi}_p' . \quad (8.3)$$

The principal components based on P are $\underline{\xi}_i'\underline{x}$ $(i = 1, 2, \ldots, p)$.

In practice, only S, the matrix of corrected s.s. and s.p. of observations based on n d.f., is available. The m.l.e. of P will then be R, where

$$S = \mathrm{diag}\left(s_{11}^{1/2}, \ldots, s_{pp}^{1/2}\right) R \,\mathrm{diag}\left(s_{11}^{1/2}, \ldots, s_{pp}^{1/2}\right). \tag{8.4}$$

We expect the eigenvalues and eigenvectors of R to estimate those of P and thus to provide the sample principal components. Analogous to the sphericity test in Section 7, we consider first the hypothesis $\bar{H}$ of equality of all the eigenvalues of P. Since all diagonal elements of P are unity, $\sum_1^p \lambda_i = p$ and hence $\bar{H}$ is equivalent to $\lambda_1 = \lambda_2 = \cdots = \lambda_p = 1$, and so from (8.3), we find that P reduces to I, if $\bar{H}$ is true. In this case the distribution of R is (see Section 3, Chapter 4)

$$\frac{\left\{\Gamma\left(\frac{n-1}{2}\right)\right\}^p}{\pi^{p(p-1)/4} \prod_{i=1}^{p} \Gamma\left(\frac{n-i}{2}\right)} |R|^{(n-p-2)/2} \, dR \tag{8.5}$$

and from this it was deduced in Chapter 4 that the hth moment of $|R|$ is

$$\frac{K(p,\, n) \cdot 2^{hp}}{K(p,\, n + 2h)} \cdot \left\{\frac{\Gamma(\frac{n}{2})}{\Gamma(h + \frac{n}{2})}\right\}^p, \tag{8.6}$$

where K(p, n) is given by (3.2.20). Following the same method used for Wilks's Λ in Section 3 of Chapter 8, Bartlett [3] showed that

$$-\left\{n - \tfrac{1}{6}(2p + 5)\right\} \log_e |R| \tag{8.7}$$

is asymptotically a χ^2 with $p(p-1)/2$ d.f. This, therefore, provides a test for the hypothesis $\bar{H}$. If this hypothesis is rejected, all the eigenvalues of P are not equal but still some of them could be, and so we set up a more general hypothesis $\bar{H}_k$, similar to H_k in Section 7.

$$\bar{H}_k: \lambda_{p-k+1} = \lambda_{p-k+2} = \cdots = \lambda_p \ (= \lambda \text{ say}). \tag{8.8}$$

The last k eigenvalues of P are equal. Since $\sum_1^p \lambda_i = p$, the common value λ, when $\bar{H}_k$ is true, is given by

$$\lambda = \frac{p - \lambda_1 - \lambda_2 - \cdots - \lambda_{p-k}}{k}, \tag{8.9}$$

and hence

$$|P| = \prod_1^p \lambda_i = \left(\prod_{i=1}^{p-k} \lambda_i\right) \cdot \left(\frac{p - \lambda_1 - \lambda_2 - \cdots - \lambda_{p-k}}{k}\right)^k. \tag{8.10}$$

If $\bar{H}_k$ is true, we, therefore, see that the ratio of $|P|$ to

$$\left(\prod_{i=1}^{p-k} \lambda_i\right) \left(p - \sum_1^{p-k} \lambda_i\right)^k k^{-k}$$

is unity. Bartlett [3, 4, 5] therefore replaced the P and the λ's in this ratio by their estimates and proposed

$$R_k = \frac{|R|\, k^k}{\left(\prod_{i=1}^{p-k} \hat{\lambda}_i\right) \left(p - \sum_1^{p-k} \hat{\lambda}_i\right)^k} \tag{8.11}$$

as the criterion for testing $\bar{H}_k$. $\hat{\lambda}_i$, here, represent the eigenvalues of R (arranged in order of magnitude). By a heuristic argument he and Lawley [27] have shown that

$$-\left[n - \tfrac{1}{6}(2p + 5) - \tfrac{2}{3}(p-k)\right]\log_e R_k \tag{8.12}$$

has an asymptotic χ^2 distribution with $\frac{1}{2}(k-1)(k+2)$ d.f.

The distribution of R, when $P \neq I$, is very complicated, and so the mathematics associated with the eigenvalues and eigenvectors of R in the non-null case is intractable. Not much work has been done in this area, and the reader may refer to Anderson [2] and Lawley [27] for some further details.

9. GEOMETRICAL INTERPRETATION OF PRINCIPAL COMPONENTS

The transformation $\underline{v} = \Gamma'\underline{x}$ of (1.3) is an orthogonal transformation, and it transforms the ellipsoid

$$\underline{x}'\Sigma^{-1}\underline{x} = \text{const} \tag{9.1}$$

into the ellipsoid

$$\underline{v}'\Delta^{-1}\underline{v} = \text{const}$$

i.e.,

$$\sum_{1}^{p} \frac{v_i^2}{\delta_i} = \text{const} . \tag{9.2}$$

This is an ellipsoid referred to its principal axes, and $\delta_i^{1/2}$ $(i = 1, 2, \ldots, p)$ are the lengths of the principal semi-axes. $\underline{\gamma}_i$ represent the direction cosines of the principal axes. Transformation to principal components is thus a transformation to new axes by rotation, so that the ellipsoid (9.1) on which the p.d.f. of $\underline{x}$ (when $\underline{x}$ has a normal distribution) is constant is turned into one referred to its principal axes. For this and various other geometrical concepts associated with principal components, reference may be made to Dempster [12].

10. PRINCIPAL COMPONENTS ANALYSIS

In Chapters 7, 8, and 9 we considered canonical analysis. In a sense, it can be called "external" analysis, as the objective is to study the relation of one vector to another vector. Principal components analysis on the other hand, is an "internal" analysis, as it is concerned with the variances and covariances of the elements of a random vector. No external relationship with any other vector is involved. As already noted in Section 2, the objective of principal components analysis is to construct new variables and replace the original variables with as few new variables as possible. In doing this, naturally, some information contained in Σ (the variance-covariance matrix of the original variables) will have to be sacrificed. Principal components are optimum in the sense that this lost information is kept to a minimum. Whether we use tr Σ or the generalized variance, i.e., $|\Sigma|$, or whether we consider the predictive ability or the norm of the difference between Σ and the corresponding matrix of the new variables, in every case the principal components provide the optimum solution. If one wants to replace the entire set of p variables by one new variable, one has to take the first principal component. If one can afford to consider more than one new variable, one has to take the desired number of principal components. Naturally, the greater the number of principal components chosen, less will be the information lost and better will be the performance of the new variables in explaining the internal relationship of the original variables. In psychology, education, anthropology, and economics, an investigator usually collects observations on a large number of variables, as he does not know, initially, which variables are more important and useful for his investigations. In fact, he tries to measure and include all possible variables which are likely to have some connection with the problem. His next problem is then to reduce these data, and here principal components analysis comes to the rescue. He can usually condense the whole information into a managable number of new variables and consider these variables in detail, without losing any vital information about the variation in the original set of variables. Principal components is thus an exploratory technique of constructing new artificial variables. The tests of significance derived in this chapter are helpful in determining the number of principal components that are necessary if all the important information is to be preserved; they also help in estimating the relative magnitude of the information lost. However these tests are based on the assumption of normality and are all large sample tests. If the assumption of normality does not hold, the technique can still be used, either by using personal discretion to determine the appropriate number of principal components or using the tests based on normality as approximations, without giving much importance to the probability levels associated with the tests.

Principal components are artificial variables and do not necessarily have any physical meaning or significance. They are linear combinations of variables that can be measured, but they themselves cannot, in general, be measured directly. They provide a different angle of viewing the original data and may disclose their nature of variation. Principal components help in interpreting and analyzing the data. Sometimes physical significance can be attached to the principal components. For example, Ahamed, (see Applied Statistics, Vol. 16, p. 17), who analyzed crimes by the method of principal components, could find a suitable meaning for his principal components. There are several other examples of such a situation in the literature.

Principal components analysis originated with Hotelling [18], who developed the technique for his work in educational psychology. The usefulness of this technique has been now exploited in many other branches of science (see, for example, Jolicoeur [22], and Spurrell [40]). It has also been used to attack problems of multicollinearity in econometrics (see Kendall [23], Farrar and Glauber [13], Haitovsky [16, 17], Massy [31], Meyer and Kraft [32]). Torgerson [44] used it in a multidimensional scaling problem, where n stimuli were to be represented in the smallest possible Euclidean space. Rao [36] gives examples of the use of principal components in determining size and shape factors in anthropometric investigations. Wernimont used the technique of principal components to detect heterogeneity.

Simonds [39] used it in connection with optical response data, while Stone [41] used it in econometrics to study interdependence of blocks of transactions.

In spite of this widespread use of principal components as a statistical technique, it must be said that the sampling theory and inference procedure associated with it are still in a very unsatisfactory state, and principal components is mostly an empirical technique. On the other hand, canonical analysis is less vague and more developed. The objectives of these two techniques are, however, entirely different. Principal components explain internal variation; canonical analysis helps to predict an unknown vector from a known one. Principal components were seen to be optimum in various ways, but it must be borne in mind that none of the optimality properties were concerned with what canonical analysis does. Thus the last principal component can be ignored, if it does not explain a significant portion of the total variation of $\underline{x}$. But if the relationship between $\underline{x}$ and $\underline{y}$ is to be studied, it can very well happen that this last principal component is the only one that predicts $\underline{y}$. One drawback with canonical analysis is that one cannot determine the canonical variables unless the sample size n is greater than p, the number of variables. It is not so with principal components. One can determine principal components even if $n < p$. Such situations arise in practice. For example, patients with rare diseases are not available in sufficient number and various characteristics such as blood pressure, temperature, etc. associated with their medical examination must be considered simultaneously. If now these patients are from different groups, a canonical analysis is necessary for studying the differences among these groups, but this cannot be carried out. Dempster [10, 11] therefore suggests that first a principal components analysis be carried out to reduce the number of variables and then a canonical analysis be carried out on the first few principal components. This is, of course, not an efficient procedure, as principal components are not meant to be useful for discriminant analysis. The principal components which were ignored might have more discriminating ability about the groups than the ones that were considered, but in the absence of any better solution, this method seems to be the only way out. In doing so, we assume that principal components will reduce the dimensionality of the problem and also that they will not lose any important information about discrimination. However a theoretical and empirical study of the efficiency of this method is needed.

Principal components analysis and factor analysis are closely related. Factor analysis has been used widely by psychologists and educationists to derive meaningful "specific" and "common" factors and to represent an observation in terms of these. We do not propose to go into the details of this. Interested readers should refer to Lawley and Maxwell's [29] excellent book on factor analysis for the statistical aspects of the problem and to Cattell [8] for an expository article on the essentials of factor analysis.

REFERENCES

[1] Anderson, T. W. (1958). Introduction to Multivariate Statistical Analysis. John Wiley and Sons, New York.

[2] Anderson, T. W. (1963). "Asymptotic theory for principal component analysis", Ann. Math. Statist., 34, p. 122.

[3] Bartlett, M. S. (1950). "Tests of significance in factor analysis", Brit. J. Psych. (Stat. Sec.), 3, p. 77.

[4] Bartlett, M. S. (1951). "The effect of standardization of a χ^2-approximation in factor analysis", Biometrika, 38, p. 337.

[5] Bartlett, M. S. (1951). "A further note on tests of significance in factor analysis", Brit. J. Psych., 4, p. 1.

[6] Bartlett, M. S. (1955). An Introduction to Stochastic Processes. Cambridge University Press.

[7] Bland, R. P., and Owen, D. B. (1966). "A note on singular normal distribution", Ann. Inst. Stat. Math., 18, p. 113.

[8] Cattell, R. B. (1965). "Factor analysis: an introduction to essentials I. The purpose and underlying models", Biometrics, 21, p. 190. (1965) ________ II. The role of factor analysis in research. Biometrics, 21, p. 405.

[9] Darroch, J. N. (1965). "An optimum property of principal components", Ann. Math. Statist., 36, p. 1579.

[10] Dempster, A. P. (1963). "Multivariate theory for general stepwise methods", Ann. Math. Statist., 34, p. 873.

[11] Dempster, A. P. (1963). "Stepwise multivariate analysis of variance based on principal variables", Biometrics, 19, p. 478.

[12] Dempster, A. P. (1969). Elements of Continuous Multivariate Analysis. Addison-Wesley Publishing Company, Massachusetts.

[13] Farrar, D. E., and Glauber, R. R. (1967). "Multicollinearity in regression analysis: the problem revisited", The Review of Economics and Statistics, 49, p. 92.

[14] Good, I. J. (1969). "Some applications of the singular decomposition of a matrix", Technometrics, 11, p. 823.

[15] Gupta, R. P. (1967). "Latent roots and vectors of a Wishart matrix", Ann. Inst. Stat. Math., 19, p. 157.

[16] Haitovsky, Y. (1966). "A note on regression on principal components", American Statistician, 20, p. 28.

[17] Haitovsky, Y. (1968). "Multicollinearity in regression analysis", Econometric Society Summer Meeting at Colorado.

[18] Hotelling, H. (1933). "Analysis of a complex statistical variable into principal components", J. Educ. Psych., 26, p. 417.

[19] James, A. T. (1960). "Distribution of the latent roots of the covariance matrix", Ann. Math. Statist., 31, p. 151.

[20] James, A. T. (1969). "Tests of equality of latent roots of the covariance matrix", Multivariate Analysis II, (edited by Krishnaiah, P. R.), Academic Press, New York.

[21] James, A. T. (1966). "Inference on latent roots by calculation of hypergeometric functions of matrix arguments", Multivariate Analysis I, (edited by Krishnaiah, P. R.), Academic Press, New York.

[22] Jolicoeur, P., and Mosimann, J. E. (1960). "Size and shape variation in the pointed turtle, a principal component analysis", Growth, 24, p. 339.

[23] Kendall, M. G. (1957). A Course in Multivariate Analysis. Charles Griffin and Company, London.

[24] Kshirsagar, A. M. (1961). "The goodness-of-fit of a single (non-isotropic) hypothetical principal component", Biometrika, 48, p. 397.

[25] Kshirsagar, A. M., and Gupta, R. P. (1965). "The goodness-of-fit of two or more principal components", Ann. Inst. Stat. Math., 17, p. 347.

[26] Kshirsagar, A. M. (1966). "The non-null distribution of a direction statistic in principal component analysis", Biometrika, 53, p. 590.

[27] Lawley, D. N. (1956). "Tests of significance for the latent roots of covariance and correlation matrices", Biometrika, 43, p. 128.

[28] Lawley, D. N. (1956). "A general method for approximating to the distribution of likelihood ratio criteria", Biometrika, 43, p. 295.

[29] Lawley, D. N., and Maxwell, A. E. (1963). Factor Analysis as a Statistical Method. Butterworths, London.

[30] Mallows, C. L. (1960). "Latent vectors of random symmetric matrices", Biometrika, 48, p. 133.

[31] Massy, W. F. (1965). "Principal components regression in exploratory statistical research", J. Ann. Stat. Assoc., 60, p. 234.

[32] Meyer, J. R., and Kroft, G. (1961). "The Evaluation of statistical costing techniques as applied in the transportation industry", Am. Economic Review, 51, p. 313.

[33] Miller, K. S. (1964). Multidimensional Gaussian Distributions. John Wiley and Sons, New York.

[34] Okamoto, M., and Kanazawa, M. (1968). "Minimization of eigenvalues of a matrix and optimality of principal components", Ann. Math. Statist., 30, p. 859.

[35] Okamoto, M. (1969). "Optimality of principal components", Multivariate Analysis II, (edited by Krishnaiah, P. R.), Academic Press, New York.

[36] Rao, C. Radhakrishna (1964). "The use and interpretation of principal component analysis in applied research", Sankhyā A, 26, p. 329.

[37] Rao, C. Radhakrishna (1965). Linear Statistical Inferences and Its Applications. John Wiley and Sons, New York.

[38] Rubin, Herman. "The topology of probability measures on a topological space", unpublished.

[39] Simonds, J. L. (1963). "Applications of characteristic vector analysis to photographic and optical response data", J. Optical Soc. of Am., 53, p. 968.

[40] Spurrell, D. J. (1963). "Some metallurgical applications of principal components", Applied Statistics, 12, p. 180.

[41] Stone, R. (1947). "An interdependence of blocks of transactions", J. Roy. Stat. Soc. (Suppl.), 9, p. 1.

[42] Sugiyama, T. (1966). "On the distribution of the largest root and the corresponding latent vector for principal component analysis", Ann. Math. Statist., 37, p. 995.

[43] Sugiyama, T. (1967). "On the distribution of the largest root of the covariance matrix", Ann. Math. Statist., 38, p. 1148.

[44] Torgenson, W. S. (1958). Theory and Methods of Scaling. John Wiley and Sons, New York.

[45] Tumura, Y. (1965). "The distribution of latent roots and vectors", Tokyo Rika University Math., 1, p. 1.

EXERCISES

1. The frequency function of the multivariate negative binomial distribution is

$$p(n_1, n_2, \ldots, n_k) = \frac{\Gamma\left(N + \sum_1^k n_i\right)}{\prod_{i=1}^{k} n_i! \;\; \Gamma(N)} q^{-N} \prod_{i=1}^{k} \left(\frac{p_i}{q}\right)^{n_i},$$

where $n_i \geq 0$, $p_i > 0$, $N > 0$, $\sum_1^k p_i = -1 + q$. ($n_1, \ldots, n_k$ are integers).

Show that the probability generating function of the distribution is

$$\left(q - \sum_1^k p_i z_i\right)^{-N}.$$

Show that the marginal distribution of $n_1, n_2, \ldots, n_m$ $(m < k)$ is

$$\frac{\Gamma\left(N + \sum_1^m n_i\right)}{\prod_1^m n_i! \; \Gamma(N)} q^{1-N} \prod_1^m \left(\frac{p_i}{q'}\right)^{n_i},$$

where $q' = 1 + \sum_1^m p_i$. Hence show that the conditional distribution of $n_{m+1}, n_{m+2}, \ldots, n_k$ given $n_1, \ldots, n_m$ is again negative binomial (multivariate). Show that

$$\text{correlation } (n_i, n_j) = p_i^{1/2} \, p_j^{1/2} / (1 + p_i)^{1/2} \, (1 + p_j)^{1/2}, \quad i \neq j$$

and

$$E(n_j | n_1, \ldots, n_m) = \left(N + \sum_1^m n_i\right) p_j / q', \quad j > m.$$

Johnson and Kotz [22]

2. $x_1, x_2, \ldots, x_k$ have the k-variate Dirichlet distribution with p.d.f.

$$\frac{\Gamma(a_1 + \ldots + a_{k+1})}{\Gamma a_1 \Gamma a_2 \ldots \Gamma a_{k+1}} x_1^{a_1-1} x_2^{a_2-1} \ldots x_k^{a_k-1}$$

$$(1 - x_1 - x_2 - \ldots - x_k)^{a_{k+1}-1}$$

$$x_i \geq 0, \quad i = 1, 2, \ldots, k$$

$$\Sigma x_i \leq 1, \quad a_i \text{ real and} > 0 .$$

Denote this p.d.f. by $P(x_1, \ldots, x_k | a_1, \ldots, a_k, a_{k+1})$.

Show that

(i) $V(x_i) = a_i(a_1+\ldots+a_{k+1} - a_i)/(a_1+\ldots+a_{k+1})^2(a_1+\ldots+a_{k+1}+1)$.

(ii) $\mathrm{Cov}(x_i, x_j) = -a_i a_j/(a_1+\ldots+a_{k+1})^2(a_1+\ldots+a_{k+1}+1)$.

(iii) The marginal distribution of $x_1, \ldots, x_m$ $(m < k)$ is the m-variate Dirichlet distribution with p.d.f. $P(x_1,\ldots,x_m|a_1,\ldots,a_m,a_{m+1}+\ldots+a_{k+1})$.

(iv) $E(x_k|x_1,\ldots,x_{k-1}) = a_k(1-x_1-\ldots-x_{k-1})/(a_k+a_{k+1})$.

(v) $V(x_k|x_1,\ldots,x_{k-1}) = a_k a_{k+1}(1-x_1-\ldots-x_{k-1})^2/\{(a_k+a_{k+1})^2 (a_k+a_{k+1}+1)\}$.

(vi) $x_1 + \ldots + x_k$ has a Beta distribution with parameters $a_1 + \ldots + a_k$ and a_{k+1}.

(vii) $\theta_1 = x_1, \theta_2 = x_2(1-x_1), \ldots, \theta_k = x_k/(1/x_1-\ldots-x_{k-1})$ are independently distributed.

Wilks [60]

3. $p_o, p_1, \ldots, p_k$ (with $\Sigma p_i = 1$) are the probabilities associated with the k+1 classes of a multinomial distribution. Let $q_i = 1-p_i$, and $s_j = 1-(p_1+\ldots+p_j)$, $s_o = 1$. If V is the variance-covariance matrix of the number of individuals falling in classes 1 to k, in a sample of size N, show that

$$V^{-1} = \frac{1}{N}\left\{\text{diag}(p_1^{-1}, p_2^{-1}, \ldots, p_k^{-1}) + s_k^{-1} E_{kk}\right\}.$$

If V = NU'U, where U' is a l.t. matrix, show that the non-zero elements of U^{-1} are

$$U^{jj} = (p_j^{-1} + s_j^{-1})^{1/2}$$

$$U^{ij} = (s_j^{-1} - s_{j-1}^{-1})^{1/2}, \quad i < j.$$

Taylor's blood serological data is given in Fisher's statistical methods for research workers, p. 290. The matrix T of total s.s. and s.p. corresponding to that data is given by Table 61.902 on p. 291. Explain, using the above result about V^{-1}, why T^{-1} has all non-diagonal elements equal.

Healy, J.M. [19]

4. Prove the following results for a set of variables $x_1, x_2, \ldots, x_p$.

(a) If all gross correlation coefficients are negative, all partial correlation coefficients of all orders of the set are also negative.

(b) If the signs of all gross correlation coefficients may be made negative by a cogredient change of signs, or if the signs of all partial correlation coefficients of highest order of the set may be made positive by a cogredient change of sign, then without any change of signs, any partial correlation of any order has the same sign as the corresponding gross correlation coefficient.

O. Riersol [52]

5. The elements of a matrix A are functions of a random variable x. If A is real, symmetric and positive definite for all values of x, show that

$$E(A^{-1}) - \{E(A)\}^{-1}$$

is positive semi-definite, provided $E(A^{-1})$ and $E(A)$ exist.

Groves and Rothenberg [17]

6. X is the $p \times N$ matrix of sample observations from a p-variate normal distribution. Show that

$$X(I - N^{-1} E_{NN})X'$$

is positive definite with probability one, if and only if $N > p$.

Dykstra [12]

7. x_1, x_2, x_3 have a trivariate normal distribution with zero means, unit variance and correlation matrix $P = [\rho_{ij}]$. Prove that

(1) $$\text{Prob}(x_1 > 0,\ x_2 > 0,\ x_3 > 0) = \frac{1}{8} + \frac{\sin^{-1}\rho_{12} + \sin^{-1}\rho_{13} + \sin^{-1}\rho_{23}}{4\pi}$$

(2) $1 + 2\rho_{12}\,\rho_{13}\,\rho_{23} \geq \rho_{12}^2 + \rho_{13}^2 + \rho_{23}^2$.

Kendall and Stuart Vol. 1 [26]

8. x, y have a bivariate normal distribution with zero means, unit standard deviations and correlation coefficient ρ. Define

$$M(h, k, \rho) = \text{Prob}(x > h, y > k); \; G(h) = \text{Prob}(x < h).$$

Prove the following results.

(1) $M(-h, -k, \rho) = G(h) + G(k) + M(h, k, \rho) - 1$,

(2) $M(-h, k, -\rho) = G(h) - M(-h, -k, \rho)$,

(3) $M(h, -k, \rho) = G(k) - M(-h, -k, \rho)$,

(4) $M(h, k, \rho) = M(h, 0, u) + M(k, 0, \nu) - \delta$, where

$$u = (\rho h - k)\,\text{sign}\,h/(h^2 - 2\rho hk + k^2)^{1/2},$$

$$\nu = (\rho k - h)\,\text{sign}\,k/(h^2 - 2\rho hk + k^2)^{1/2},$$

$$\text{sign}\,h = 1 \text{ if } h \geq 0 \text{ and } = -1 \text{ if } h < 0,$$

$$\delta = \begin{cases} 0 \text{ if } hk > 0 \text{ or } hk = 0 \text{ but } h + k \geq 0 \\ 1/2 \quad \text{otherwise.} \end{cases}$$

Owen [47]

9. In a certain manufacturing process, the observed measurement S consists of two parts P, the true value of the product which has the $N(\mu, \sigma_p^2)$ distribution and T, the error of measurement which has the $N(\lambda, \sigma_t^2)$ distribution and is independent of P. The performance specification limits are $\mu + k_1\sigma_p$, $\mu - k_2\sigma_p$ and so the consumer's loss CL is

$$\text{Prob}\ [P > \mu + k_1\sigma_p \text{ or } P < \mu - k_2\sigma_p \text{ and}$$
$$\mu + \lambda - (k_2\sigma_2 - b_2\sigma_t) < S < \mu + \lambda + (k_1\sigma_p - b_1\sigma_t)]$$

where b_1, b_2, k_1, k_2 are suitable constants. Show that this can be expressed in terms of the $M(h, k, \rho)$ function of the previous exercise, as

$$CL = M(k_1, -q_2, \rho) - M(k_1, q_1, \rho) - M(k_2, q_2, \rho) + M(k_2, -q_1, \rho)$$

where

$$q_1 = (k_1\sigma_p - b_1\sigma_t)/(\sigma_p^2 + \sigma_t^2)^{1/2} ,$$
$$q_2 = (k_2\sigma_p - b_2\sigma_t)/(\sigma_p^2 + \sigma_t^2)^{1/2} .$$

Similarly, the producer's loss PL is

$$\text{Prob}\ [\mu - k_2\sigma_p < P < \mu + k_1\sigma_p \text{ and } S < \mu + \lambda - (k_2\sigma_p - b_2\sigma_t)$$
$$\text{or } S > \mu + \lambda + (k_1\sigma_p - b_1\sigma_t)].$$

Show that PL can be expressed as

$$CL + G(k_1) + G(k_2) - G(q_1) - G(q_2) ,$$

where the function G is defined in the previous exercise.

Owen [48]

10. $X_1, \ldots, X_n$ have an n-variate normal distribution with zero means and variance-covariance matrix

$$(1-\rho)I + \rho\, E_{nn} .$$

Define $F_n(x|\rho)$ as Prob $(X_1 < x_1, \ldots, X_n < x_n)$ and $F_o(x|\rho) \equiv 1$. $G(x)$ is the c.d.f. of a $N(0, 1)$ variable. Show that

$$F_n(t|\rho) = \int_{-\infty}^{\infty} G^n\left(\frac{t - \sqrt{\rho}\, y}{\sqrt{1-\rho}}\right) G'(y)dy .$$

Dunnet and Sobel [11]

11. Let $x' = [x_1, \ldots, x_p]$, $y' = [y_1, \ldots, y_p]$, $z = [z_1, \ldots, z_k]$. $\underline{x}$ has the $N_p(\underline{x}|\underline{0}|\Sigma)$ distribution; $\underline{y}$ has the $N_p(\underline{y}|0|I)$ distribution and is independent of $\underline{z}$, which has the $N_k(\underline{z}|0|I)$ distribution. Suppose it is possible to express Σ as $C^2 + BB'$, where $C = \text{diag}(C_1, \ldots, C_p)$ and $B = [b_{ij}]$ all $C_i > 0$. By expressing $\underline{x}$ as $C\underline{y} - B\underline{z}$, show that the p-fold integral

$$\int_{a_1}^{\infty} \cdots \int_{a_p}^{\infty} N_p(\underline{x}|\underline{0}|\Sigma)\,d\underline{x}$$

reduces to the k-fold integral

$$(2\pi)^{-k/2} \int_{-\infty}^{\infty} \cdots \int_{-\infty}^{\infty} \prod_{i=1}^{p} P\left\{(a_i + \sum_{j=1}^{k} b_{ij}z_j)/C_i\right\} e^{-z'z/2}\, d\underline{z}$$

where

$$P(a) = \text{Prob}\,(y_1 \geq a).$$

Obtain sufficient conditions for the existence of the matrices B, C satisfying $\Sigma = C^2 + BB'$. Express

$$\begin{bmatrix} 10 & 1 & 3 & 7 & 6 \\ 1 & 11 & -5 & 0 & -3 \\ 3 & -5 & 10 & 5 & 6 \\ 7 & 0 & 5 & 12 & 9 \\ 6 & -3 & 6 & 9 & 10 \end{bmatrix}$$

as $C^2 + BB'$.

Webster [58]

12. X is the $p \times n$ matrix of n sample observations from the $N_p(\underline{x}|0|\Sigma)$ distribution and L is the likelihood of the sample observations. The elements of Σ are σ_{ij} and those of Σ^{-1} are σ^{ij}. The information matrix U is defined as the $p(p+1)/2 \times p(p+1)/2$ matrix

$$- E\left[\frac{\partial^2 \log L}{\partial \sigma_{ij} \quad \partial \sigma_{rt}}\right].$$

Show that

(i) $U = -\frac{n}{2}\left[\frac{\partial^2 \log|\Sigma|}{\partial\sigma_{rt} \ \partial\sigma_{ij}}\right]$,

(ii) $U^{-1} = -\frac{1}{n} K \Delta$,

where $\Delta = \left[\frac{\partial\sigma_{ij}}{\partial\sigma^{rt}}\right]$ and K is a diagonal matrix of order $\frac{p(p+1)}{2}$

defined by

$$K = \text{diag}(k_{11}, k_{12}, \ldots, k_{1p}, \ldots, k_{p1}, \ldots, k_{pp})$$

where

$$k_{ij} = 2 \quad \text{if} \quad i = j \quad \text{and} \quad 1 \quad \text{if} \quad i \neq j \ .$$

Smith and Hocking [55]

13. f is the p.d.f. of a p-variate normal distribution with means $\underline{\mu}$ and variance-covariance matrix $\Sigma = [\sigma_{ij}]$. Show that

$$\frac{\partial f}{\partial\sigma_{jk}} = \frac{\partial^2 f}{\partial x_j \ \partial x_k}\left(1 - \frac{\delta_{jk}}{2}\right),$$

where

$$\delta_{jk} = 1 \quad \text{if} \quad j = k \quad \text{and zero otherwise} .$$

14. $\underline{x}$ has the $N_p(\underline{x}|\underline{\mu}|\Sigma)$ distribution. Show that the

$$\text{Prob}(x_1 > a_1, \ x_2 > a_2, \ \ldots, \ x_p > a_p)$$

increases with an increase in any off diagonal element of Σ.

Webster [58]

15. If ϕ denotes the p.d.f. of $\underline{x}$ which has the $N_k(\underline{x}|\underline{0}|A^{-1})$ distribution and $M(\underline{\alpha})$ denotes the moment generating function

(m.g.f.) of $\underline{x}$, show that $\alpha_i M$ is the m.g.f. of $\frac{-\partial\phi}{\partial x_i}$ and $\frac{\partial M}{\partial \alpha_i}$ is the m.g.f. of $x_i\phi$ (α_i, x_i are respectively the ith elements of $\underline{\alpha}$ and $\underline{x}$). Hence show that

$$\alpha_i M = a_{1i}\frac{\partial M}{\partial \alpha_1} + a_{2i}\frac{\partial M}{\partial \alpha_2} + \dots + a_{ki}\frac{\partial M}{\partial \alpha_k}, \quad (i=1,2,\dots,k)$$

where a_{ij} are the elements of A. Solving these k equations in $\frac{1}{M}\frac{\partial M}{\partial \alpha_r}$, $(r = 1, 2, \dots, k)$, show that

$$M = \exp\left(\frac{1}{2}\underline{\alpha}' A^{-1} \underline{\alpha}\right).$$

Steyn [56]

16. The p.d.f. of the joint distribution of $x_1, \dots, x_n$ is

$$f(\underline{x}) = (2\pi)^{-n/2} \exp\left\{-\frac{1}{2}\underline{x}'\underline{x}\right\}\cdot\left\{1 + \prod_1^n\left(x_i e^{-x_i^2/2}\right)\right\},$$

where $\underline{x}$ is the column vector of $x_1, \dots, x_n$. Define X(j) as the vector of all the $x_1, \dots, x_n$ excluding x_j $(j = 1, 2, \dots, n)$. The p.d.f. of X(j) is denoted by $g_j(X(j))$. Show that

$$g_j(X(j)) = \int_{-\infty}^{\infty} f(\underline{x})\, dx_j = I_1 + I_2$$

where

$$I_2 = \int_{-\infty}^{\infty} (2\pi)^{-n/2} e^{-\underline{x}'\underline{x}/2} \left\{\prod_1^n \left(x_i e^{-x_i^2/2}\right)\right\} dx_j = 0.$$

Hence show that any proper subset of $x_1, \dots, x_n$ are mutually independent and jointly normal, yet $(x_1, \dots, x_n)$ together are not normally distributed, nor are independent. [Note that this result proves that the converse of Theorem 2, Chapter 2 is not true.]

Pierce and Dykstra [50]

17. The variance-covariance matrix of v variables $\underline{x}$ with mean μE_{v1} is $\sigma^2 NN'$, where N is the incidence matrix of a symmetric balanced incomplete block design with b = v blocks, v treatments, r replications and k = r plots in each block. Find the best estimate of μ, an unbiased estimate of σ^2 and v linear combinations of the elements of $\underline{x}$ such that their variance-covariance matrix is $\sigma^2 I$.

18. x_1, x_2 have a bivariate normal distribution with means μ_1, μ_2, variances σ_{11}, σ_{22} respectively and covariance σ_{12}. A random sample of size n_{12} is available from this distribution. In addition, an independent random sample of size n_1 from the distribution of x_1 only and a random sample of size n_2 from the distribution of x_2 only is available. Using all these observations, obtain optimum estimates of the parameters μ_1, μ_2, σ_{11}, σ_{12}, σ_{22}.

Smith [54]

19. $\underline{x}$ has the $N_p(\underline{x}|\underline{\mu}|\sigma^2 R)$ distribution, where the diagonal elements of R are all unity. $u = f\, s^2/\sigma'^2$ has a χ^2 distribution with f d.f. and is independent of $\underline{x}$. Show that the distribution of $\underline{t} = s^{-1}\underline{x}$ is

$$\frac{(\sigma'/\sigma)^p \exp(-\underline{\mu}'R^{-1}\underline{\mu}/2\sigma^2)}{(\pi f)^{p/2}\,|R|^{1/2}\,\Gamma(f/2)} \sum_{\alpha=0}^{\infty} \frac{(\sigma'/\sigma)^{\alpha}\, 2^{\alpha/2}\, (\underline{t}'R^{-1}\underline{\mu})^{\alpha}\,\Gamma\frac{1}{2}(f+p+\alpha)\, d\underline{t}}{\sigma^{\alpha} f^{\alpha/2} \alpha! \left(1+\dfrac{\sigma'^2}{\sigma^2}\cdot\dfrac{\underline{t}'R^{-1}\underline{t}}{f}\right)^{(f+p+\alpha)/2}}$$

This is known as the noncentral multivariate $\underline{t}$ distribution. The central distribution is obtained by putting $\sigma' = \sigma$ and $\underline{\mu} = \underline{0}$.

Kshirsagar [31]

20. B has the $W_p(B|n|I)$ distribution. C is a l.t. matrix with elements C_{ij} such that $B = CC'$. Define

$$\nu_{11}^2 = C_{11}^2, \text{ and, } \nu_{ii}^2 = C_{ii}^2 + \sum_{j=1}^{i-1} C_{ij}^2 \quad (i = 1, 2, \ldots, p).$$

Further, let

$$C_{i,i-1} = \left(\nu_{ii}^2 - C_{i1}^2 - \cdots - C_{i,i-2}^2\right)^{1/2} \frac{t_{i,i-1}/\sqrt{n-i+1}}{1+t_{i,i-1}^2/(n-i+1)}$$

$$\vdots$$

$$C_{i,i-2} = \left(\nu_{ii}^2 - C_{i1}^2 - \cdots - C_{i,i-3}^2\right)^{1/2} \frac{t_{i,i-2}/\sqrt{n-i+2}}{1+t_{i,i-2}^2/(n-i+2)}$$

$$C_{i1} = \nu_{ii} \cdot \frac{t_{i1}/\sqrt{(n-1)}}{1+t_{i1}^2/(n-1)}.$$

Obtain the joint density of $\nu_{11}^2, \nu_{22}^2, \ldots, \nu_{pp}^2, t_{21}, t_{31}, \ldots, t_{p1}, t_{32}, \ldots, t_{p2}, \ldots, t_{p,p-1}$. Hence show that the ν_{ii}^2 are all χ^2 variables with n d.f. while the t_{ij}'s are student's t variables and that all these variables are independent. Compare this decomposition of B with Bartlett's decomposition.

Kabe [24]

21. X represents the p x n matrix of the sample observations, when a random sample of size n is drawn from the $N_p(\underline{x}|\underline{\mu}|\Sigma)$ distribution. A is an n x n symmetric matrix of constant elements, with

$$A = \lambda_1 \underline{q}_1 \underline{q}_1' + \cdots + \lambda_n \underline{q}_n \underline{q}_n'$$

as its spectral decomposition ($\underline{q}$'s are orthogonal and unit). Show that the moment generating function of XAX' (defined as

$$E\left(e^{\operatorname{tr}\theta XAX'}\right),$$

where θ is a symmetric matrix of order p) is

$$\prod_{i=1}^{t} |I - 2\lambda_i\Sigma\theta|^{-1/2} \exp\left\{\sum_{i=1}^{t} \lambda_i \underline{q}_i' E_{n1}\underline{\mu}'\theta(I - 2\lambda_i\Sigma\theta)^{-1}\underline{\mu}E_{1n}\underline{q}_i\right\}$$

t is the rank of A. Show further that the cumulant generating function of XAX' is

$$\sum_{s=0}^{\infty} 2^s \operatorname{tr} \underline{\mu}E_{1n}A^{s+1} E_{n1}\underline{\mu}'\,\theta(\Sigma\theta)^s + \sum_{s=1}^{\infty} 2^{s-1} \operatorname{tr} A^s \operatorname{tr}(\Sigma\theta)^s/s.$$

Khatri [28]

22. S has the $W_p(s|n|\sigma^2R)$ distribution, where the diagonal elements of R are all unity. $u = fs^2/\sigma^2$ has a χ^2 distribution with f d.f. and is independent of S. Show that the distribution of $T = s^{-2}S$ is

$$\frac{\Gamma\left(\frac{np+f}{2}\right) |R|^{-n/2} |T|^{(n-p-1)/2} \left\{1 + \frac{1}{f}\operatorname{tr} R^{-1}T\right\}^{(np+f)/2} dT}{f^{np/2} \pi^{p(p-1)/4} \Gamma\left(\frac{f}{2}\right) \prod_{i=0}^{p-1} \Gamma\left(\frac{n-i}{2}\right)}.$$

Cornish [10]

23. S_o and S_1 have independent Wishart distributions $W_p(S_o|n_o|I)dS_o$ and $W_p(S_1|n_1|I)dS_1$ respectively. $S_o^{1/2}$ is any matrix such that $(S_o^{1/2}) = S_o$ and T is a l.t. matrix such that $S_o = TT'$. Prove that the distribution of $V = S^{-1/2} S_1 S_o^{-1/2}$ is

$$\text{Const. } |V|^{(n_1-p-1)/2} |I+V|^{-(n_o+n_1)/2} dV, \quad (V > 0)$$

while that of $U = T^{-1}S_1T'^{-1}$ is

$$\text{Const. } |U|^{(n_1-p-1)/2} |I+U|^{-(n_o+n_1+p+1)/2} \prod_{j=1}^{p} |I_j + U_j| dU,$$
$$(U > 0)$$

where

U_j is the matrix of the first j rows and columns of U.

Olkin and Rubin [44]

24. The k+1 matrices S_j (j = 0, 1, ..., k) of order p have independent Wishart distributions $W_p(S_j|n_j|I)dS_j$. Matrices $S_o^{1/2}$ and T are defined as in the previous exercise, while M is defined as an upper triangular matrix such that $S_o = MM'$. Further let (for j = 1, 2, ..., k)

$$V_j = S_o^{-1/2}\, S_j\, S_o^{-1/2} ,$$

$$U_j = T^{-1}\, S_j\, T'^{-1} ,$$

$$L_j = M^{-1}\, S_j\, M'^{-1} ,$$

$U_{j\alpha}$ = the matrix of the first α rows and columns of U_j,

$L^*_{j\alpha}$ = the matrix of the last α rows and columns of L_j.

Show that the joint density of the V_j's is,

$$\text{Const.} \prod_{j=1}^{k} |V_j|^{(n_j-p-1)/2} \left|I + \sum_{j=1}^{k} V_j\right|^{-n/2} ,\ V_j > 0$$

that of the U_j's is

$$\text{Const.} \prod_{j=1}^{k} |U_j|^{(n_j-p-1)/2} |I + \Sigma U_j|^{-(n-p-1)/2} \prod_{\alpha=1}^{p} \left|I + \sum_{j=1}^{k} U_{j\alpha}\right|^{-1},$$

while the density of the L_j's is

$$\text{Const.} \prod_{j=1}^{k} |L_j|^{(n_j-p-1)/2} \left|I + \sum_{j=1}^{k} L_j\right|^{-(n-p-1)/2} \prod_{\alpha=1}^{p} \left|I + \sum_{j=1}^{k} L^*_{j\alpha}\right|^{-1}$$

while $n = n_o + n_1 + \dots + n_k$.

Olkin and Rubin [44]

25. $S_o, S_1, \ldots, S_k$ are independently distributed as $W_p(S_j|n_j|\Sigma)dS_j$ $(j = 0, 1, \ldots, k)$. Let

$$W_j = \left(\sum_{o}^{k} S_i\right)^{-1/2} S_j \left(\sum_{o}^{k} S_i\right), \quad j = 1, 2, \ldots, k$$

$$V_j = S_o^{-1/2} S_j S_o^{-1/2}, \quad j = 1, 2, \ldots, k$$

$$Z_j = \left(I + \sum_{1}^{k} V_j\right)^{-1/2} V_j \left(I + \sum_{1}^{k} V_j\right)^{-1/2}, \quad j = 1, 2, \ldots, k.$$

Prove that the joint distribution of the W_j's is

$$\text{Const.} \prod_{j=1}^{k} |W_j|^{(n_j-p-1)/2} \left|I - \sum_{1}^{k} W_j\right|^{(n_o-p-1)/2} \pi dW_j,$$

and that the joint distribution of the z_j's is the same as above, with W_j replaced by z_j.

Olkin and Rubin [44]

26. S_1 and S_2 are independently distributed as $W_p(S_j|n_j|\Sigma)dS_j$ $(j = 1, 2)$. $S_2 = TT'$, where T is l.t. The distribution of

$$Y = T'(S_1 + S_2)^{-1}T$$

is

$$\text{Const.} \frac{|Y|^{n_2/2} |I - Y|^{(n_1-p-1)/2} dY}{\prod_{\alpha=1}^{p} |Y^*_\alpha|}$$

where Y^*_α is the matrix of the last α rows and columns of Y.

If T is not l.t. but upper triangular, show that the distribution of Y above remains the same, except that Y^*_α is replaced by Y_α, which is the matrix of the first α rows and columns of Y.

Olkin and Rubin [44]

27. A sample of size $k \leq p$ is drawn from a $N_p(\underline{x}|\underline{0}|\Sigma)$ population and X, which is a $p \times k$ matrix represents these sample observations. V has an independent $W_p(V|n|\Sigma)$ distribution and $V = TT'$ where T is l.t. Also $V = MM'$, where M is upper triangular. Show that the distribution of $G = X'V^{-1}X'$ is

$$\text{Const. } |G|^{(p-k-1)/2} \; |I+G|^{-(n+k)/2} dG.$$

Show also that the distribution of $H = X'T'^{-1}$ is

$$\text{Const. } |(I+H'H)|^{-(n+k-p-1)/2} \prod_{\alpha=1}^{p} |I+(H'H)_\alpha|^{-1}$$

where $(H'H)_\alpha$ is the matrix of the first α rows and columns of H'H.

Show that the distribution of $H = X'M'^{-1}$ is the same as above, except that $(H'H)_\alpha$ is replaced by $(H'H)^*_\alpha$, which is the matrix of the last α rows and columns of H'H.

Olkin and Rubin [44]

28. S_j $(j = 1, 2, \ldots, k)$ are independently distributed as $W_p(S_j|n_j|\Sigma)dS_j$. Let T_j be a l.t. matrix such that

$$S_1 + \ldots + S_j = T_j T_j' \quad (j = 1, 2, \ldots, k-1).$$

Show that

$$W_j = T_j^{-1} \; S_{j+1} \; T_j'^{-1} \quad (j = 1, 2, \ldots, k-1)$$

are independently distributed.

Khatri [27]

29. S_1, S_2 are independently distributed as $W(S_j|n_j|I)dS_j$ $(j = 1, 2)$. Let $Z = S_1 + S_2$, $W = S_1^{-1/2} S_2 S_1^{-1/2}$, where

$$(S_1^{1/2})^2 = S_1 .$$

Show that Z and W are not independent, but if $W = T^{-1} S_2 T'^{-1}$, where T is l.t. and $S_1 = TT'$, then Z and W are independent.

Olkin and Rubin [44]

30. $S_1, S_2, \ldots, S_k$ are independently distributed as $W_p(S_j|n_j|\Sigma)dS_j$ $(j = 1, 2, \ldots, k)$. Prove that

$$Y_j = (S_1 + \ldots + S_{j+1})^{-1/2} S_{j+1} (S_1 + \ldots + S_{j+1})^{-1/2}$$

are independently distributed, $j = 1, 2, \ldots, k-1$.

Olkin and Rubin [44]

31. Y is a $k \times n$ matrix, D is a $(p-k) \times n$ matrix $(n > p)$. Show that

$$\int\limits_{\substack{YY'=G \\ YD'=V}} dY = 2^{-k} \prod_{i=1}^{k} C(n-p+i) \, |DD'|^{-p/2} |G-V(DD')^{-1}V'|^{\frac{n-p-1}{2}}$$

where C(n) is the surface area of a unit n-dimensional sphere.

Prove further that (if $\underline{d}$ and $\underline{v}$ are column vectors)

$$\int\limits_{\substack{(Y-Q)A(Y-Q)'=G \\ Y\underline{d}=\underline{v}}} dY = 2^{-k} \prod_{i=1}^{k} c(n-p-1+i)(\underline{d}'A^{-1}\underline{d})^{-k/2}|A|^{-k/2} .$$

$$\left|G - \frac{(\underline{v} - Q\underline{d})(\underline{v} - Q\underline{d})'}{\underline{d}'A^{-1}\underline{d}}\right|^{(n-k-2)/2} .$$

Show also that ($\underline{x}$ is a $p \times 1$ vector)

$$\int_{-\infty}^{\infty} \exp\{-\tfrac{1}{2}(\underline{x}-\underline{v})'\, M^{-1}(\underline{x}-\underline{v}) - \tfrac{1}{2}\,\underline{v}'\, H^{-1}\,\underline{v}\}\, d\underline{v}$$

$$= (2\pi)^{p/2}\, |M^{-1} + H^{-1}|^{-1/2} \exp\{-\tfrac{1}{2}\, x'[M^{-1}$$

$$- M^{-1}(M^{-1} + H^{-1})^{-1} M^{-1}]\, \underline{x}\}$$

Kabe [25]

32. $z = x + iy$ (where $i = \sqrt{-1}$) is said to be a complex normal variable if x and y have a bivariate normal distribution. The vector $\underline{\xi}' = (z_1, z_2, \ldots, z_p)$ is said to be a p-variate complex normal variable if the vector of real and imaginary parts has a 2p-variate normal distribution. Suppose $z_j = x_j + iy_j$ and $E(z_j) = 0$, $j = 1, 2, \ldots, p$. Let

$$E\begin{bmatrix} x_j x_k & x_j y_k \\ y_j x_k & y_j y_k \end{bmatrix} = \frac{1}{2}\begin{bmatrix} 1 & 0 \\ 0 & 1 \end{bmatrix} \sigma_k^2 , \quad \text{if } j = k$$

$$= \frac{\sigma_j \sigma_k}{2}\begin{bmatrix} \alpha_{jk} & -\beta_{jk} \\ \beta_{jk} & \alpha_{jk} \end{bmatrix}, \quad \text{if } j \neq k .$$

Show that the Hermitian positive definite complex covariance matrix of the vector $\underline{\xi}$ is $\Sigma = [\sigma_{jk}]$, of order $p \times p$, where

$$E(z_j \bar{z}_k) = \sigma_{jk} = \sigma_k^2 , \quad \text{if } j = k$$

$$= (\alpha_{jk} + i\,\beta_{jk})\sigma_j \sigma_k , \quad \text{if } j \neq k.$$

($\bar{z}_k$ denotes the complex conjugate of z_k).

Show further that the p.d.f. of the p-variate complex normal distribution is

$$\pi^{-p} |\Sigma|^{-1} \exp(-\bar{\underline{\xi}}' \Sigma^{-1} \underline{\xi}).$$

$\underline{\xi}_1, \underline{\xi}_2, \ldots, \underline{\xi}_n$ is a random sample of size n from this complex normal distribution. Prove that the p.d.f. of the distinct elements of the complex Wishart matrix

$$A = \sum_{j=1}^{n} \underline{\xi}_j \bar{\underline{\xi}}_j'$$

is

$$\frac{1}{\pi^{p(p-1)/2}\, \Gamma(n)\Gamma(n-1) \ldots \Gamma(n-p+1) \cdot |\Sigma|^n} |A|^{n-p} e^{-\operatorname{tr} \Sigma^{-1} A}.$$

T is a complex upper triangular matrix, with positive real diagonal elements, T_{jj}, such that $\bar{T}'T = A$. Prove that the p.d.f. of the nonzero elements of T is

$$\text{const.} \prod_{j=1}^{p} (T_{jj})^{2n-(2j-1)} \exp(-\operatorname{tr} \Sigma^{-1} \bar{T}'T).$$

Goodman [16]

33. A matrix S has the $W_p(S|n|\Sigma)$ distribution and a variable u has an independent $B(u|\frac{f}{2}, \frac{n-f}{2})$ distribution, where $n > f$. Define $T = uS$. Show that, for any vector $\underline{a}$ of constant elements,

(i) $\underline{a}'T\underline{a}/\underline{a}'\Sigma\underline{a}$ has the χ^2 distribution with f d.f. if $\underline{a}'\Sigma\underline{a} \neq 0$

and (ii) $\underline{a}'T\underline{a} = 0$ with probability one, if $\underline{a}'\Sigma\underline{a} = 0$.

Show further that

(iii) $E(T) = f\,\Sigma$.

Verify that T cannot have a Wishart distribution. (Hint: If at all T has a Wishart distribution, it must be, by (iii) $W_p(T|f|\Sigma)$. In that case $r_{ij} = t_{ij}/(t_{ii}t_{jj})^{1/2}$ has the distribution of a sample correlation coefficient, based on a sample of size f+1, but r_{ij} is also $= s_{ij}/(s_{ii}s_{jj})^{1/2}$ and it has the distribution of a sample correlation coefficient obtained from a sample of size n+1 hence etc.). This result shows that the converse of Lemma 2 of Chapter 3 is not true.

Mitra [41]

34. S is a symmetric $p \times p$ matrix with the following properties

(i) If for any vector $\underline{a}$, $\underline{a}'\Sigma\underline{a} = 0$, then $\underline{a}'S\underline{a} = 0$ with probability one.

(ii) For every matrix L (not necessarily square) satisfying $L\Sigma L' = I$, the diagonal elements of LSL' are distributed as independent χ^2 variables with f d.f.

Show that the distribution of S is $W_p(S|f|\Sigma)dS$.

Mitra [41]

35. x and y have a bivariate normal distribution with zero means, unit variances and correlation coefficient ρ. Obtain an unbiased estimate of

$$\text{Prob}(x > 0,\ y > 0)$$

from a sample of size n. If r is the sample correlation

coefficient, based on a sample of size n, show that

$$\text{Prob}(r > 0) = \text{Prob}(t > -\sqrt{n-1} \cdot \rho)$$

where t is a student's t variable with n-1 d.f.. Instead of the usual definition of r viz.

$$r = \frac{\sum_1^n (x_i - \bar{x})(y_i - \bar{y})}{\left\{\sum_1^n (x_i - \bar{x})^2 \sum_1^n (y_i - \bar{y})^2\right\}^{1/2}},$$

a new definition is employed (as means of x, y are known to be zero)

$$r' = \sum_1^n x_i y_i / (\Sigma x_i^2)^{1/2} (\Sigma y_i^2)^{1/2} .$$

In what way does the distribution of r' differ from that of r?

36. $S = [s_{ij}]$ is the matrix of corrected s.s. and s.p. of sample observations in a sample of size N from a bivariate normal population. Obtain the joint distribution of the two studentized regression coefficients

$$u = \left\{\frac{(N-2)s_{22}}{(1-r^2)s_{11}}\right\}^{1/2}(b_1 - \beta_1), \quad v = \left\{\frac{(N-2)s_{11}}{(1-r^2)s_{22}}\right\}^{1/2}(b_2 - \beta_2)$$

where $b_1 = s_{12}/s_{22}$, $b_2 = s_{12}/s_{11}$, $r = s_{12}/\sqrt{s_{11}\, s_{22}}$ and β_1, β_2 are the true regression coefficients of the two variables, in their regressions on each other.

Mohn [43]

37. Show that the unconditional or marginal distribution of the matrix B of regression coefficients, in the regression of

one vector on another, which was defined in (4.4.2) is

$$\text{const. } |\Sigma_{11}^{-1} + (B-\beta)'\Sigma_{22\cdot 1}^{-1}(B-\beta)|^{-(n+p-k-1)/2}\, dB .$$

Kshirsagar [31]

38. R is the correlation matrix of p random variables. Show that

$$|R| \leq \{1 + (p-1)\,\bar{r}\}\,(1-\bar{r})^{p-1}$$

where

$$\bar{r} = 2 \sum_{j}\sum_{\substack{k \\ j\neq k}} r_{jk}/p(p-1) .$$

Show that the equality sign occurs if and only if $r_{jk} = \bar{r}$, for all j, k.

Aitkin, M.A., et.al. [1]

39. Let R be the sample correlation matrix calculated from a sample of size n+1 drawn from a p-variate normal population, in which the true correlation matrix is P. Show that the limiting distribution of

$$n^{1/2}(|R| - |P|)$$

is normal with mean 0 and variance

$$2|P|^2\,(\text{tr } P^2 - p).$$

r_{ij}, ρ_{ij} are typical elements of R and P.

Show further that the limiting covariance between $n^{1/2}r_{ij}$ and $n^{1/2}r_{k\ell}$ is

$$\tfrac{1}{2}\,\rho_{ij}\rho_{k\ell}(\rho_{ik}^2 + \rho_{i\ell}^2 + \rho_{jk}^2 + \rho_{j\ell}^2) + \rho_{ik}\rho_{j\ell} + \rho_{i\ell}\rho_{jk}$$
$$-(\rho_{ij}\rho_{ik}\rho_{i\ell} + \rho_{ji}\rho_{jk}\rho_{j\ell} + \rho_{ki}\rho_{kj}\rho_{k\ell} + \rho_{\ell i}\rho_{\ell j}\rho_{\ell k}).$$

Olkin and Siotoni [45]

40. $\bar{\underline{x}}$ and S represent the sample mean vector and the matrix of the corrected s.s. and s.p. of observations in a sample of size n from $N_p(\underline{x}|\underline{\mu}|\Sigma)$. Hotelling's T^2 for testing the hypothesis $\underline{\mu} = \underline{\mu}_o$ is

$$T^2 = n(n-1)(\bar{\underline{x}}-\underline{\mu}_o)' \, S^{-1}(\bar{\underline{x}}-\underline{\mu}_o) \; .$$

A statistician, by mistake, used

$$T'^2 = n(n-1)(\bar{\underline{x}}-\underline{\mu}_o)' \, A^{-1}(\bar{\underline{x}}-\underline{\mu}_o)$$

where A is the matrix of the s.s. and s.p. of the sample observations measured about the true means $\underline{\mu}_o$. Find the relation between T'^2 and T^2 and hence derive the distribution of T'^2.

41. Show that Hotelling's T^2, obtained in (5.4.63), for testing the significance of contrasts, can be expressed as

$$\frac{T^2}{n-1} = n\left\{\bar{\underline{x}}' \, S^{-1} \, \bar{\underline{x}} - \frac{(\bar{\underline{x}}'S^{-1}E_{p1})^2}{E_{1p} \, S^{-1} \, E_{p1}}\right\}$$

and is thus independent of the choice of the contrast matrix C. Hence show that the test can be generalized to test the hypothesis $\underline{\mu} = \gamma \, \underline{a}$, where $\underline{a}$ is a given vector but γ is unspecified, by replacing E_{p1} above by $\underline{a}$, (see 5.5.58 also)

Williams, E. J. [61]

42. Samples of sizes $n_1, n_2, \ldots, n_q$ from q independent p-variate normal populations $N_p(\underline{x}|\underline{\mu}_\alpha|\Sigma)$, ($\alpha$ = 1, 2, ..., q) are drawn. $\bar{\underline{x}}_\alpha$ is the vector of the sample means and S_α is the matrix of the corrected s.s. and s.p. of observations for the αth

sample. Show that the hypothesis

$$H:\ \beta_1\underline{\mu}_1 + \beta_2\underline{\mu}_2 + \dots + \beta_q\underline{\mu}_q = \underline{\mu}$$

(where $\beta_1, \dots, \beta_q$ and $\underline{\mu}$ are specified) can be tested by using Hotelling T^2 defined by

$$\frac{T^2}{\sum_1^q n_\alpha - q} = (\sum_\alpha \beta_\alpha^2/n_\alpha)^{-1}\ (\sum_\alpha \beta_\alpha \bar{\underline{x}}_\alpha - \underline{\mu})'\ (\sum_\alpha S_\alpha)^{-1}$$

$$\times\ (\sum_\alpha \beta_\alpha \bar{\underline{x}}_\alpha - \underline{\mu}) .$$

Prove that, when the hypothesis is true,

$(f-p+1)T^2/fp$ has the F distribution with p and f-p+1 d.f.

where $f = \sum_1^\alpha n_\alpha - q$. Obtain the power function of this test. If the variance-covariance matrices of the q populations are not the same and if n_1 is the smallest of $n_1, n_2, \dots, n_q$, we define

$$y_{ir} = \beta_1 x'_{ir} + \sum_{\alpha=2}^{q} \beta_\alpha \sqrt{\frac{n_1}{n_\alpha}} \left(x_{ir}^\alpha - \frac{1}{n_1}\sum_{s=1}^{n_1} x_{is}^\alpha + \frac{1}{\sqrt{n_1 n_\alpha}} \sum_{s=1}^{n_\alpha} x_{is}^\alpha\right)$$

where x_{ir}^α is the rth observation on the ith variate in the αth population. Let Y be the $p \times n_1$ matrix of all the quantities y_{ir} $(i = 1, \dots, p;\ r = 1, \dots, n_1)$. Let

$$\bar{\underline{y}} = Y\, E_{n_1 1}/n_1 , \quad \text{and} \quad S_y = Y(I - n_1^{-1} E_{n_1 n_1})Y' .$$

Define

$$T^2 = n_1(n_1-1)(\bar{\underline{y}} - \underline{\mu})'\ S_y^{-1}(\bar{\underline{y}} - \underline{\mu}) .$$

Prove that

$$\frac{(n_1-p)T^2}{p(n_1-1)} \text{ has the F distribution with p and } n_1-p \text{ d.f.}$$

if the hypothesis H is true.

Anderson [2]

43. Samples of size N are available from three independent p-variate normal populations $N_p(\underline{x}|\underline{\mu}_\alpha|\Sigma_\alpha)$, ($\alpha$ = 1, 2, 3). Let $\underline{x}_{\alpha r}$ be the vector of the rth observation from the αth population (r = 1, 2, ..., N; α = 1, 2, 3). The hypothesis H: $\underline{\mu}_1 = \underline{\mu}_2 = \underline{\mu}_3$ is to be tested. Consider

$$\underline{y}_r = a_1\underline{x}_{1r} + a_2\underline{x}_{2r} + a_3\underline{x}_{3r},$$

$$\underline{z}_r = b_1\underline{x}_{1r} + b_2\underline{x}_{2r} + b_3\underline{x}_{3r},$$

where $a_1 + a_2 + a_3 = 0$, $b_1 + b_2 + b_3 = 0$ and the vectors (a_1, a_2, a_3), (b_1, b_2, b_3) are linearly independent. Show that H is equivalent to the hypothesis

$$E(\underline{y}_r) = 0, \quad E(\underline{z}_r) = 0 \quad (r = 1, 2, \ldots, N).$$

Hence show that H can be tested by the Hotelling's T^2 test using

$$T^2 = N(N-1)\,[\bar{\underline{y}}'|\bar{\underline{z}}']S^{-1}\left[\frac{\bar{\underline{y}}}{\bar{\underline{z}}}\right]$$

where $\bar{\underline{y}} = \sum_r \underline{y}_r/N$, $\bar{\underline{z}} = \sum_r \underline{z}_r/N$ and S is the matrix of the corrected s.s. and s.p. of the sample observations $\underline{y}_r$, $\underline{z}_r$.

Anderson [4]

44. S has the $W_p(S|f|\Sigma)$ distribution and $\underline{x}$ has an independent $N_p(\underline{x}|\underline{\mu}|\Sigma)$ distribution. Show that the distribution of

$$\underline{t} = S^{-1/2} (\underline{x} - \underline{\mu}) f^{1/2}$$

is

$$\frac{\Gamma(\frac{f+1}{2})}{(\pi f)^{p/2} \Gamma(\frac{f-p+1}{2})} \frac{d\underline{t}}{|I + \underline{t}\underline{t}'/f|^{(f+1)/2}} .$$

Transforming from $\underline{t}$ to $\underline{t}'\underline{t}$ and θ_1, θ_2, ..., θ_{p-1}, by the spherical polar transformation and integrating out the θ's, derive the distribution of Hotelling's T^2, given by

$$\frac{T^2}{f} = (\underline{x} - \underline{\mu})' S^{-1} (\underline{x} - \underline{\mu}) .$$

For the spherical polar transformation see p. 19 of Lancaster [36].

45. Using the notation in Section 3 of Chapter 5 and the conditional distribution of $(T_p^2 - T_k^2)/(f + T_k^2)$, derived in Theorem 3 of that section, obtain the marginal distribution of Rao's U statistic, when $\lambda_p^2 \neq \lambda_k^2$. Obtain also the marginal distribution of $(T_p^2 - T_k^2)$.

Subrahmaniams [57]

46. S has a $W_p(S|n|\Sigma)$ distribution; $\underline{x}_i$ (i = 1, 2, ..., m) have independent $N_p(\underline{x}|\underline{\mu}_i|\Sigma)$ distributions which are independent of S also. Let

$$\underline{x}_i' = [\underline{x}_{1i}'|\underline{x}_{2i}'] , \quad S = \begin{bmatrix} S_{11} & S_{12} \\ S_{21} & S_{22} \end{bmatrix} \begin{matrix} q \\ p-q \end{matrix} , \quad \underline{\mu}_i' = [\underline{\mu}_{1i}'|\underline{\mu}_{2i}']$$
$$\quad q \quad p-q \qquad\qquad q \quad\quad p-q \qquad\qquad\qquad q \quad p-q$$

$$\Sigma = \left[\begin{array}{c|c} \Sigma_{11} & \Sigma_{12} \\ \hline \Sigma_{21} & \Sigma_{22} \end{array}\right]\begin{matrix} q \\ p-q \end{matrix}, \quad \Delta_{qi}^2 = \underline{\mu}_{1i}' \Sigma_{11}^{-1} \underline{\mu}_{1i}$$
$$\qquad q \qquad p-q \qquad\qquad \Delta_{pi}^2 = \underline{\mu}_i' \Sigma^{-1} \underline{\mu}_i$$

$$R_i = \left\{1 + \underline{x}_{1i}'\left(S_{11} + \sum_{j=1}^{i-1} x_{1j}x_{1j}'\right)^{-1} \underline{x}_{1i}\right\} \Big/ \left\{1 + \underline{x}_i'\left(S + \sum_{j=1}^{i-1} \cdot \underline{x}_i\underline{x}_i'\right)^{-1} \underline{x}_i\right\} \qquad (i = 1, 2, \ldots, m).$$

If $\Delta_{pi}^2 = \Delta_{qi}^2$ for each i, show that R_i $(i = 1, \ldots, m)$ are independently distributed as const. $R_i^{(n-p+i-2)/2}$ $(1 - R_i)^{(p-q-2)/2}\, dR_i$. Show that the R_i's are independent of $S + \sum_1^m \underline{x}_j\underline{x}_j'$.

Khatri [27]

47. A $p \times p$ matrix B has the $W_p(B|f|I)$ distribution and the vectors $\underline{x}$ and $\underline{y}$ have independent $N_p(\underline{x}|\underline{0}|I_p)$ and $N_p(\underline{y}|\underline{0}|I_p)$ distributions, independent of B also. Show that the joint distribution of

$$\underline{u} = A^{-1/2}\underline{x} \quad \text{and} \quad \underline{v} = A^{-1/2}\underline{y}$$

where $A = B + \underline{x}\underline{x}' + \underline{y}\underline{y}'$ is

$$\frac{\Gamma(f+1)}{(2\pi)^p\, \Gamma(f-p+1)} \left|I - \underline{u}\underline{u}' - \underline{v}\underline{v}'\right|^{(f-p-1)/2} d\underline{u}\, d\underline{v}\,.$$

Hence show that the p.d.f. of the joint distribution of the two correlated Hotelling's T^2, viz. $1 - r = \underline{x}'A^{-1}\underline{x}$ and $1 - s = \underline{y}'A^{-1}\underline{y}$ is

$$\frac{\Gamma(f+1)}{4\pi\, \Gamma(p-1)\, \Gamma(f-p+1)} \int_{-\pi/2}^{\pi/2} \{rs - (1-r)(1-s)\sin^2\theta\}^{(f-p-1)/2} \times \cos^{p-2}\theta\, d\theta\,.$$

Show further that $\underline{u}$ and $\underline{z}$, where

$$\underline{z} = (I - \underline{u}\underline{u}')^{-1/2} \underline{v}$$

are independently distributed. Hence show that the covariance between r and s is $E(r) - E(r)\, E(\underline{z}'\underline{z}) + E\{r(\underline{z}'\underline{u})^2\} - \{E(r)\}^2$, which reduces to

$$\frac{-2p(f-p+2)}{(f+1)(f+2)^2(f+4)} .$$

Kshirsagar and Young [33]

48. In Chapter 6, Section 2, the sample discriminant function was obtained by maximizing the ratio of the between population s.s. to the within population s.s. If instead, the ratio of between population s.s. to the total s.s. is maximized, show that the discriminant function so obtained is proportional to the other discriminant function.

Healy [20]

49. $\underline{\mu}_1$ and $\underline{\mu}_2$ are the mean vectors of two multivariate normal distributions, with the same variance-covariance matrix Σ. Show that the discriminant function $\{\Sigma^{-1}(\underline{\mu}_1 - \underline{\mu}_2)\}'\underline{x}$ is good not only for these two populations, but for any population whose mean is of the form $a_1\underline{\mu}_1 + a_2\underline{\mu}_2$, where $a_1 + a_2 = 1$, and variance-covariance matrix Σ. This proves the sufficiency property of the discriminant function for the set of alternatives $a_1\underline{\mu}_1 + a_2\underline{\mu}_2$, which are means collinear with $\underline{\mu}_1$, $\underline{\mu}_2$.

Rao [51]

50. $\underline{\delta}' = [\delta_1, \delta_2, \delta_3]$ is the difference between the means of two trivariate normal populations, having the same correlation matrix $[\rho_{ij}]$. Show that the increase in the squared distance between the two populations, when the third variate is added to a discriminant function containing only the first two variates is

$$\frac{\left(1 - \rho_{12}^2\right)\left\{\delta_3 - \frac{\rho_{13}(\delta_1 - \rho_{12}\delta_2)}{1 - \rho_{12}^2} - \frac{\rho_{23}(\delta_2 - \rho_{12}\delta_1)}{1 - \rho_{12}^2}\right\}^2}{1 - \rho_{12}^2 - (\rho_{13} - \rho_{23})^2 - 2\,\rho_{13}\rho_{23}(1 - \rho_{12})} .$$

When is this quantity $> \delta_3^2$?

Cochran [9]

51. $\underline{x}$ is the vector of p multinormal variables, with variance-covariance matrix $(1 - \rho_1)I + \rho_1 E_{pp}$. $\underline{\delta}$ is the difference in the values of $E(\underline{x})$ in two populations, with the same variance-covariance matrix and Δ_p^2 is the squared (Mahalanobis) distance between them. If the $\underline{x}$'s are all uncorrelated, the distance is $(\underline{\delta}'\underline{\delta})^{1/2}$. Show that $\Delta_p^2 > \underline{\delta}'\underline{\delta}$ always, when $\rho_1 < 0$. But if $\rho_1 > 0$, $\Delta_p^2 < \underline{\delta}'\underline{\delta}$, unless

$$\rho_1 > \frac{(E_{1p}\underline{\delta})^2 - \underline{\delta}'\underline{\delta}}{(p-1)\,\underline{\delta}'\underline{\delta}} .$$

q more variables $\underline{y}$, such that $V(\underline{y}) = (1 - \rho_2)I + \rho_2 E_{qq}$, $\text{cov}(\underline{x}, \underline{y}) = \rho_3 E_{pq}$ are considered. $\underline{d}$ is the difference in the means of $\underline{y}$ in the two populations. Δ_{p+q}^2 is the squared distance between the two populations when all the p+q variates $\underline{x}$ and $\underline{y}$ are considered. Show that

$$\Delta^2_{p+q} - \Delta^2_p, \text{ where } \rho_3 = 0$$

$$= \frac{q\,\rho_3^2(\underline{\delta}'\,E_{p1})^2}{\{1+(p-1)\rho_1\}r} + \frac{p\,\rho_3^2(\underline{d}'\,E_{q1})^2}{\{1+(q-1)\rho_2\}r} - \frac{2\rho_3(E_{1p}\,\underline{\delta})(E_{1q}\,\underline{d})}{r}$$

where $r = \{1+(p-1)\rho_1\}\{1+(q-1)\rho_2\} - pq\,\rho_3^2$.

Hence show that the cross correlation ρ_3 is helpful in discrimination if ρ_3 is negative but is harmful if ρ_3 is positive, unless

$$\rho_3 > \frac{2(E_{1p}\,\underline{\delta})(E_{1q}\,\underline{\delta})}{\dfrac{q(E_{1p}\,\underline{\delta})^2}{1+(p-1)\rho_1} + \dfrac{p(E_{1q}\,\underline{d})^2}{1+(q-1)\rho_2}} .$$

Cochran [9]

52. Consider two p-variate normal populations with different means and the same variance-covariance matrix. Samples of sizes n_1, n_2 are available from these two populations and $\underline{b}'\underline{x}$ is the sample discriminant function constructed in the usual way, using Fisher's method. Show that any other linear function $\underline{a}'\underline{x}$ will also be good enough as a discriminant function and will not differ significantly (level of significance $100\alpha\,\%$) from $\underline{b}'\underline{x}$, if its correlation (estimated) with $\underline{b}'\underline{x}$ is at least as large as F^*, where

$$F^{*2} = 1 - \frac{F'_o}{R^2(1+F'_o)} ,$$

$$F'_o = \frac{p-1}{n_1+n_2-p-1}\,F_o ,$$

F_o = 100α % point of the F distribution with p-1 and $n_1 + n_2 - p - 1$ d.f.

and

$$\frac{(n_1 + n_2 - 2)(n_1 + n_2)R^2}{n_1 n_2 (1 - R^2)} = \text{Mahalanobis's } D_p^2 \text{ for the two populations.}$$

Fisher [13]

53. $\underline{d}$ denotes the vector of the difference between the sample means of two p-variate normal populations and W is the pooled sample estimate of the common variance-covariance matrix of the two populations. $G = [\underline{g}_1 | \underline{g}_2 | \ldots | \underline{g}_k]$ is a $p \times k$ matrix of rank k. $\underline{\ell}$ is any vector satisfying the constraints $\underline{\ell}'W\underline{\ell}$ = constant, $\underline{\ell}'G = 0$. Show that

$$\frac{(\underline{\ell}'\underline{d})^2}{\underline{\ell}'W\underline{\ell}} \leq \frac{(\underline{\ell}^{*\prime}\underline{d})^2}{\underline{\ell}^{*\prime}W\underline{\ell}^*},$$

when $\underline{\ell}^* = W^{-1}\{I - G(G'W^{-1}G)^{-1}G'W^{-1}\}\underline{d}$.

Use this result to show that $\underline{\ell}^{*\prime}\underline{x}$ is a discriminant function whose value does not change, if the point $\underline{x}$ is moved in any of the directions $\underline{g}_1, \ldots, \underline{g}_k$. If the point $\underline{x}$ is moved in any other direction, the change in the value of $\underline{\ell}^{*\prime}\underline{x}$ is the maximum possible under the stated constraints.

Burnaby [6]

54. Using the notation of Section 6 of Chapter 6, show that the "Jackknifed" discriminant function $W_r^*(\underline{x}_r)$ can be expressed in terms of the "whole" discriminant function $W(\underline{x}_r)$ (see

6.6.1) as $8f(n_1 - 1)^2\, W_r^*(\underline{x}_r)$

$$= \frac{(f-1)a\,(\lambda-1)^2\,[W^2(\underline{x}_r) - \gamma_r^2 D_p^2] - 4\lambda\gamma_r^2 - \lambda D_p^2 + 2(\lambda^2+1)W(\underline{x}_r)}{1 - a\,\gamma_r^2 - a\,D_p^2/4 + a\,W(\underline{x}_r)}$$

where

$$\gamma_r^2 = f\,\underline{z}_r'\,S^{-1}\,\underline{z}_r\,, \quad \underline{z}_r = \underline{x}_r - \tfrac{1}{2}(\underline{\bar{x}} + \underline{\bar{y}})$$

$$\lambda = 2n_1 - 1\,, \qquad a = n_1/f(n_1 - 1)\ .$$

Hence show that $\underline{x}_r$ will be misclassified by $W_r^*(\underline{x})$ i.e. $W_r^*(x_r) \le 0$, when

$$W(\underline{x}_r) \le -\frac{(\lambda^2+1)}{a(\lambda-1)^2} + \left[\left\{\frac{\lambda^2+1}{a(\lambda-1)^2}\right\}^2 + \gamma_r^2 D_p^2 + \frac{4\lambda\gamma_r^2}{a(\lambda-1)^2} + \frac{\lambda D_p^2}{a(\lambda-1)^2}\right]^{1/2} = A,\ \text{say.}$$

Hence show that the proportion of misclassified observations in the first sample is not m_1/n_1, as $W(x)$ incorrectly shows but is in fact $(m_1+m_1')/n_1$, where m_1 is the number of observations in the first sample for which $W(\underline{x}) \le 0$ and m_1' is the additional number of observations for which $W(\underline{x}) > 0$ but $\le A$, given above.

Hills [21]

55. Using Anderson's classification statistic $W(\underline{x})$ as defined in (6.6.1), to classify a future observation $\underline{x}$ independent of the preliminary samples from which W is calculated and using the same notation as in Chapter 6, show that $(i = 1, 2)$

$$E(W(\underline{x})|\bar{\underline{x}}, \bar{\underline{y}}, S, \underline{x} \in \pi_i) = W(\underline{\mu}(i))$$

and

$$V(W(\underline{x})|\bar{\underline{x}}, \bar{\underline{y}}, S, \underline{x} \in \pi_i) = V_1(W(\underline{x})) = f^2 \underline{d}' S^{-1} \Sigma S^{-1} \underline{d} .$$

From these, derive the following unconditional expectations

$$E(W(\underline{x})|\underline{x} \in \pi_i) = \frac{f}{2(f-p-1)}\left[(-1)^{i+1} \Delta_p^2 - \frac{p(n_2-n_1)}{n_1 n_2}\right]$$

and

$$E\{V_1(W(\underline{x}))\} = \frac{f^2(f-1)}{(f-p)(f-p-1)(f-p-3)}\left[\Delta_p^2 + p\left(\frac{1}{n_1} + \frac{1}{n_2}\right)\right] .$$

Hint: Use (4.7.12)].

Lachenbruch [35]

56. Using the delta technique of obtaining the large sample standard errors of functions of random variables, show that the asymptotic variance covariance matrix of the vector of discriminant function coefficients viz. $\underline{\ell} = f S^{-1} \underline{d}$ (using notation of Section 5, Chapter 6) is

$$\left(\frac{1}{n_1} + \frac{1}{n_2}\right) \Sigma^{-1} + \frac{1}{f} \Delta_p^2 \Sigma^{-1} + \frac{1}{f} \Sigma^{-1} \underline{\delta\delta}' \Sigma^{-1} .$$

Compare this with the exact expression (6.5.17).

Kendall and Stuart [26]

57. If $\underline{x}$ and $\underline{y}$ are two p and q component random vectors, show that the addition of extra variates to either $\underline{x}$ or $\underline{y}$ can never decrease any of the canonical correlations between $\underline{x}$ and $\underline{y}$, in absolute value.

Chen [8]

58. The variance-covariance matrix of p + q variables $\underline{x}$, $\underline{y}$ is

$$\left[\begin{array}{c|c} (r_1 - \lambda_{11})I_p + \lambda_{11}E_{pp} & \lambda_{12}\, E_{pq} \\ \hline \lambda_{12}\, E_{qp} & (r_2 - \lambda_{22})I_q + \lambda_{22}E_{qq} \end{array}\right]\begin{matrix} p \\ q \end{matrix} .$$

$$\qquad p \qquad\qquad\qquad q$$

Obtain the canonical correlations between $\underline{x}$ and $\underline{y}$ and also obtain the canonical variables.

59. Consider the design of experiments model

$$y_{ijk} = (\mu + \alpha_i + \beta_j + e_{ijk}), \quad i = 1, 2, \ldots, v; \quad j = 1, 2, \ldots, b$$

$$n_{ij} \text{ is an integer} \geq 0 \qquad k = 1, 2, \ldots, n_{ij}$$

where μ, α_i, β_j are fixed, and e_{ijk} are normal independent variables with zero means and a common variance σ^2. Define $N = [n_{ij}]$ of order $v \times b$.

$$T_i = \sum_j \sum_k y_{ijk}, \quad B_j = \sum_i \sum_k y_{ijk}, \quad r_i = \sum_j n_{ij}, \quad k_j = \sum_i n_{ij},$$

$$\underline{T}' = [T_1, \ldots, T_v], \quad \underline{B}' = [B_1, \ldots, B_b],$$

$$D_1 = \text{diag}(r_1, \ldots, r_v), \quad D_2 = \text{diag}(k_1, \ldots, k_b),$$

$$\underline{Q} = \underline{T} - N D_2^{-1} \underline{B}, \quad \underline{P} = \underline{B} - N' D_1^{-1} \underline{T} .$$

Show that the number of nonzero canonical correlations between $\underline{Q}$, $\underline{P}$ is equal to the rank of the matrix $C D_1^{-1} N$, where

$$C = D_1 - N D_2^{-1} N' .$$

Show further that if rank of C is v - 1, all the canonical correlations between $\underline{Q}$ and $\underline{P}$ will be zero, if and only if $n_{ij} = r_i k_j$ for all i, j.

Chakrabarti [7]

60. r_1, r_2 are the sample canonical correlations in a sample of size N, between $\underline{x}$ and $\underline{y}$, where $\underline{x}$ has two elements and $\underline{y}$ has p elements. Let $q = r_1 r_2$ and $z = (1 - r_1^2)(1 - r_2^2)$. Obtain the double Mellins transform of the p.d.f. of q and z and by inverting it, derive the distribution of q and z.

Kabe [23]

61. $r_1^2, r_2^2, \ldots, r_p^2$ are the squares of the sample canonical correlations between a p-vector and an independent q-vector $(p < q)$. Let $\phi_i = r_i^2/(1 - r_i^2)$. Show that the joint distribution of the elementary symmetric functions of ϕ_i viz.

$$\eta_1 = \Sigma\, \phi_i, \quad \eta_2 = \Sigma\, \phi_i \phi_j, \quad \eta_3 = \Sigma\, \phi_i \phi_j \phi_k, \ldots, \eta_p = \phi_1 \phi_2 \cdots \phi_p$$

is (under the assumption of normality of the vectors)

$$\text{const. } \eta_p^{(q-p-1)/2} (1 + \eta_1 + \ldots + \eta_p)^{-n/2}\, d\eta_1 \ldots d\eta_p\,,$$

n being the d.f. on which the matrix of the corrected s.s. and s.p. of sample observations is based.

62. The sample canonical correlations $r_1 > \ldots > r_p$ between two vectors $\underline{x}$ and $\underline{y}$ of p and q components $(p \leq q)$ respectively correspond to the roots of the determinantal equation in r^2,

$$|-(1 - r^2)S_{22} + S_{22\cdot 1}| = 0$$

where S_{11}, S_{12}, S_{22} etc. are all defined in Section 3 of Chapter 7. The residual canonical correlations between $\underline{x}$, $\underline{y}$, when x_1 is eliminated are defined as $\phi_1 > \ldots > \phi_{p-1}$ where the ϕ's are the roots of the determinantal equation

$$| - (1-\phi^2) S_{22\cdot x_1} + S_{22.1}| = 0 .$$

Let $\underline{u}$, $\underline{v}$ be the sample canonical variables corresponding to $\underline{x}$ and $\underline{y}$. Show by using (7.3.11), that $\underline{x} = D'\underline{u}$, where D is defined by (7.3.11). Transforming from $\underline{x}$, $\underline{y}$ to $\underline{u}$ and $\underline{v}$, show that the above determinantal equation, yielding residual canonical correlations, reduces to

$$\sum_{i=1}^{p} \frac{d_i^2(1-r_i^2)}{r_i^2 - \phi^2} = 0 ,$$

where d_i are the elements of the first column of D. Hence or otherwise show that

(a) $$d_i^2 = \frac{s(1-\eta)}{(1-r_i^2)} \cdot \frac{\prod_{k=1}^{p-1} (r_i^2 - \phi_k^2)}{\prod_{j \neq i} (r_i^2 - r_j^2)} ,$$

where $\eta = \sum_1^p d_i^2 \, r_i^2 / \sum_1^p d_i^2$, $s = \sum_1^p d_i^2$.

(b) $$(1-\eta) \prod_k (1-\phi_k^2) = \prod_{i=1}^{p} (1-r_i^2) .$$

(c) $$\sum_k \phi_k^2 = \sum_i r_i^2 - (\eta - \sum_i d_i^2 \, r_i^4)/(1-\eta) .$$

Kshirsagar [30]
Williams [61]

63. With r_i, d_i, ϕ_i defined as in the previous exercise, show that the Jacobian of transformation from $d_1, \ldots, d_p$ to $\phi_1^2, \ldots, \phi_{p-1}^2$ and s is the absolute value of

$$s(p/2)^{-1} \frac{\prod_i (1 - r_i^2)^{(p-1)/2} \prod_{h \neq k} (\phi_h^2 - \phi_k^2)}{\prod_k (1 - \phi_k^2)^{p/2} \prod_{i=1}^{p} \prod_{k=1}^{p-1} (r_i^2 - \phi_k^2)^{1/2}} .$$

The distribution of the matrix D is given by (7.6.8). Using (7.6.8) with $\Sigma = I$, show that the conditional distribution of $\phi_1^2, \ldots, \phi_{p-1}^2$ when r_i^2 are fixed is

$$\frac{\text{const.} \prod (1 - r_i^2)^{(p-1)/2} \prod_{h \neq k} (\phi_h^2 - \phi_k^2) \, \pi \, d\phi_k^2}{\pi (1 - \phi_k^2)^{p/2} \, \pi (r_i^2 - \phi_k^2)^{1/2}} .$$

Williams [61]

64. The $p \times n$ matrix X represents a random sample of size n from a p-variate population with mean vector $\underline{\mu} = \underline{0}$. $\bar{\underline{x}}$ is the vector of the sample means and $S = [s_{ij}]$ is the matrix of corrected s.s. and s.p. $\underline{u}$ is the column vector of all the $p(p+1)/2$ distinct elements of S and $\lambda_1, \lambda_2, \ldots, \lambda_p$ are the canonical correlations between $\bar{\underline{x}}$ and $\underline{u}$. β_{1p} is defined as $2 \sum_1^p \lambda_i^2$. Obtain β_{1p} in terms of the moments of the distribution of $\underline{x}$, neglecting cumulants of order higher than 3, of $\underline{x}$ and conside g second order moments of $\bar{\underline{x}}$ and $\underline{u}$ only to order n^{-1}. Explain how β_{1p} can be regarded as a measure of multivariate skewness. Define

$$\beta_{2p} = E\{(\underline{x} - \underline{\mu})' \Sigma^{-1} (\underline{x} - \underline{\mu})\}^2 .$$

Show how it measures the kurtosis of the multivariate distribution of $\underline{x}$. Show that $\beta_{1p} = 0$ and $\beta_{2p} = p(p+2)$ for a p-variate normal distribution. Define b_{1p} and b_{2p} as sample quantities corresponding to β_{1p}, β_{2p}. When the p.d.f. of $\underline{x}$ is $N_p(\underline{x}|\underline{\mu}|\Sigma)$, prove the following results

(i) $n\, b_{1p}/6$ is asymptotically a χ^2 with $p(p+1)(p+2)/6$ d.f.

(ii) $E(b_{2p}) = p(p+2)(n-1)/(n+1)$

(iii) $V(b_{2p}) = 8\, p(p+2)/n + O(n^{-2})$

(iv) $\{b_{2p} - E(b_{2p})\}/\{V(b_{2p})\}^{1/2}$ is asymptotically $N(0,1)$.

Mardia [38]

65. A $p \times p$ symmetric matrix L has the distribution (multivariate Beta) $f(L)dL = \text{const.}\ |L|^{(q-p-1)/2}\ |I - L|^{(n-q-p-1)/2}\ dL$.

Let $L = TT'$, where T is l.t. and has elements t_{ij} $(i < j)$. Let T be partitioned as

$$\begin{bmatrix} t_{11} & 0 \\ \underline{t} & T_2 \end{bmatrix} \begin{matrix} 1 \\ p-1 \end{matrix}$$
$$\begin{matrix} 1 & p-1 \end{matrix}$$

and let $I - T_2T_2' = MM'$, where M is again l.t.

Show that T_2, t_{11} and $\underline{z}_{(1)} = (1 - t_{11}^2)^{-1/2}\, M^{-1}\underline{t}$ are independently distributed. Show further that T_2 has the same p.d.f. as T with n replaced by n-1 and p replaced by p-1, t_{11}^2 has a beta distribution, while the distribution of $\underline{z}_{(1)}$ is

$$\text{const.}\ (1 - \underline{z}_{(1)}'\underline{z}_{(1)})^{(q-p-1)/2}\ d\underline{z}_{(1)}\ .$$

From $\underline{z}_{(1)}' = (z_{21}, z_{31}, \ldots, z_{p1})$, transform to

z_{21} and $y_{31} = z_{31}/(1-z_{21}^2)^{1/2}, \ldots, y_{p1} = z_{p1}/(1-z_{21}^2)^{1/2}$ and hence show that z_{21} is independent of $y_{31}, \ldots, y_{p1}$ and has a beta distribution.

Continue this decomposition of the distribution of L, by repeating the above for T_2 and $(y_{31}, \ldots, y_{p1})$. Finally, when this decomposition is completed, observe that f(L)dL can be written as

$$\prod_{i=1}^{p} B\left(t_{ii}^2 \middle| \frac{n-(i-1)-q}{2}, \frac{q}{2}\right)dt_{ii}^2 \prod_{i=1}^{p-1} \prod_{r=i+1}^{p} B\left(\xi_{ri}^2 \middle| \frac{1}{2}, \frac{q-(r-i)}{2}\right)d\xi_{ri}^2 .$$

This is the decomposition of the multivariate beta distribution of L into separate independent univariate beta distributions. $(B(z|\ell, m))$ is defined by (4.2.25).

66. A symmetric matrix L of order $p \times p$, such that both L and $I - L$ are positive definite has the p.d.f.

$$G_p(L|q|n-q) = C(p, q, n)\ |L|^{(q-p-1)/2}|I-L|^{(n-q-p-1)/2},$$

where $C(p, q, n)$ is the constant in the p.d.f. and $n > p+q$, $p < q$.

(a) Show that

$$C(p, q, n) = \prod_{i=1}^{p} \Gamma\left(\frac{n-i+1}{2}\right)/\pi^{p(p-1)/4} \prod_{i=1}^{p} \left\{\Gamma\left(\frac{q-i+1}{2}\right)\Gamma\left(\frac{n-q-i+1}{2}\right)\right\}.$$

(b) If $M = nL$, show that the asymptotic distribution of M is (as $n \to \infty$) $W_p(M|q|I)dM$.

(c) The p.d.f. of $I - L$ is $G_p(I - L|n - q|q)$.

(d) For a fixed vector $\underline{a}$, $\underline{a}'L\underline{a}/\underline{a}'\underline{a}$ has the beta distribution with parameters q/2 and (n-q)/2.

(e) If S has the $W_p(s|f|\Sigma)$ distribution and is independent of L, then $T_1 = S^{1/2}\, L(S^{1/2})'$ and $T_2 = S^{1/2}(I - L)(S^{1/2})'$ are independently distributed as $W_p(T_1|q|\Sigma)$ and $W_p(T_2|n-q|\Sigma)$ respectively.

(f) If L* has the p.d.f. $G_p(L^*|\, n\, |m)$ and is independent of L, then

$$M = (L^*)^{1/2}\, L(L^{*1/2})'$$

has the distribution $G_p(M|q|n+m-q)$.

(g) Partition L as

$$\left[\begin{array}{c|c} L_{11} & L_{12} \\ \hline L_{21} & L_{22} \end{array}\right]\begin{array}{l} p-m \\ m \end{array} \qquad \text{and} \quad q \geq p-m\,.$$
$$\begin{array}{cc} p-m & m \end{array}$$

Also let $L_{22\cdot 1} = L_{22} - L_{21}L_{11}^{-1}L_{12}$. Show that the p.d.f. of $L_{22\cdot 1}$ is $G_m(L_{22\cdot 1}|q-p+m|n-q)$ and that it is independently distributed of L_{11} which has the p.d.f. $G_{p-m}(L_{11}|q|n-q)$.

(h) Let L_i be the matrix of the first i rows and columns of L, with $|L_0|$ defined as 1. Show that $|L_i|/|L_{i-1}|$, (i = 1, 2, ..., p) are independent beta variables with parameters (q+1-i)/2 and (n-q)/2.

(i) A is any p x p orthogonal matrix, of constant elements. Show that the distribution of M = ALA' is $G_p(M|q|n-q)dM$.

(j) For a fixed vector $\underline{a}$, show that $\underline{a}'\underline{a}/\underline{a}'L^{-1}\underline{a}$ has the beta distribution with parameters $(q-p+1)/2$ and $(n-q)/2$.

Mitra [42]

67. Prove (9.2.16) explicitly. It gives the relation between the vectors $\underline{m}$ and $\underline{g}$.

[Hint: (7.3.5) shows that $\underline{m}$ is proportional to $C_{yy}^{-1}C_{yx}\underline{\ell}$.
(9.2.15) shows that $\underline{\ell} = A^{-1}\underline{z} = A^{-1}U\underline{g}$.
Use (9.1.34) for C_{yy}^{-1}, C_{yx}.
From (9.2.15), $(I - \underline{h}\underline{h}')U'A^{-1}U\underline{g} = \theta\underline{g}$.
Putting all these together, and using

$$\sqrt{n_1}\, g_1 + \dots + \sqrt{n_k}\, g_k = 0,$$

the required result follows.]

68. Consider k groups of individuals and p+1 attributes A_o, A_1, ..., A_p. We record for each individual, whether he possesses the attribute A_j or not ($j = 0, 1, 2, \dots, p$). Define a variable x_i ($i = 1, 2, \dots, p$) which takes the value 1, if an individual possesses the attribute A_i and is zero, otherwise. Let n_{ij} be the number of individuals in the ith group, possessing the attribute A_j. A multivariate analysis of variance table giving "between groups", within groups and "total" matrices of s.s. and s.p. is constructed for the p variables $x_1, \dots, x_p$. Show that the between groups matrix is $B = [brs]$, where

$$brs = \sum_{i=1}^{k} \frac{n_{ir}n_{is}}{n_{i.}} - \frac{n_{.r}n_{.s}}{N} \qquad (r, s = 1, \dots, p)$$

where

$$N = \sum_{i=1}^{k} \sum_{j=0}^{p} n_{ij} \quad \text{and} \quad n_{i.} = \sum_{j=0}^{p} n_{ij}, \quad n_{.j} = \sum_{i=1}^{k} n_{ik}.$$

Similarly show that the total matrix is $A+B$, where $A=[a_{rs}]$ and

$$a_{rs} + b_{rs} = n_{.r} - n_{.r}^2/N, \quad \text{if } r = s,$$

$$= -n_{.r}n_{.s}/N, \quad \text{if } r \neq s.$$

When $k=2$, define $f_r = n_{1r}n_{2.} - n_{2r}n_{1.}$ $(r=1, 2, \ldots, p)$ and let $\underline{f}$ be the column vector of these elements. Show that the optimum scores to be assigned to the attributes A_o, A_1, ..., A_p are 0, h_1, ..., h_p where the vector $\underline{h}$ of the elements h_1, ..., h_p is proportional to $(A+B)^{-1}\underline{f}$. Hence show that we may assign the scores

$$\frac{n_{1r}n_{2.} - n_{2r}n_{1.}}{n_{.r}} \qquad (r = 0, 1, 2, \ldots, p)$$

to A_o, A_1, ..., A_p respectively (see Section 6 of Chapter 9).

69. Consider a Markov chain of order 1 and s states. Define a variable $N_u(t)$ as

$$N_u(t) = 1, \quad \text{if the process is in state u, at time t}$$

$$= 0, \quad \text{otherwise}, \qquad (u = 1, 2, \ldots, s).$$

Determine optimum scores β_1, ..., β_s to be assigned to the s states, by maximizing the correlation between

$$x(t) = \underline{\beta}'\underline{N}(t) \quad \text{and} \quad x(t+1) = \underline{\beta}'\underline{N}(t+1),$$

where

$$\underline{\beta}' = [\beta_1, \ldots, \beta_s] \quad \text{and} \quad \underline{N}'(t) = [N_1(t), \ldots, N_s(t)].$$

Show that the optimum scores are provided by the eigenvector of the transition probability matrix of the Markov chain, corresponding to the largest eigenvalue, excluding the eigenvalue that is unity.

Patel [49]

70. $S = [s_{ij}]$ is the matrix of corrected s.s. and s.p. of N observations from a bivariate normal population with mean vector $\underline{\mu}$ and variance-covariance matrix $\Sigma = (\sigma_{ij})$. ρ is the population correlation coefficient. If it is given that $\sigma_{11} = \sigma_{22}$, show that the likelihood ratio criterion to test the null hypothesis $\rho = \rho_o$, is a function of the statistic

$$u = \frac{2\, s_{12}}{s_{11} + s_{22}} .$$

Obtain the distribution and moments of u, when $\rho \neq 0$ and $\sigma_{11} \neq \sigma_{22}$.

Mehta and Gurland [40]

71. Obtain the likelihood ratio criterion for testing the hypothesis that the correlation matrix of $\underline{x}$ (which has the $N_p(\underline{x}|\underline{\mu}|\Sigma)$ distribution) is

$$(1 - \rho)I + \rho\, E_{pp} ,$$

where ρ is not specified. What is the m.l.e. of ρ?

Aitken, M. A. et.al. [1]

72. $\underline{x}$ has a p-variate normal distribution with mean $\underline{\theta}$ and variance-covariance matrix $\Sigma = \sigma^2(1-\rho)I + \rho E_{pp}$. Define the regions

$$\omega_1 = \{\underline{\theta}, \Sigma: \theta_1 = \dots = \theta_p, \Sigma > 0\}$$

$$\omega_2 = \{\underline{\theta}, \Sigma: \theta_1 = \dots = \theta_k, -\infty < \theta_j < \infty, \quad j = k+1, \dots, p, \quad \Sigma > 0\}$$

$$\omega_3 = \{\underline{\theta}, \Sigma: -\infty < \theta_j < \infty, \quad j = 1, 2, \dots, p, \quad \Sigma > 0\},$$

where θ_j are the elements of $\underline{\theta}$ and $\Sigma > 0$ means Σ is positive definite. Obtain likelihood ratio statistics and their central and noncentral distributions, for the following hypotheses:

H_{13}: $(\theta, \Sigma) \in \omega_1$ versus ω_3

H_{12}: $(\theta, \Sigma) \in \omega_1$ versus ω_2

H_{23}: $(\theta, \Sigma) \in \omega_2$ versus ω_3.

Wilks [59]

Olkin and Shrikhande [46]

73. Show that the likelihood ratio criterion $-2 \log_e \lambda_b$ of (11.7.23) can be expressed as

$$nk \log_e(n/e) - n \log |G_1| + \mathrm{tr}(G_1 + G_2)$$

where G_1 and G_2 have independent Wishart distributions with $n - (p-k)$ and $(p-k)$ d.f. respectively. Hence obtain the characteristic function of $-2 \log_e \lambda_b$. From the characteristic function, deduce that

$$E(-2 \log_e \lambda_b) = k + f_a + n(\mathrm{tr}\, \Lambda - k - \log_e |\Lambda|) + O(n^{-1})$$

where

$$\Lambda = \Delta_1^{-1/2} (\Sigma_{11}^* - \Sigma_{12}^* \Sigma_{22}^{*-1} \Sigma_{21}^*)\Delta_1^{-1/2} ,$$

$$\Sigma^* = \begin{array}{c} \left[\begin{array}{c|c} \Sigma_{11}^* & \Sigma_{12}^* \\ \hline \Sigma_{21}^* & \Sigma_{22}^* \end{array}\right] \begin{array}{l} k \\ p-k \end{array} \\ \quad k \qquad p-k \end{array} = \left[\begin{array}{c|c} \Gamma_1' \Sigma \Gamma_1 & \Gamma_1' \Sigma \Gamma_2 \\ \hline \Gamma_2' \Sigma \Gamma_1 & \Gamma_2' \Sigma \Gamma_2 \end{array}\right] .$$

Mallows [37]

74. A sample of size N is drawn from a multivariate normal population whose variance-covariance matrix is

$$\Sigma = \Psi + \sigma^2 I_p ,$$

where Ψ is a positive semi-definite matrix of rank m. Obtain the maximum likelihood estimate of σ^2.

Anderson [3]

75. The variance-covariance matrix of $u = p(p-1)/2$ variables is $\sigma^2 I_u + P$, where $P = [p_{rs}]$ and the rows and columns of P are numbered by pairs of integers. For example,

$$r = (i, j);\ i, j = 1, 2, \ldots, p;\ i < j$$

$$s = (k, \ell);\ k, \ell = 1, 2, \ldots, p;\ k < \ell .$$

$p_{rs} = 1$, if i, j, k, ℓ are all distinct and zero otherwise. Show that P satisfies the equation,

$$P^3 - \left\{\binom{p-2}{2} - p + 4\right\} P^2 - \left\{\binom{p-2}{2}(p-4) + (p-3)\right\} P + (p-3)\binom{p-2}{2} I = 0 .$$

Hence obtain the variances of all the principal components of the variables. Show further that the total of all the variables is a principal component.

Bartlett [5]

76. Show that the distribution of the matrix U defined in (11.6.2), when $\delta_1 = \dots = \delta_p = \lambda$ is

$$\text{const. } \exp\left\{-\frac{1}{4\lambda^2} \text{tr } UU'\right\} dU .$$

Hence show that the distribution of the eigenvalues $h_1 > \dots > h_p$ of U is

$$\frac{1}{2^{p(p+3)/4} \lambda^{p(p+1)/2} \prod_{i=1}^{p} \Gamma\left(\frac{p+1-i}{2}\right)} \prod_{i<j} (h_i - h_j)$$

$$\times \exp\left(-\frac{1}{4\lambda^2} \text{tr } H^2\right) dH$$

where $H = \text{diag}(h_1, \dots, h_p)$.

Anderson [3]

77. S is the matrix of corrected s.s. and s.p. of sample observations when a sample of size $N = n+1$ is drawn from a p-variate normal population whose variance-covariance matrix is Σ. The largest eigenvalue of Σ is δ_1, the corresponding unit eigenvector is $\underline{\gamma}$. All other eigenvalues of Σ are less than δ_1 but are all equal to δ. Show that $\delta^{-1}(\text{tr } S - \underline{\gamma}' S \underline{\gamma})$ is a χ^2 with $n(p-1)$ d.f. Show further that this χ^2 can be broken up into independent χ^2 variables with p-1 and (p-1)(n-1) d.f., the χ^2 with p-1 d.f. being

$$\eta = \delta^{-1}\left\{\frac{\underline{\gamma}' S^2 \underline{\gamma}}{\underline{\gamma}' S \underline{\gamma}} - \underline{\gamma}' S \underline{\gamma}\right\} .$$

If $\underline{\gamma}'\underline{x}$ is expressed as $e_1u_1 + \dots + e_pu_p$, where $u_1, \dots, u_p$ are the sample principal components, show that η can be expressed also as

$$\eta = \sum_{1}^{p} f_i - \sum_{1}^{p-1} \phi_k - \lambda^2$$

where $f_1, \ldots, f_p$ are the eigenvalues of S, $\lambda^2 = \underline{\gamma}' S \underline{\gamma}$ and the ϕ_k's are the roots of the determinantal equation in ϕ

$$|f_i \delta_{ij} - \frac{1}{\lambda^2} e_i e_j f_i f_j - \phi \delta_{ij}| = 0$$

δ_{ij} being the Kronecker delta. Show that the ϕ_k's can be interpreted as the residual eigenvalues of S when $\underline{\gamma}'\underline{x}$ is eliminated.

Kshirsagar [32]

78. S is the matrix of corrected s.s. and s.p. of the sample observations when a sample of size n from a p-variate normal population with variance-covariance matrix diag($\sigma_{11}, \sigma_{22}, \ldots, \sigma_{pp}$). Q is a l.t. matrix with elements $q_{ij} (i \leq j)$ such that S = QQ'. If R_i is the sample multiple correlation coefficient between x_i and $x_1, \ldots, x_{i-1}$ $(i = 2, 3, \ldots, p)$, show that

$$R_i^2 = \sum_{j=1}^{i-1} q_{ij}^2 \Big/ \sum_{j=1}^{i} q_{ij}^2 \ .$$

Write down the distribution of Q and by making a transformation from q_{ii} to R_i^2 $(i = 2, 3, \ldots, p)$, obtain the joint distribution of all the R_i^2's and the q_{ij}'s $(i < j)$. By integrating out the q_{ij}'s, show that the R_i^2's are independently distributed as beta variables.

Roy, S. N., and Bargemann [53]

79. R is the correlation matrix obtained from a sample of size n drawn from a p-variate normal population whose variance-covariance matrix is I. R_i is the matrix of the first i rows and columns of R. Show that $|R_i|/|R_{i-1}|$ (i = 2, 3, ..., p) are independently distributed as beta variables. In the notation of Chapter 8, show that $|R_i|/|R_{i-1}|$ has the $\Lambda(n, 1, i-1)$ distribution [where $\Lambda(n, p, q)$ is Wilks's Λ]. Hence show that $a_i \log_e|R_i|/|R_{i-1}|$ has approximately a χ^2 distribution with (i-1) d.f. (i = 2, 3, ..., p), where $a_i = -[n - \frac{1}{2}(i+1)]$ is Bartlett's correction factor. To combine all these results, add all $\log_e|R_i|/|R_{i-1}|$ and choose an overall correction factor by taking the weighted average of the factors a_i, the weights being the d.f. (i-1). Show that this leads to the result that

$$-\left\{n - \frac{2p+5}{6}\right\} \log_e|R|$$

has an asymptotic χ^2 distribution with p(p-1)/2 d.f. [This is an alternative heuristic derivation of (11.8.7)].

80. X is the n x p matrix of n observations on p variables $\underline{x}$, and $\underline{y}$ is the vector of the corresponding n observations on a variable y. The usual method of predicting y from $\underline{x}$ is to consider the regression of y on $\underline{x}$ and this leads to (assuming zero means for all the variables)

$$\hat{\underline{y}} = X(X'X)^{-1}X'\underline{y}$$

as the predicted value. Instead of this procedure, the

principal components of $\underline{x}$ are first obtained by finding the eigenvalues and eigenvector of the matrix X'X. The observations on the principal components are then given by the matrix V such that $V'V = I$, $V'(X'X)V = \text{diag}(\lambda_1, \ldots, \lambda_p)$ where the λ's are the eigenvalues of X'X. We then define factor scores S by the relation $S = XV$. Now consider the regression of y on the principal components to predict y. This gives the predicted value as

$$\underline{y}^* = S(S'S)^{-1}S'\underline{y} \ .$$

Show that this method leads to the same predicted value of y or that $\hat{\underline{y}} = \underline{y}^*$.

Haitovsky [18]

81. X represents the $p \times N$ matrix of sample observations from a p-variate singular normal distribution with zero means and variance-covariance matrix Σ of rank $r < p$. The matrix $S = XX'$ is the Wishart matrix and is said to have a singular Wishart distribution of N d.f. and associated matrix Σ. Show that the characteristic function of such a singular Wishart distribution is

$$\phi(\theta) = |I_p - 2\sqrt{-1}\,\Sigma\,\theta|^{-N/2} ,$$

where θ is the symmetric $p \times p$ matrix of elements θ_{ij} and $\phi(\theta)$ is defined as the expected value of

$$\exp\left[\sqrt{-1}\left\{\sum_i \theta_{ii}s_{ii} + 2\sum_{i<j} \theta_{ij}s_{ij}\right\}\right] ,$$

s_{ij} being the elements of S.

Mitra [41]

82. X is an $n \times p$ matrix of random variables which are all independent and N(0, 1). Show that there exists an orthogonal matrix $P = [p_{ij}]$ of order n and a matrix $B = [b_{ij}]$, which is upper triangular, i.e., $b_{ij} = 0$ $(j < i)$ and $b_{ii} > 0$, such that X = PB. Let $\hat{P}$ denote the set of those elements p_{ij} of P for which simultaneously $j < i$ and $j \leq p$. Show that the Jacobian of transformation from X to B and P is

$$\prod_{i=1}^{m} b_{ii}^{n-i} \cdot \prod_{j=1}^{m} |P_j|^{-1}$$

where $m = \min(n, p)$ and P_j is the matrix of the first j rows and columns of P. Hence show that B and $\hat{P}$ are independently distributed, the distribution of $\hat{P}$ being,

$$\text{const.} \prod_{j=1}^{m} |P_j|^{-1} \prod_{i=1}^{n} \prod_{\substack{j<i \\ j<P}} d\, p_{ij} ,$$

where $|\hat{P}_j|$ should be expressed in terms of P.

Mauldon [39]

83. S is a $p \times p$ symmetric matrix with eigenvalues $f_1 > f_2 > \cdots > f_p$ (> 0). Then there exists an orthogonal matrix $E = [e_{ij}]$ of order p, such that $S = EFE'$, where $F = \text{diag}(f_1, \ldots, f_p)$. Since $EE' = I$, all the p^2 elements of E are not functionally independent; only $p(p-1)/2$ are. We can choose these to be e_{ij} $(j = i+1, \ldots, p;\ i = 1, 2, \ldots, p-1)$. The transformation $S = EFE'$ determines E and F uniquely if we impose $f_1 > \cdots > f_p$ and $e_{12}, \ldots, e_{1p}$ all > 0. Show that the

Jacobian of transformation from S to F and $e_{ij}(j > i,\ i = 1, \ldots, p-1)$ is

$$\prod_{i<j} (f_i - f_j) \cdot \prod_{i=1}^{p-1} |E_i|^{-1}$$

where E_i is the matrix of the first i rows and columns of E. [Hint: $J(S \to E, F) = J(dS \to dE, dF)$ where dS is matrix of differentials. $dS = EF(dE') + E(dF)E' + (dE)FE'$ or $W = FG' + dF + GF$, where $W = E'(dS)E$, $G = E'(dE)$. Note that $G + G' = 0$. Observe that $J(dS \to dE, dF) = J_1(dS \to W)\ J_2(W \to G, dF) \cdot J_3(G \to dE)$, $J_1 = |E|^{-(p+1)} = 1$. Then

$$w_{ii} = df_i,\ w_{ij} = (f_j - f_i)g_{ij} \quad (i \neq j).$$

Hence

$$J_2 = \prod_{i<j} (f_i - f_j).$$

To obtain J_3, we use $dE = EG$, hence

$$\begin{bmatrix} de_{21} \\ \vdots \\ de_{p1} \end{bmatrix} = \begin{bmatrix} e_{22} & \cdots & e_{2p} \\ \\ e_{p2} & \cdots & e_{pp} \end{bmatrix} \begin{bmatrix} g_{21} \\ \vdots \\ g_{p1} \end{bmatrix}, \quad \text{as } g_{11} = 0 .$$

Therefore

$$J(de_{21}, \ldots, de_{p1} \to g_{21}, \ldots, g_{p1}) = \begin{vmatrix} e_{22} & \cdots & e_{2p} \\ e_{p2} & \cdots & e_{pp} \end{vmatrix} = e_{11}$$

by Jacobi's theorem. Putting similar results together

$$J_3 = \prod_{i=1}^{p-1} |E_i|^{-1}$$

and hence the result.]

84. If the matrix S of the previous exercise has the $W_p(S|n|I)$ distribution, show that the matrix E has the distribution

$$\frac{\pi^{p(p+1)/4}}{\prod_{i=1}^{p} \Gamma\left(\frac{p+1-i}{2}\right)} \prod^{p-1} |E_i|^{-1} \prod_{i=1}^{p-1} \prod_{j>i} de_{ij} .$$

Observe that this is an alternative form of the Haar invariant distribution obtained in (11.5.7). In the above expression the density is not expressed in terms of the variables $e_{ij}(j > i)$ and that can be done theoretically at least by using $EE' = I$. Also compare this distribution with the distribution of exercise 82, when $n = p$.

85. X is a $p \times n$ $(n > p)$ matrix of random variables and possesses a density with respect to a Lebgesque measure. Suppose, for any orthogonal matrix K of order n, the distribution of XK does not depend on the elements of K, then XX' and $H = (XX')^{-1/2}X$ are independently distributed, the distribution of H being the Haar invariant distribution.

Khatri [29]

REFERENCES

[1] Aitken, M. A., Nelson, W. C., and Reinfurt, Karen (1968). "Tests for correlation matrices", Biometrika, 55, p. 327.

[2] Anderson, T. W. (1958). Introduction to Multivariate Statistical Analysis. John Wiley and Sons, New York.

[3] Anderson, T. W. (1963). "Asymptotic theory for principal component analysis", Ann. Math. Statist., 34, p. 122.

[4] Anderson, T. W. (1963). "A test for equality of means when covariance matrices are unequal", Ann. Math. Statist., 34, p. 673.

[5] Bartlett, M. S. (1951). "The effect of standardization on a χ^2-approximation in factor analysis", Biometrika, 38, p. 337. (Appendix by Ledermann.)

[6] Burnaby, T. P. (1966). "Growth-invariant discriminant functions and generalized distances", Biometrics, 22, p. 96.

[7] Chakrabarti, M. C. (1962). Mathematics of Design and Analysis of Experiments. Asia Publishing House, Bombay.

[8] Chen, C. W. (1971). "On some problems in canonical correlation analysis", Biometrika, 58, p. 399.

[9] Cochran, W. G. (1964). "On the performance of the linear discriminant function", Technometrics, 6, p. 179.

[10] Cornish, E. A. (1955). "The sampling distribution of statistics derived from the multivariate t distribution", Austral. J. Physics, 8, p. 193.

[11] Dunnett, and Sobel, M. (1955). "Approximations to the probability integral and certain percentage points of a multivariate analogue of Student's t distribution", Biometrika, 42, p. 258.

[12] Dykstra, R. L. (1970). "Establishing the positive definiteness of the sample covariance matrix", Ann. Math. Statist., 41, p. 2153.

[13] Fisher, R. A. (1950). Contributions to Mathematical Statistics. John Wiley and Sons, New York.

[14] Fisk, P. R. (1970). "A note on a characterization of the multivariate normal distribution", Ann. Math. Statist., 41, p. 486.

[15] Gilbert, E. S. (1969). "The effect of unequal variance-covariance matrices on Fisher's linear discriminant function", Biometrics, 25, p. 505.

[16] Goodman, N. R. (1963). "Statistical analysis based on a certain multivariate complex Gaussian distribution (An Introduction)", Ann. Math. Statist., 34, p. 152.

[17] Groves, T., and Rothenberg, T. (1969). "A note on the expected value of an inverse matrix", Biometrika, 56, p. 690.

[18] Haitovsky, Y. (1966). "A note on regression on principal components", Am. Statistician, 20, p. 28.

[19] Healy, M. J. R. (1964). "A property of the multinomial distribution and the determination of appropriate scores", Biometrika, 51, p. 265.

[20] Healy, M. J. R. (1965). "Computing a discriminant function from within-sample dispersions", Biometrics, 21, p. 1011.

[21] Hills, M. (1966). "Allocation rules and their error rates", J. Roy. Stat. Soc. B., 28, p. 1.

[22] Johnson, N. L., and Kotz, S. (1969). Continuous Distributions. Vol. 1 and 2, Houghton Mifflin Company, Boston.

[23] Kabe, D. G. (1958). "Some applications of Meijer-G functions to distribution problems in statistics", Biometrika, 45, p. 578.

[24] Kabe, D. G. (1964). "Decomposition of Wishart distribution", Biometrika, 51, p. 267.

[25] Kabe, D. G. (1965). "On the noncentral distribution of Rao's U statistic", Ann. Inst. Stat. Math., 17, p. 75.

[26] Kendall, M. G., and Stuart, A. (1968). The Advanced Theory of Statistics. Hafner Publishing Company, New York.

[27] Khatri, C. G. (1959). "On the mutual independence of certain statistics", Ann. Math. Statist., 30, p. 1258.

[28] Khatri, C. G. (1959). "On conditions for the forms of the type XAX' to be distributed independently or to obey Wishart distribution", Cal. Stat. Assoc. Bull., 8, p. 162.

[29] Khatri, C. G. (1970). "A note on Mitra's paper, a density free approach to the matrix variate beta distribution", Sankhyā A, 32, p. 311.

[30] Kshirsagar, A. M. (1960). "A note on the derivation of some exact multivariate tests", Biometrika, 47, p. 480.

[31] Kshirsagar, A. M. (1961). "Some extensions of the multivariate t-distribution and the multivariate generalization of the distribution of the regression coefficient", Proc. Camb. Phil. Soc., 57, p. 80.

[32] Kshirsagar, A. M. (1961). "Goodness-of-fit of a single (non-isotropic) hypothetical principal component", Biometrika, 48, p. 397.

[33] Kshirsagar, A. M., and Young, John C. (1971). "Correlation between two Hotelling's T^2", Themis Technical Report, Department of Statistics, Southern Methodist University, Dallas.

[34] Kullback, S. (1967). "On testing correlation matrices", Applied Statistics, 16, p. 80.

[35] Lachenbruch, P. A. (1968). "On expected probabilities of misclassification in discriminant analysis, necessary sample size, and a relation with the multiple correlation coefficient", Biometrics, 24, p. 823.

[36] Lancaster, H. O. (1969). _The Chi-Squared Distribution_. John Wiley and Sons, New York.

[37] Mallows, C. L. (1960). "Latent vectors of random symmetric matrices", Biometrika, 48, p. 133.

[38] Mardia, K. V. (1970). "Measures of multivariate skewness and kurtosis with applications", Biometrika, 57, p. 519.

[39] Mauldon, J. G. (1955). "Pivotal quantities for Wishart's and related distributions and a paradox in fiducial theory", J. Roy. Stat. Soc. B, 17, p. 79.

[40] Mehta, J. S., and Gurland, J. (1969). "Some properties and an application of a statistic arising in testing correlation", Ann. Math. Statist., 40, p. 1736.

[41] Mitra, S. K. (1970). "Some characteristic and non-characteristic properties of the Wishart distribution", Sankhyā A, 32, p. 19.

[42] Mitra, S. K. (1970). "A density-free approach to the matrix variate beta distribution", Sankhyā A, 32, p. 81.

[43] Mohn, Erik (1968). "The joint distribution of two studentized regression coefficients", Biometrika, 55, p. 424.

[44] Olkin, I., and Rubin, H. (1964). "Multivariate Beta distributions and independence properties of the Wishart distribution", Ann. Math. Statist., 35, p. 261.

[45] Olkin, I., and Siotani (1964). "Asymptotic distribution of functions of a correlation matrix", Technical Report No. 6, Department of Statistics, Stanford University, Stanford.

[46] Olkin, I., and Shrikhande, S. S. (1970). "An extension of Wilks's test for the equality of means", Ann. Math. Statist., 41, p. 683.

[47] Owen, D. B. (1956). "Tables for computing bivariate normal probability", Ann. Math. Statist., 27, p. 1075.

[48] Owen, D. B., and Wiessen, J. M. (1959). "A method of computing bivariate normal probability", Sandia Corporation Report, Bell System Technical Journal, 38, No. 2.

[49] Patel, H. I. (1970). "The choice of optimum scores in a Markov chain of order one", Ann. Inst. Stat. Maths., 22, p. 535.

[50] Pierce, David A., and Dykstra, R. L. (1969). "Independence and the multivariate normal distribution", American Statistician, 23, p. 39.

[51] Rao, C. Radhakrishna (1965). Linear Statistical Inference and Its Applications. John Wiley and Sons, Inc., New York.

[52] Reiersol, O. (1956). "A note on the signs of gross correlation coefficients and partial correlation coefficients", Biometrika, 43, p. 480.

[53] Roy, S. N., and Bargmann, R. E. (1958). "Tests of multiple independence and the associated confidence bounds", Ann. Math. Statist., 29, p. 493.

[54] Smith, W. B. (1968). "Bivariate samples with missing data", Answer to Query, Technometrics, 10, p. 867.

[55] Smith, W. B., and Hocking, R. R. (1968). "A simple method for obtaining the information matrix for a multivariate normal distribution", American Statistician, 22, p. 18.

[56] Steyn, S. (1951). "The Wishart distribution derived by solving simultaneous linear differential equations", Biometrika, 38, p. 470.

[57] Subrahmaniam, Kathleen, and Subrahmaniam, S. (1971). "On the distribution of $(D^2_{p+q} - D^2_q)$ statistic; percentage points and the power of the test", Technical Report No. 7, Department of Statistics, University of Manitoba, Canada.

[58] Webster, J. T. (1970). "On the application of the method of Das in evaluating a multivariate normal integral", Biometrika, 57, p. 657.

[59] Wilks, S. S. (1946). "Sample criteria for testing equality of means, equality of variances and equality of covariances in a normal multivariate distribution", Ann. Math. Statist., 17, p. 257.

[60] Wilks, S. S. (1964). <u>Mathematical Statistics</u>. John Wiley and Sons, New York.

[61] Williams, E. J. (1952). "Some exact tests in multivariate analysis", Biometrika, 39, p. 17.

[62] Williams, E. J. (1970). "Comparing means of correlated variates", Biometrika, 57, p. 459.

APPENDIX

JACOBIANS OF CERTAIN TRANSFORMATIONS

	Old Variables	New Variables	Transformation	Jacobian (Absolute value of)
(1)	$\underline{y}$	$\underline{x}$	$\underline{y} = A\underline{x}$ A is pxp, nonsingular	$\lvert A \rvert$
(2)	Y pxq	X pxq	Y = A X A is pxp, nonsingular	$\lvert A \rvert^q$
(3)	Y pxq	X pxq	Y = A X B A is pxp, B is qxq both nonsingular	$\lvert A \rvert^q \lvert B \rvert^p$
(4)	L pxp symmetric	M pxp symmetric	L = AMA', A is pxp nonsingular	$\lvert A \rvert^{p+1}$
(5)	L pxp l.t.	M pxp l.t.	L = AM, A, pxp l.t. of elements a_{ij}	$\prod_{i=1}^{p} a_{ii}^{i}$
(6)	V pxp symmetric	T pxp l.t. of elements t_{ij}	V = TT'	$2^p \prod_{i=1}^{p} t_{ii}^{p+1-i}$
(7)	A, B pxp	$\phi_1 > \phi > \ldots > \phi_p > 0$ roots of $\lvert A - \phi B \rvert = 0$ and $\underset{pxp}{W} = [w_{ij}]$, $w_{1i} > 0$	A = W diag $(\phi_1, \ldots, \phi_p)$W' B = WW'	$\lvert W \rvert^{p+2} . 2^p$ $\prod_{i<j}^{p} (\phi_i - \phi_j)$
(8)	V pxp	W	$W = V^{-1}$	$\lvert W \rvert^{-2p}$

ERRATA

Page	Line	For	Read
26	12	compartment	component
79	6 eq. 8.3	$W_p(D\|n\|I) = e^{-\lambda^2/2} \ldots$	$W_p(D\|n\|I) \cdot e^{-\lambda^2/2} \ldots$
122	9	$\underline{\mu} \neq 0$	$\mu = 0$
122	10	$\underline{\mu} = 0$	$\underline{\mu} \neq 0$
138	21	add equation no. (3.18)	which is missing
293	18 eq. 2.5	$\prod_{i=1}^{p} \left\{ \frac{\Gamma\left(\frac{f_1 + f_2 + 1 - i}{2}\right)}{\Gamma\left(\frac{f_1 + 1 - i}{2}\right) \Gamma\left(\frac{f_2 + 1 - i}{2}\right)} \right\}^{-1}$	-1 is missing
373	20	after $\xi_2, \ldots, \xi_p$, add provided we define $y_6 = 1$ and not 0 as in (1.28)	
447	eq. 7.7	$(f_1 f_2 \ldots f_k)$	$(f_1 f_2 \ldots f_{p-k})$
447	eq. 7.7	$\operatorname{tr} S - f_1 - \ldots - f_k$	$\operatorname{tr} S - f_1 - \ldots - f_{p-k}$

AUTHOR INDEX

Underlined numbers give the page on which the complete reference is listed.

E

F

G

H

I

J

K

SUBJECT INDEX